Image-Based Computational and Experimental Biomedical Flows

Image-Based Computational and Experimental Biomedical Flows

Editor

Huidan (Whitney) Yu

MDPI

Basel • Beijing • Wuhan • Barcelona • Belgrade • Novi Sad • Cluj • Manchester

Editor
Huidan (Whitney) Yu
Purdue University in
Indianapolis
Indianapolis, IN
USA

Editorial Office
MDPI AG
Grosspeteranlage 5
4052 Basel, Switzerland

This is a reprint of articles from the Special Issue published online in the open access journal *Fluids* (ISSN 2311-5521) (available at: https://www.mdpi.com/journal/fluids/special_issues/Imaged_Biomedical_Flows).

For citation purposes, cite each article independently as indicated on the article page online and as indicated below:

Lastname, A.A.; Lastname, B.B. Article Title. *Journal Name* **Year**, *Volume Number*, Page Range.

ISBN 978-3-7258-2417-5 (Hbk)
ISBN 978-3-7258-2418-2 (PDF)
doi.org/10.3390/books978-3-7258-2418-2

Contents

About the Editor vii

Preface ix

Huidan (Whitney) Yu
Image-Based Computational and Experimental Biomedical Flows
Reprinted from: *Fluids* **2024**, *9*, 227, doi:10.3390/fluids9100227 1

Maria Sabrina Souza, Andrews Souza, Violeta Carvalho, Senhorinha Teixeira, Carla S. Fernandes, Rui Lima and João Ribeiro
Fluid Flow and Structural Numerical Analysis of a Cerebral Aneurysm Model
Reprinted from: *Fluids* **2022**, *7*, 100, doi:10.3390/fluids7030100 5

Augusto Fava Sanches, Suprosanna Shit, Yigit Özpeynirci and Thomas Liebig
CFD to Quantify Idealized Intra-Aneurysmal Blood Flow in Response to Regular and Flow Diverter Stent Treatment
Reprinted from: *Fluids* **2022**, *7*, 254, doi:10.3390/fluids7080254 21

Maria Antonietta Boniforti, Roberto Magini and Tania Orosco Salinas
Hemodynamic Investigation of the Flow Diverter Treatment of Intracranial Aneurysm
Reprinted from: *Fluids* **2023**, *8*, 189, doi:10.3390/fluids8070189 34

Maria Antonietta Boniforti, Maria Chiara Cesaroni, Roberto Magini, Edoardo Pasqui and Gianmarco de Donato
Image-Based Numerical Investigation in an Impending Abdominal Aneurysm Rupture
Reprinted from: *Fluids* **2022**, *7*, 269, doi:10.3390/fluids7080269 52

Francesco Duronio and Andrea Di Mascio
Blood Flow Simulation of Aneurysmatic and Sane Thoracic Aorta Using OpenFOAM CFD Software
Reprinted from: *Fluids* **2023**, *8*, 272, doi:10.3390/fluids8100272 68

Pablo Jeken-Rico, Aurèle Goetz, Philippe Meliga, Aurélien Larcher, Yigit Özpeynirci and Elie Hachem
Evaluating the Impact of Domain Boundaries on Hemodynamics in Intracranial Aneurysms within the Circle of Willis
Reprinted from: *Fluids* **2023**, *9*, 1, doi:10.3390/fluids9010001 83

Jana Korte, Thomas Rauwolf, Jan-Niklas Thiel, Andreas Mitrasch, Paulina Groschopp, Michael Neidlin, et al.
Hemodynamic Assessment of the Pathological Left Ventricle Function under Rest and Exercise Conditions
Reprinted from: *Fluids* **2023**, *8*, 71, doi:10.3390/fluids8020071 99

Abdulaziz Al Baraikan, Krzysztof Czechowicz, Paul D. Morris, Ian Halliday, Rebecca C. Gosling, Julian P. Gunn, et al.
Modelling The Hemodynamics of Coronary Ischemia
Reprinted from: *Fluids* **2023**, *8*, 159, doi:10.3390/fluids8050159 114

Safia Ihsan Ali, David Patton, Kimberley A. Myers and Julio Garcia
Repaired Tetralogy of Fallot Pressure Assessment: Insights from 4D-Flow Pressure Mapping
Reprinted from: *Fluids* **2023**, *8*, 196, doi:10.3390/fluids8070196 147

Huidan Yu, Monsurul Khan, Hao Wu, Chunze Zhang, Xiaoping Du, Rou Chen, et al.
Inlet and Outlet Boundary Conditions and Uncertainty Quantification in Volumetric Lattice Boltzmann Method for Image-Based Computational Hemodynamics
Reprinted from: *Fluids* **2022**, *7*, 30, doi:10.3390/fluids7010030 . . . . **170**

Weichen Hong, Huidan Yu, Jun Chen, John Talamantes, Dave M. Rollins, Xin Fang, et al.
A Mock Circulation Loop to Characterize In Vitro Hemodynamics in Human Systemic Arteries with Stenosis
Reprinted from: *Fluids* **2023**, *8*, 198, doi:10.3390/fluids8070198 . . . . **185**

Andrey Yukhnev, Ludmila Tikhomolova, Yakov Gataulin, Alexandra Marinova, Evgueni Smirnov, Andrey Vrabiy, et al.
V Flow Measurements of Pulsatile Flow in Femoral-Popliteal Bypass Proximal Anastomosis Compared with CFD Simulation
Reprinted from: *Fluids* **2024**, *9*, 64, doi:10.3390/fluids9030064 . . . . **200**

John F. LaDisa, Jr., Arash Ghorbannia, David S. Marks, Peter Mason and Hiromasa Otake
Advancements and Opportunities in Characterizing Patient-Specific Wall Shear Stress Imposed by Coronary Artery Stenting
Reprinted from: *Fluids* **2022**, *7*, 325, doi:10.3390/fluids7100325 . . . . **215**

About the Editor

Huidan (Whitney) Yu

Dr. Huidan (Whitney) Yu is a Professor of Mechanical Engineering at Purdue University in Indianapolis and a Research Professor of Vascular Surgery at Indiana University School of Medicine. Her expertise lies in image-based computational fluid dynamics (ICFD) and its medical applications, with a focus on hemodynamics and cardiovascular flows. Dr. Yu's research integrates advanced computational tools, including the lattice Boltzmann method and GPU parallel computing, with experimental techniques such as a mock circulation loop to study complex vascular behaviors. She has made significant contributions to patient-specific modeling of cardiovascular diseases, enhancing personalized treatment strategies and diagnostic capabilities for conditions such as arterial stenosis and ischemic strokes. Dr. Yu is a prolific researcher and has been widely recognized for her innovative work at the intersection of engineering and medicine. In addition to her research, Dr. Yu is a dedicated educator and mentor, guiding students in computational modeling and biomedical applications. Her leadership in organizing the Special Issue *Image-Based Computational and Experimental Biomedical Flows* reflects her commitment to advancing the field through collaboration across disciplines.

Preface

It is with great pleasure that I present this reprint, "Image-Based Computational and Experimental Biomedical Flows", a curated collection of thirteen groundbreaking research papers that focus on the intersection of medical imaging data and fluid dynamics in cardiovascular flows. The primary aim of this reprint is to provide readers with a comprehensive overview of the latest techniques in computational modeling and experimental measurements, enabling the quantification of 4-D (space+time) hemodynamics in human vascular systems derived from patient-specific imaging data. These works represent the forefront of innovation in the field, addressing complex cardiovascular pathologies such as intracranial aneurysms, coronary artery disease, and myocardial ischemia, with the goal of advancing diagnostic accuracy and improving treatment outcomes through personalized medicine.

The motivation behind this special collection stems from the increasing need for precise, patient-specific approaches in cardiovascular disease management. The studies featured herein bridge the gap between in silico simulations, in vitro experiments, and in vivo data, demonstrating the potential of these methodologies to revolutionize how cardiovascular conditions are diagnosed, treated, and understood. By leveraging tools such as computational fluid dynamics (CFD), the lattice Boltzmann method, and advanced experimental setups, these papers provide valuable insights into hemodynamic phenomena that are otherwise difficult to capture.

This reprint contains an editorial. It is intended for researchers, clinicians, and engineers who are working at the intersection of medical imaging, fluid dynamics, and cardiovascular research. It serves as both a reference for those already immersed in the field and a resource for those new to these cutting-edge technologies.

I would like to express my sincere gratitude to the authors for their contributions and to the reviewers for their invaluable feedback. Special thanks go to the editorial team at Fluids for their unwavering support throughout the preparation of this Special Issue. Without their collective efforts, this reprint would not have been possible.

Huidan (Whitney) Yu
Editor

fluids

MDPI

Editorial

Image-Based Computational and Experimental Biomedical Flows

Huidan (Whitney) Yu

Purdue University in Indianapolis, Indianapolis, IN 46202, USA; yu597@purdue.edu

Fluids is pleased to present a Special Issue named "Image-Based Computational and Experimental Biomedical Flows", a curated collection of thirteen featured research papers that explore the integration between medical imaging data and 4-D (space + time) fluid dynamics for patient-specific cardiovascular flows. This Issue highlights recent developments in both computational and experimental methodologies, delivering valuable insights into the intricate hemodynamic complexities within human vascular systems. By bridging the gap between medical imaging and advanced fluid dynamics, these studies open new avenues for more accurate diagnosis, personalized treatment strategies, and the potential to revolutionize cardiovascular disease management. The featured articles dive deep into a diverse range of topics, including patient-specific modeling of blood flow in major arteries, the hemodynamics of intracranial aneurysms, and the intricate behavior of coronary artery stents. Each paper not only showcases state-of-the-art techniques but also emphasizes the real-world applications of these technologies in understanding and predicting vascular behavior under both healthy and diseased conditions. A recurring theme throughout this Issue is the synergy between computational fluid dynamics (CFD), image-based hemodynamic simulations, and innovative experimental setups, designed to replicate human arterial systems. By leveraging advanced computational tools like the lattice Boltzmann method and OpenFoam and experimental setting such as mock circulation loops, along with physical devices like flow diverter stents, these studies provide highly accurate simulations of blood flow, pressure distributions, and vascular responses.

The contributions in this Special Issue cover a wide range of topics.

Aneurysm Studies: A significant portion of this Issue is devoted to the study of aneurysms, reflecting the critical importance of this pathology in cardiovascular research [1,2]. Souza et al. [3] present a comprehensive analysis of cerebral aneurysms that goes beyond traditional flow studies. By coupling CFD with structural simulations, they investigate not only the flow behavior but also the biomechanical response of the aneurysm wall. Their results, spanning various Reynolds numbers and rheological models, reveal valuable insights into how flow patterns influence wall stresses and deformations, potentially aiding in rupture risk assessment. The cutting-edge topic of flow diverter stents for treating intracranial aneurysms is addressed by both Sanches et al. [4] and Boniforti et al. [5]. Sanches et al. [4] use CFD simulations to quantify the dramatic changes in intra-aneurysmal hemodynamics induced by these devices. They report significant reductions in wall shear stress and flow velocity, coupled with increased turnover time, suggesting a higher likelihood of thrombotic occlusion and reduced rupture risk. Boniforti et al. [5] expand on this work by conducting a detailed numerical investigation of flow diverter treatment, offering insights into optimal device selection based on porosity and other parameters. In a pioneering study, Boniforti et al. [6] apply CFD techniques to investigate an abdominal aortic aneurysm caught in the act of rupturing. By analyzing CT images of the rupturing aneurysm and creating a virtual pre-rupture model, they identify hemodynamic parameters associated with rupture risk. Their findings highlight the potential of CFD as a predictive tool for aneurysm management. Duronio and Di Mascio [7] utilize the open-source CFD software OpenFOAM to simulate blood flow in both healthy and aneurysmal thoracic aortas. Their

Citation: Yu, H. Image-Based Computational and Experimental Biomedical Flows. *Fluids* **2024**, *9*, 227. https://doi.org/10.3390/fluids9100227

Received: 20 September 2024
Accepted: 27 September 2024
Published: 30 September 2024

work not only provides insights into the hemodynamic changes associated with aortic aneurysms but also demonstrates the capabilities of open-source tools for patient-specific cardiovascular modeling. Jeken-Rico et al. [1] address a critical methodological challenge by evaluating the impact of domain boundaries on hemodynamic simulations of intracranial aneurysms. Their work provides valuable guidance for researchers and clinicians on how to set up computational domains for accurate and reliable aneurysm simulations, particularly in the complex anatomical context of the Circle of Willis.

Cardiac Function and Disease: Several papers in this Issue focus on cardiac function and specific cardiac pathologies [8–10]. Korte et al. [11] advance the research by analyzing left ventricular hemodynamics in patients with mitral valve insufficiency under both rest and exercise conditions. Their echocardiography-based simulations reveal intriguing differences in kinetic energy patterns between rest and exercise states, as well as between different stages of valve insufficiency. This work demonstrates the potential of computational modeling to provide new diagnostic and prognostic indicators for cardiac function. Baraikan et al. [12] present an ambitious multi-scale modeling framework to quantify myocardial ischemia. By integrating various levels of cardiovascular modeling, from zero-dimensional models to CFD simulations, their approach offers a comprehensive tool for personalizing and predicting ischemic burden, with potential implications for clinical decision-making in coronary artery disease. Ihsan Ali et al. [13] leverage 4-D flow MRI to assess pressure changes in the repaired tetralogy of Fallot patients. This non-invasive imaging technique offers a unique window into the hemodynamics of congenital heart disease, providing valuable data for both clinical assessment and computational model validation.

Arterial Flow and Stenosis: This Issue also addresses the critical topic of vascular flows, particularly in the context of stenosis. Yu et al. [14] make significant strides in improving the accuracy of image-based computational hemodynamics. They develop physiological inlet and outlet boundary conditions based on patient-specific medical data and integrate them into a volumetric lattice Boltzmann method. Their approach, validated on six human aortorenal arterial systems, shows excellent agreement with medical measurements.

Hong et al. [15] develop an innovative mock circulation loop for in vitro hemodynamic measurements in stenosed arteries. This experimental setup bridges the gap between computational and experimental approaches, allowing for the validation of numerical models and the investigation of flow phenomena that may be challenging to capture in silico. Yukhnev et al. [16] present a study that exemplifies the power of combining advanced experimental techniques with computational modeling. They compare ultrasound vector flow measurements to CFD simulations for pulsatile flow in femoral-popliteal bypass grafts, validating computational approaches and showcasing the capabilities of advanced ultrasound imaging in capturing complex flow patterns.

Coronary Artery Interventions: LaDisa et al. [17] provide a comprehensive review of recent advancements in modeling wall shear stress changes induced by coronary artery stents. This paper synthesizes state-of-the-art computational approaches, image-based reconstruction methods, and novel boundary condition implementations, offering a roadmap for future patient-specific modeling efforts in interventional cardiology.

Methodological Advancements: Several papers in this Issue focus on methodological advancements that have broad implications for the field of cardiovascular flow modeling. Yu et al. [14] demonstrate the potential of the lattice Boltzmann method [18–22] for hemodynamics simulations, offering an alternative to traditional Navier–Stokes-based approaches. Jeken-Rico et al. [1] provide crucial insights into the impact of domain boundaries on CFD simulations, addressing a key challenge in the computational modeling of vascular flows. Duronio and Di Mascio [7] showcase the capabilities of open-source CFD software (OpenFOAM) for cardiovascular simulations, potentially democratizing access to advanced modeling tools.

Experimental Techniques: While computational methods dominate this Issue, several papers highlight the crucial role of experimental approaches in validating and complementing numerical simulations. Hong et al. [15] develop a mock circulation loop that allows for detailed in vitro studies of stenosed arteries. Yukhnev et al. [16] demonstrate the power of advanced ultrasound vector flow measurements in capturing complex arterial flow patterns. Ihsan Ali et al. [13] showcase the potential of 4D flow MRI for non-invasive assessment of cardiovascular hemodynamics.

Multi-scale and Integrated Approaches: A trend towards more comprehensive, multi-scale modeling approaches is evident in several papers. Souza et al. [3] combine fluid dynamics and structural analysis to provide a more complete picture of aneurysm biomechanics. Baraikan et al. [12] present a multi-scale framework that integrates various levels of cardiovascular modeling to study coronary ischemia.

This Special Issue represents the cutting edge of image-based cardiovascular flow modeling and analysis, showcasing the incredible strides being made in this rapidly evolving field. From refining fundamental simulation techniques to tackling highly complex clinical challenges, these studies reveal the immense potential of computational and experimental approaches to transform our understanding of cardiovascular physiology and pathology. By seamlessly integrating computational models, experimental methods, and clinical data, these papers illustrate the remarkable synergy between these domains, driving forward the science of cardiovascular flows. As medical imaging technologies continue to advance and computational power reaches unprecedented heights, the fusion of in silico, in vitro, and in vivo approaches is set to revolutionize the future of cardiovascular medicine [23,24]. The promise of personalized cardiovascular care is no longer a distant goal, but an emerging reality. Patient-specific simulations will soon guide clinical decision-making, enabling tailored treatment strategies that optimize outcomes for each individual. With this confluence of technology and medicine, we stand on the brink of a new era—one where cardiovascular disease can be understood, treated, and potentially prevented with unparalleled precision and insight.

Conflicts of Interest: The authors declare no conflict of interest.

References

1. Jeken-Rico, P.; Goetz, A.; Meliga, P.; Larcher, A.; Özpeynirci, Y.; Hachem, E. Evaluating the Impact of Domain Boundaries on Hemodynamics in Intracranial Aneurysms within the Circle of Willis. *Fluids* **2023**, *9*, 1. [CrossRef]
2. Berg, P.; Saalfeld, S.; Voß, S.; Beuing, O.; Janiga, G. A review on the reliability of hemodynamic modeling in intracranial aneurysms: Why computational fluid dynamics alone cannot solve the equation. *Neurosurg. Focus* **2019**, *47*, E15. [CrossRef]
3. Souza, M.S.; Souza, A.; Carvalho, V.; Teixeira, S.; Fernandes, C.S.; Lima, R.; Ribeiro, J. Fluid Flow and Structural Numerical Analysis of a Cerebral Aneurysm Model. *Fluids* **2022**, *7*, 100. [CrossRef]
4. Sanches, A.F.; Shit, S.; Özpeynirci, Y.; Liebig, T. CFD to Quantify Idealized Intra-Aneurysmal Blood Flow in Response to Regular and Flow Diverter Stent Treatment. *Fluids* **2022**, *7*, 254. [CrossRef]
5. Boniforti, M.A.; Magini, R.; Salinas, T.O. Hemodynamic Investigation of the Flow Diverter Treatment of Intracranial Aneurysm. *Fluids* **2023**, *8*, 189. [CrossRef]
6. Boniforti, M.A.; Cesaroni, M.C.; Magini, R.; Pasqui, E.; de Donato, G. Image-Based Numerical Investigation in an Impending Abdominal Aneurysm Rupture. *Fluids* **2022**, *7*, 269. [CrossRef]
7. Duronio, F.; Di Mascio, A. Blood Flow Simulation of Aneurysmatic and Sane Thoracic Aorta Using OpenFOAM CFD Software. *Fluids* **2023**, *8*, 272. [CrossRef]
8. Huang, J.; Wang, Y.; Lin, L.; Li, Z.; Shan, Z.; Zheng, S. Comparison of dynamic changes in aortic diameter during the cardiac cycle measured by computed tomography angiography and transthoracic echocardiography. *J. Vasc. Surg.* **2019**, *69*, 1538–1544. [CrossRef]
9. Hathaway, Q.A.; Roth, S.M.; Pinti, M.V.; Sprando, D.C.; Kunovac, A.; Durr, A.J.; Cook, C.C.; Fink, G.K.; Cheuvront, T.B.; Grossman, J.H.; et al. Machine-learning to stratify diabetic patients using novel cardiac biomarkers and integrative genomics. *Cardiovasc. Diabetol.* **2019**, *18*, 78. [CrossRef] [PubMed]
10. Taylor, C.A.; Fonte, T.A.; Min, J.K. Computational Fluid Dynamics Applied to Cardiac Computed Tomography for Noninvasive Quantification of Fractional Flow Reserve: Scientific Basis. *J. Am. Coll. Cardiol.* **2013**, *61*, 2233–2241. [CrossRef]
11. Korte, J.; Rauwolf, T.; Thiel, J.-N.; Mitrasch, A.; Groschopp, P.; Neidlin, M.; Schmeißer, A.; Braun-Dullaeus, R.; Berg, P. Hemodynamic Assessment of the Pathological Left Ventricle Function under Rest and Exercise Conditions. *Fluids* **2023**, *8*, 71. [CrossRef]
12. Al Baraikan, A.; Czechowicz, K.; Morris, P.D.; Halliday, I.; Gosling, R.C.; Gunn, J.P.; Narracott, A.J.; Williams, G.; Garg, P.; Malawski, M.; et al. Modelling The Hemodynamics of Coronary Ischemia. *Fluids* **2023**, *8*, 159. [CrossRef]

13. Ali, S.I.; Patton, D.; Myers, K.A.; Garcia, J. Repaired Tetralogy of Fallot Pressure Assessment: Insights from 4D-Flow Pressure Mapping. *Fluids* **2023**, *8*, 196. [CrossRef]
14. Yu, H.; Khan, M.; Wu, H.; Zhang, C.; Du, X.; Chen, R.; Fang, X.; Long, J.; Sawchuk, A.P. Inlet and Outlet Boundary Conditions and Uncertainty Quantification in Volumetric Lattice Boltzmann Method for Image-Based Computational Hemodynamics. *Fluids* **2022**, *7*, 30. [CrossRef]
15. Hong, W.; Yu, H.; Chen, J.; Talamantes, J.; Rollins, D.M.; Fang, X.; Long, J.; Xu, C.; Sawchuk, A.P. A Mock Circulation Loop to Characterize In Vitro Hemodynamics in Human Systemic Arteries with Stenosis. *Fluids* **2023**, *8*, 198. [CrossRef]
16. Yukhnev, A.; Tikhomolova, L.; Gataulin, Y.; Marinova, A.; Smirnov, E.; Vrabiy, A.; Suprunovich, A.; Khubulava, G. V Flow Measurements of Pulsatile Flow in Femoral-Popliteal Bypass Proximal Anastomosis Compared with CFD Simulation. *Fluids* **2024**, *9*, 64. [CrossRef]
17. LaDisa, J.F.; Ghorbannia, A.; Marks, D.S.; Mason, P.; Otake, H. Advancements and Opportunities in Characterizing Patient-Specific Wall Shear Stress Imposed by Coronary Artery Stenting. *Fluids* **2022**, *7*, 325. [CrossRef]
18. Zhang, X.; Gomez-Paz, J.; Chen, X.; McDonough, J.M.; Islam, M.; Andreopoulos, Y.; Zhu, L.; Yu, H. Volumetric lattice Boltzmann method for wall stresses of image-based pulsatile flows. *Sci. Rep.* **2022**, *12*, 1697. [CrossRef]
19. An, S.; Yu, H.; Yao, J. GPU-accelerated volumetric lattice Boltzmann method for porous media flow. *J. Pet. Sci. Eng.* **2017**, *156*, 546–552. [CrossRef]
20. An, S.; Yu, H.; Wang, Z.; Kapadia, B.; Yao, J. Unified mesoscopic modeling and GPU-accelerated computational method for image-based pore-scale porous media flows. *Int. J. Heat Mass Transf.* **2017**, *115*, 1192–1202. [CrossRef]
21. Wang, Z.; Zhao, Y.; Sawchuck, A.P.; Dalsing, M.C.; Yu, H. GPU acceleration of Volumetric Lattice Boltzmann Method for patient-specific computational hemodynamics. *Comput. Fluids* **2015**, *115*, 192–200. [CrossRef]
22. Yu, H.; Chen, X.; Wang, Z.; Deep, D.; Lima, E.; Zhao, Y.; Teague, S.D. Mass-conserved volumetric lattice Boltzmann method for complex flows with willfully moving boundaries. *Phys. Rev. E* **2014**, *89*, 063304. [CrossRef]
23. An, S.; Yu, H.; Islam, M.; Zhang, X.; Zhan, Y.; Olivieri, J.J.; Ambati, J.; Yao, J.; Gelfand, B.D. Effects of donor-specific microvascular anatomy on hemodynamic perfusion in human choriocapillaris. *Sci. Rep.* **2023**, *13*, 22666. [CrossRef]
24. Yu, H.; Khan, M.; Wu, H.; Du, X.; Chen, R.; Rollins, D.M.; Fang, X.; Long, J.; Xu, C.; Sawchuk, A.P. A new noninvasive and patient-specific hemodynamic index for the severity of renal stenosis and outcome of interventional. *Int. J. Numer. Methods Biomed. Eng.* **2022**, *38*, e3611. [CrossRef]

fluids

MDPI

Article

Fluid Flow and Structural Numerical Analysis of a Cerebral Aneurysm Model

Maria Sabrina Souza [1], Andrews Souza [2], Violeta Carvalho [2,3], Senhorinha Teixeira [3], Carla S. Fernandes [1], Rui Lima [2,4,*] and João Ribeiro [1,5]

1 Instituto Politécnico de Bragança, ESTiG, C. Sta. Apolónia, 5300-253 Bragança, Portugal; sabrinasouza680@gmail.com (M.S.S.); cveiga@ipb.pt (C.S.F.); jribeiro@ipb.pt (J.R.)
2 Mechanical Engineering and Resource Sustainability Center (MEtRICs), University of Minho, Campus de Azurém, 4800-058 Guimarães, Portugal; andrews.va.souza@alunos.ipb.pt (A.S.); violeta.carvalho@dps.uminho.pt (V.C.)
3 ALGORITMI Research Centre, University of Minho, Campus de Azurém, 4800-058 Guimarães, Portugal; st@dps.uminho.pt
4 Transport Phenomena Research Center (CEFT), Faculdade de Engenharia da Universidade do Porto (FEUP), R. Dr. Roberto Frias, 4200-465 Porto, Portugal
5 Centro de Investigação de Montanha (CIMO), Instituto Politécnico de Bragança, 5300-252 Bragança, Portugal
* Correspondence: rl@dem.uminho.pt

Abstract: Intracranial aneurysms (IA) are dilations of the cerebral arteries and, in most cases, have no symptoms. However, it is a very serious pathology, with a high mortality rate after rupture. Several studies have been focused only on the hemodynamics of the flow within the IA. However, besides the effect of the flow, the development and rupture of the IA are also associated with a combination of other factors such as the wall mechanical behavior. Thus, the objective of this work was to analyze, in addition to the flow behavior, the biomechanical behavior of the aneurysm wall. For this, CFD simulations were performed for different Reynolds numbers (1, 100, 500 and 1000) and for two different rheological models (Newtonian and Carreau). Subsequently, the pressure values of the fluid simulations were exported to the structural simulations in order to qualitatively observe the deformations, strains, normal stresses and shear stress generated in the channel wall. For the structural simulations, a hyperelastic constitutive model (5-parameter Mooney–Rivlin) was used. The results show that with the increase in the Reynolds number (Re), the recirculation phenomenon is more pronounced, which is not seen for Re = 1. The higher the Re, the higher the strain, displacement, normal and shear stresses values.

Keywords: CFD; structural simulation; intracranial aneurysm; cerebral aneurysm; hemodynamics

Citation: Souza, M.S.; Souza, A.; Carvalho, V.; Teixeira, S.; Fernandes, C.S.; Lima, R.; Ribeiro, J. Fluid Flow and Structural Numerical Analysis of a Cerebral Aneurysm Model. *Fluids* **2022**, 7, 100. https://doi.org/10.3390/fluids7030100

Academic Editor: Huidan (Whitney) Yu

Received: 3 January 2022
Accepted: 28 February 2022
Published: 7 March 2022

1. Introduction

Intracranial aneurysms (IA) are local dilations of the intracranial arteries; their occurrence in the world population varies from 0.5% to 6% [1,2]. About 0.25% of these IAs rupture and present subarachnoid hemorrhage [3]. The mortality rate from these hemorrhages reaches 60% [4]. The literature indicates how a combination of multiple factors can lead to the onset and development of IA: genetic diseases, fungal infection, hypertension, cigarette smoke inhalation, alcohol intake and just aging [5,6]. These factors result in an alteration of the original arterial wall properties and flow behavior. Factors such as weakening of the vessel wall, hemodynamic changes and inflammatory processes can also trigger the appearance of the aneurysm, which can become a stable dilatation or rupture [7–9]. The interaction of these factors and complex processes that lead to aneurysm formation still need to be better understood. Thus, it is essential to study the biomechanical behavior of the blood vessel together with the behavior of blood flow to determine regions more susceptible to rupture.

An important characteristic of blood vessels is the existence of soft tissues, which have a behavior called hyperelasticity [10]. Hyperelastic materials have a mechanical behavior characterized by a high deformation before reaching the breaking strength [11,12]. Due to the risks involved, it is very difficult to analyze in vivo the biomechanical behavior of blood vessels, particularly in the region of the aneurysm [13]. However, it is possible to manufacture flow phantom biomodels for experimental tests with compatible properties and geometries [14,15]. Flow phantoms basically consist of models that mimic vascular geometry for flow analysis (rigid flow phantoms) and flow analysis and arterial wall deformation (flexible flow phantoms). These biomodels were produced in silicone rubber, as they allow optical access to the flow and also reproduce the physiological compliance of the artery [16,17]. In its manufacture, materials whose biomechanical behavior is close to the real biological material were used [18]. A frequently used material is polydimethylsiloxane (PDMS) [19,20]. PDMS has been widely used in different kinds of research fields [21–27] due to its excellent properties and ability to replicate reliable geometries, in addition to its high flexibility, low glass transition temperature, good chemical resistance, excellent thermal stability and biocompatibility [28].

In addition to the experimental approach, in recent years, there has been an increase in numerical hemodynamic studies using the computational fluid dynamics tool CFD [29–34]. This tool has been effective in investigating the relationship of vascular hemodynamics with the mechanisms of initiation, growth and rupture [9,35–37]. CFD is also used to evaluate endovascular treatments [38,39] and validate new endovascular devices [40,41]. Most of these studies use Newtonian flow models; however, blood flow does not have this type of behavior, and there is no consensus on which type of viscosity model should be adopted to model blood flow in IA [37,42]. Some studies consider that blood viscosity follows a Newtonian behavior in cerebral arteries, as shear rates are considered to be higher than the range necessary for non-Newtonian properties to become effective [37,43]. Other studies report that, for medium and large arteries, blood viscosity is practically constant, which justifies, in blood flow simulations, considering it as Newtonian [44]. However, some studies claim that Newtonian viscosity results in excessive predictions of aneurysm wall shear stress, which could compromise the effectiveness of predicting the risk of rupture [37]. Given these controversies about which type of viscosity to use, it is important to compare the type of regime for each application. In this study, numerical simulations were performed using different Reynold numbers (Re), and comparisons were made between Newtonian and non-Newtonian flow (Carreau fluid model).

Although the numerical study using the CFD tool has achieved some advances in the understanding of IAs, they have some limitations, most of these studies use rigid vessel wall aneurysm models, and in reality, there is an interaction of blood flow with the vessel wall. Endothelial cells present in vessel walls detect blood flow through a variety of mechanical cell surfaces and intracellular sensors, any change in blood flow. Endothelial cells initiate different biological processes [45,46] that lead to a dynamic cycle of wall degeneration and subsequent remodeling. Studies have also shown that endothelial cells respond to WSS [47]. Thus, recent studies reported the importance of taking into account the interaction between blood flow and the vessel wall in the model of IA formation, growth and rupture [18,48]. However, there are few studies that precisely characterize this phenomenon of interaction; as such, in order to fill the little information reported in scientific publications, the aim of this work was to numerically determine the state of displacement and stress on the IA wall caused by different flow conditions. Thus, in order to characterize the mechanical behavior of biological tissues under different flow conditions, numerical simulations were performed using the Mooney–Rivlin constitutive model that mathematically describes hyperelastic materials [49].

In this work, numerical simulations were performed by combining both fluid flow and structural analysis [50,51] using Ansys® software. The fluid flow analysis simulates the blood flow inside the vessels, and the resulting pressure in the vessel wall was analyzed for mechanical deformations occurring in the vessel wall as well as in the aneurysm. The existence or absence of the recirculation phenomenon in different Reynolds numbers (Re) was studied, and it was possible to visualize and analyze the development of the fluid recirculation phenomenon inside the aneurysm. Finally, the numerical characterization of the biomechanical behavior of aneurysms was carried out.

2. Problem Description

In the present study, numerical simulations of blood flow and biomechanical wall's behavior of an in vitro channel with aneurysm, represented in Figure 1, were performed, resorting to Ansys® software. The geometry and dimensions of the idealized IA model are based on the work developed by Parlea et al. [52].

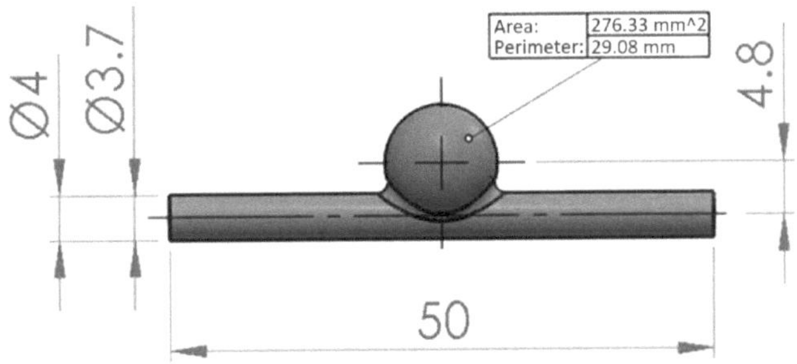

Figure 1. Studied IA model.

Initially, fluid dynamics simulations were performed in order to observe hemodynamic patterns such as flow recirculation inside the aneurysm and pressure gradients for different flow velocities. The fluid flow analysis also allows obtaining the pressure distribution on the channel walls, which are later used as boundary conditions in structural simulations. Structural simulations were carried out in order to analyze the biomechanical behavior of the wall through the values of displacement, strain and stress.

2.1. Fluid Flow Simulations

For flow simulations, both continuity and Navier–Stokes equations were solved using the finite volume commercial software Fluent, Ansys 2020 R2.

The steady and incompressible regime and rigid walls were prescribed. Simulations were carried out considering and discarding the non-Newtonian properties of blood, which permits evaluating the impact of non-Newtonian characteristics of blood in the properties of studied flows. As constitutive equations, Newtonian and Carreau models were used, the last one being described by Equation (1):

$$\eta = \eta_\infty + (\eta_0 - \eta_\infty)\left[1 + (\lambda\dot{\gamma})^2\right]^{(n-1)/2} \quad (1)$$

where η is the viscosity, η_0 is the zero viscosity, η_∞ is the infinite viscosity, λ is the characteristic time (the inverse of the shear rate for which pseudoplastic behavior begins), n is the power index and $\dot{\gamma}$ is the shear rate. For blood, parameters of Equation (1) assume the values η_0 = 0.056 Pa·s, η_∞ = 0.00345 Pa·s, λ = 3.313 s and n = 0.3568 [26]. In the case of the Newtonian model, blood viscosity considered was 0.00345 Pa s and the density 1050 kg/m^3 [26].

2.1.1. Geometrical Domain and Mesh

Initially, the idealized geometry of the aneurysm based on other studies [11] was drawn in the SolidWorks® CAD software, converted to Parasolid format (.x_t) and then exported to Ansys®.

For fluid flow simulation, only the fluid domain was considered (Figure 1). The analyzed geometry is a cylindrical channel with 3.7 mm in diameter and 50 mm in length, with a central sphere representing an artery and an aneurysm idealized saccular, respectively.

A mesh consisting predominantly of hexahedral elements was generated, with a total of 170,820 elements and 174,332 nodes (Figure 2). Before choosing the mesh, a preliminary mesh test was performed. A mesh with twice the number of elements was created, keeping the same spatial distribution, and then the simulations were performed. The maximum velocities values obtained by both meshes were observed, and a difference of less than 1% was found in the results, which shows that the results are independent of the mesh used.

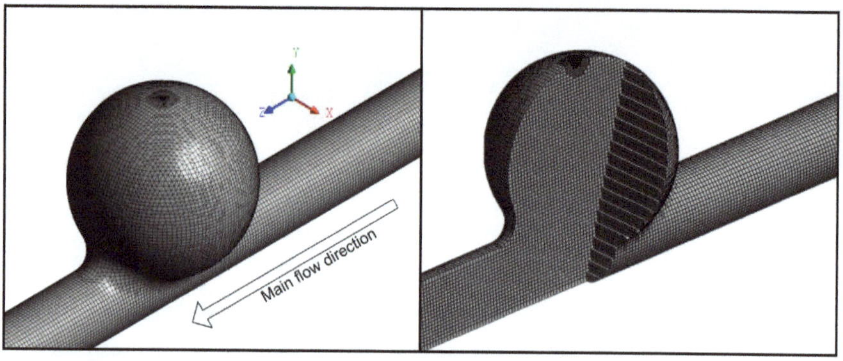

Figure 2. Detail of the mesh used in CFD calculations.

2.1.2. Boundary Conditions and Solver

The simulations were carried out considering atmospheric pressure as the operating pressure. As boundary conditions, four distinct mean velocities were imposed in the inlet of the channel: 0.9 mm/s, 88.8 mm/s, 444 mm/s and 888 mm/s. These velocities were chosen in order to obtain the corresponding Reynolds numbers (Re): 1, 100, 500 and 1000, respectively, for Newtonian flows. These Re values were chosen in order to verify the occurrence or not of fluid recirculation and if this phenomenon would increase according to the growth of Re values. At the structural level, observe whether the deformations were proportional to the increase in Re.

The wall of the channel was assumed rigid, and as the wall velocity is zero, the non-slip condition was applied at the wall. Additionally, the gauge pressure (0 Pa) was imposed in the outlet of the channel.

The pressure-based solver was used to solve numerical integrations, as this is the most suitable for problems with incompressible fluids. Speed–pressure coupling was performed using the PISO scheme. The spatial discretization of the pressure equation was performed with the PRESTO! approximation, while the moment equations were discretized with the QUICK scheme.

2.1.3. Model Validation

In order to validate CFD calculations, simulations were carried out for the conditions referred before in a channel similar to the one presented above but lengthier in the section before the aneurysm to guarantee a fully developed flow in that region. Two tests in that section were performed. First, velocity profiles were compared with the analytical solution for a Newtonian fully developed flow in a circular pipe which can be determined by [53]:

$$v(r) = 2u\left(1 - \left(\frac{r}{R}\right)^2\right) \tag{2}$$

where v is the velocity, u is the mean velocity, R is the radius of the channel and r is the radial position.

In Figure 3, it is possible to verify the good agreement between numerical results and analytic profile (mean relative error range between 2.4% and 2.6% for the three velocities), which concludes that the numerical model predicts the local properties of the studied flow well.

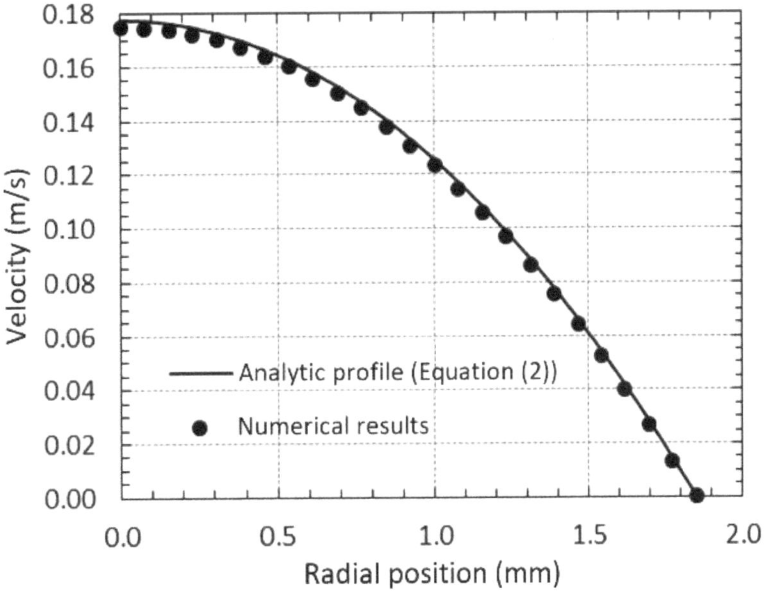

Figure 3. Velocity profile for fully developed Newtonian flow in the region before the aneurysm, u = 88.8 mm/s.

The second test consists of the comparison between the obtained pressure drops and predictions from the Haggen–Poiseuille equation [54]:

$$\Delta P = \frac{8\eta L u}{R^2} \Leftrightarrow \frac{\Delta P}{L} = \frac{8\eta u}{R^2} \quad (3)$$

where ΔP represents the pressure drop and L is the channel length. Once again, a good agreement between pressure drop obtained by Equation (3) and numerical results was found, as shown in Table 1.

Table 1. Pressure drop for fully developed Newtonian flows in the region before the aneurysm.

Re	$(\Delta P/L)_{\text{Equation (3)}}$ (Pa/m)	$(\Delta P/L)_{\text{numerical}}$ (Pa/m)	Relative Error (%)
100	716.11	713.79	0.33
500	3580.54	3575.83	0.13
1000	7161.08	7364.60	2.84

2.2. Structural Simulations

The structural simulations were performed using the static structure of the Ansys software. Initially, new material was inserted into the Engineering Data. The material considered was Polydimethylsiloxane (PDMS) Sylgard 184, as it is a hyperelastic material, as are the arteries. A stress/strain curve of the PDMS obtained by tensile testing was then inserted into the properties and approximated by a hyperelastic model previously defined in Ansys, Mooney–Rivlin parameter 5 model, which is a constitutive model adequate for this kind of simulation [55]. The Mooney–Rivlin constitutive law is described by Equation (4):

$$\Psi = C_{10}(I_1 - 3) + C_{01}(I_2 - 3) + C_{20}(I_1 - 3)^2 + C_{11}(I_1 - 3)(I_2 - 3) + C_{02}(I_2 - 3)^2 \quad (4)$$

where Ψ is the strain energy density function; I_1 and I_2 are the strain constants; and C_{10}, C_{01}, C_{20}, C_{11} and C_{02} are the material parameters, which are determined using the data obtained experimentally by tensile test.

In Figure 4, the approximation of the mentioned curve can be seen.

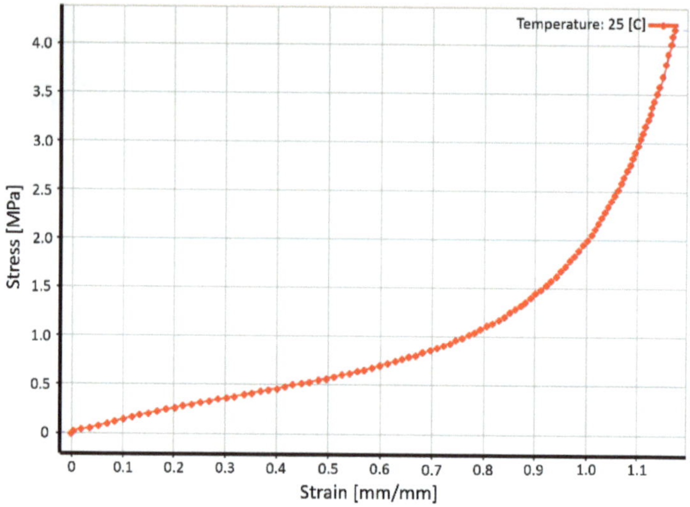

Figure 4. Approximation of the curve stress/strain by the Mooney–Rivlin 5 Parameter model.

The geometry for this simulation represents only the channel wall that has 0.15 mm of thickness.

As the geometry for the flow simulations, this geometry was previously drawn in Solidworks and later exported to Ansys. A tetrahedral mesh with an element size of 0.5 mm was generated, with a total of 48,931 nodes and 24,370 elements. This mesh can be seen in Figure 5a. Next, the boundary conditions were defined, fixed support was applied to the sides of the model, and then the pressure imported from fluid simulations in Fluent was applied to the inner wall of the channel. Figure 5b illustrates the pressure applied to the channel wall.

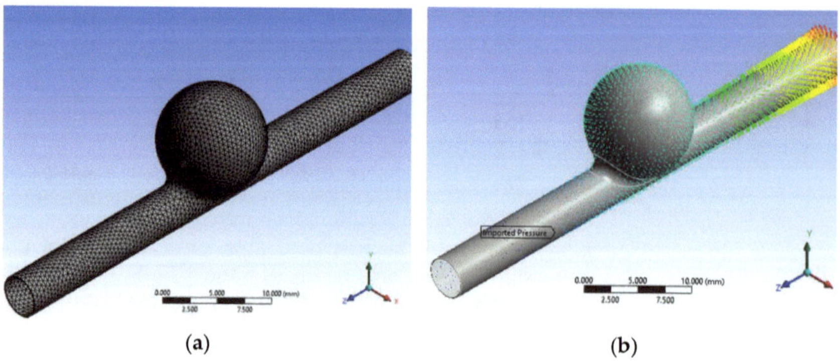

Figure 5. (**a**) Mesh used for the outer part of the channel (wall); (**b**) pressure imported from fluent for the structural static.

Finally, the software was configured to calculate the solutions for displacement, strain, the normal and shear stress.

3. Results

In this section, the results of the fluid flow simulations for different velocities and the results of the structural simulations are presented.

3.1. Fluid Flow Analysis

Fluid flow simulations were carried out considering the positive z-axis as the main flow direction, as shown in Figure 2.

3.1.1. Flow Patterns

The hemodynamic pattern was analyzed, and similarities were found between all the studied flows (distinct inlet velocities and different rheological models). Figure 6 illustrates the flow behavior for the four Re analyzed in the central plane at x position (x = 0). It is clear that for Re = 1, the flow makes a smooth curvature, bypassing the geometry of the aneurysm wall; in this case, there is no recirculation. However, for the other Re, which has much higher flow velocities, recirculation inside the aneurysm is noted, which is more pronounced as the number of Re increases.

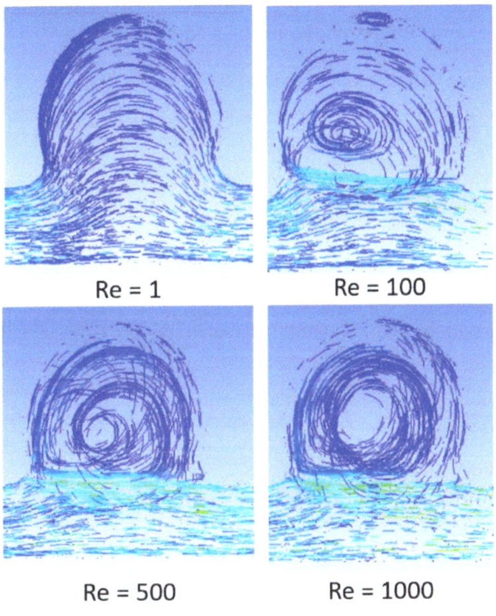

Figure 6. Streamlines for different Re.

By observing the streamlines (Figure 7), two distinct sections can be considered concerning flow behavior. In a region close to the inlet and outlet of the channel, the typical behavior of laminar flow in a circular pipe was observed. However, when we move closer to the bulge, the flow suffers a deviation in the direction of protuberance and vortex formation inside it was observed. The last phenomenon is more pronounced for the higher velocities, as already shown.

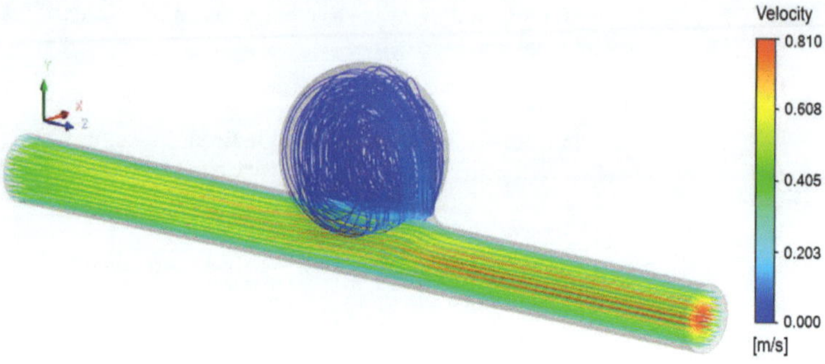

Figure 7. Streamlines for Newtonian flow and Re = 500.

In Figure 8, it is also possible to verify the behavior described before. Lines 1 and 3 exhibits the typical shape of velocity profiles in a circular pipe, that is, symmetric profiles about the center of the channel. The shape of these two profiles differs since the region before the aneurysm is clearly a region of developing flow, and after the aneurysm, the parabolic profile appears since flow is closer to a developed flow. This behavior is the reason for using a lengthier channel for the model validation. It should be noted that for lower velocity (88.8 mm/s), the difference presented and discussed in this paragraph was not found, and two parabolic profiles were observed.

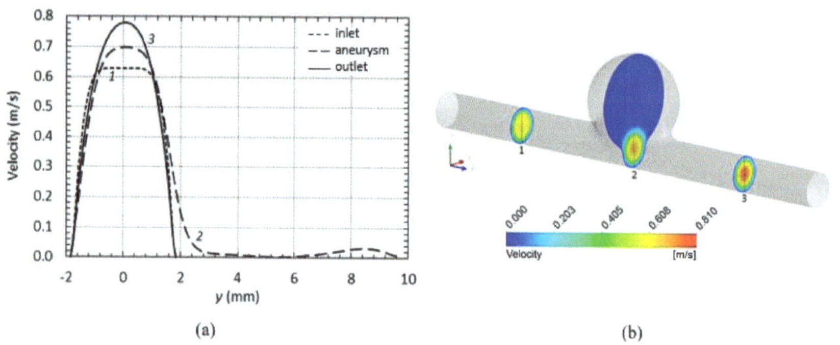

Figure 8. (**a**) Velocity profiles in three different regions of the channel for Newtonian model and u = 444 mm/s. (**b**) Three cutting planes corresponding to the curves of (**a**), respectively.

The profile represented by line 2 in Figure 8 illustrates well the lower velocities in the aneurysm. Moreover, it is possible to verify a little shift of the maximum velocity in the bulge direction.

By comparing the obtained results for the same velocity inlet, considering and discarding the non-Newtonian properties of the blood, the differences observed in velocity's magnitude could be neglected, which is the reason why the authors decide to present only an example of profiles and streamlines.

3.1.2. Pressure

As observed for velocity, pressure distribution was also revealed to be almost independent of non-Newtonian properties of the blood, as shown in Figure 9b. However, pressure is always higher when blood rheology is described by the Carreau model.

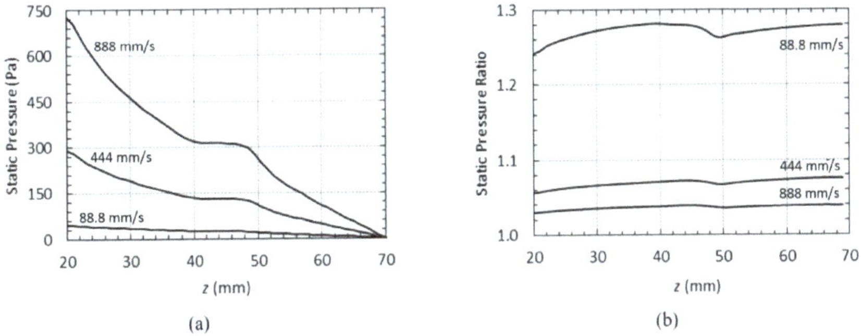

Figure 9. (**a**) Static pressure along the channel for Carreau model and different velocities. (**b**) Ratio between static pressure for Carreau and Newtonian models.

In Figure 9a, static pressure along the channel is presented for the Carreau model. As expected, pressure decrease in the flow direction, except in the region of the aneurysm where pressure remains almost constant due to the increase of the cross-section of the channel and velocity close to zero.

Pressure drop for all studied flows was determined and presented in Table 2. Analyzing the ratio between ΔP obtained, using Carreau model and Newtonian model as constitutive models, a reduction in this ratio with the increase in Re, although they are all very close to 1.

Table 2. Pressure drop for both rheological models.

Re	$\Delta P_{\text{Carreau}}$ (Pa)	$\Delta P_{\text{Newtonian}}$ (Pa)	ΔP_{ratio}
100	46.73	37.64	1.24
500	290.41	274.81	1.06
1000	726.12	704.84	1.03

The influence of mean velocity in the ratios of pressure drop, as well as pressure, could be explained by blood viscosity obtained with the Carreau model. As shown in Figure 10, blood viscosity increase with the decrease in mean velocity and assume values higher than the Newtonian viscosity (0.00345 Pa.s), which leads to higher pressure drops.

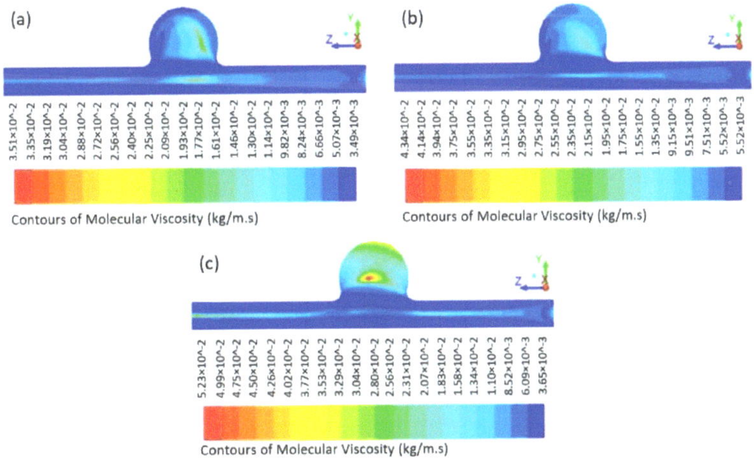

Figure 10. Viscosity in the central plane (x = 0) for non-Newtonian flows and distinct velocities. (**a**) u = 888 mm/s; (**b**) u = 444 mm/s; (**c**) u = 88.8 mm/s.

As referred before, flow simulations were performed in order to find the blood pressure in the wall (Figure 11) and use it as a boundary condition in the wall structural analysis.

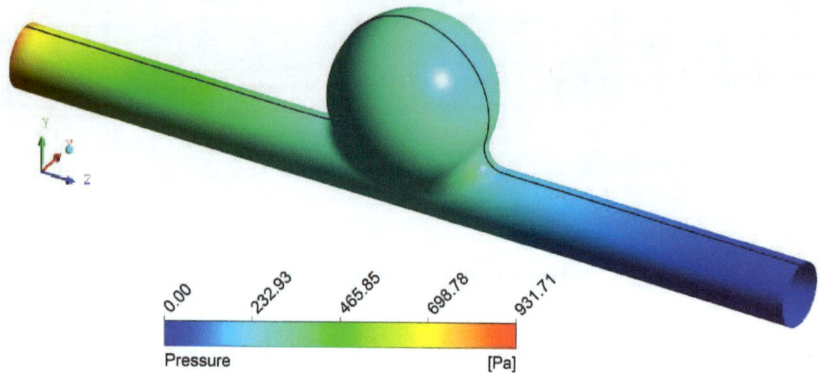

Figure 11. Wall pressure distribution for Carreau model and u = 888 mm/s.

In Figure 12, static pressure obtained along the black line of Figure 11 is presented. Once again, the decrease in pressure in the main flow direction was observed, except in the region of the aneurysm. Here the pressure remains constant along with the bulge and exhibits pronounced increases/decreases in the transition of the circular channel to the aneurysm.

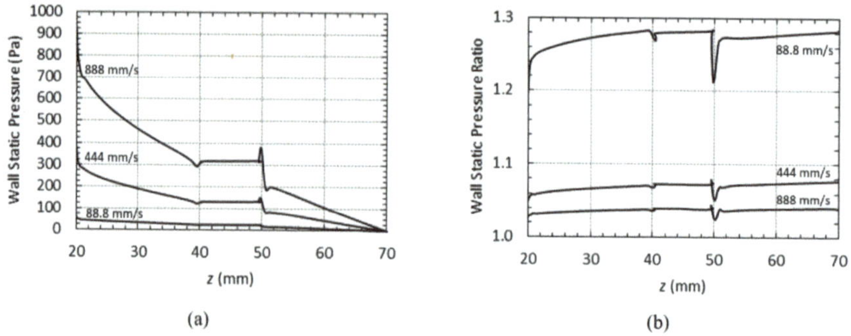

Figure 12. (**a**) Static pressure along the wall for Carreau model and different velocities. (**b**) Ratio between wall's static pressure for Carreau and Newtonian models.

The minimums observed in Figure 12 are in the transition from the circular duct to the bulge, where blood viscosity obtained with the Carreau model moves closer to the Newtonian viscosity, as could be verified in Figure 10.

3.2. Structural Analysis

The results of the channel wall structural simulations are presented below.

By observing Figure 13 below, it is possible to observe that the displacement along the channel occurs in a similar way; in a qualitative perspective, this effect can also be justified by the fixation support applied on the sides of the geometry during the modeling of the problem, which causes that the total displacement values are null at the ends, but do not affect the results in the aneurysm. Displacement values increase as the Reynolds number increases.

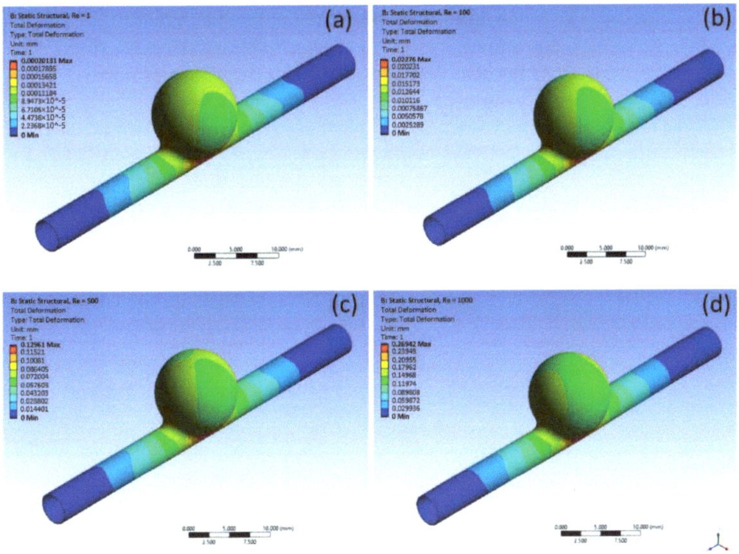

Figure 13. Total displacement: (**a**) Re = 1; (**b**) Re = 100; (**c**) Re = 500; (**d**) Re = 1000.

However, it is clear that at the dome of the aneurysm, there is a change in the behavior of total displacement for each Re. The total displacement becomes more and more intense on the sides of the aneurysm (areas of great effort) and is reduced at the top of the aneurysm.

Figure 14 shows the strain values. For these results, it is notorious both in the channel and in the dome of the aneurysm the increase of strain as the inflow velocity is increased. The highest strain values are concentrated in the transition zones, from channel to aneurysm and from the aneurysm to channel.

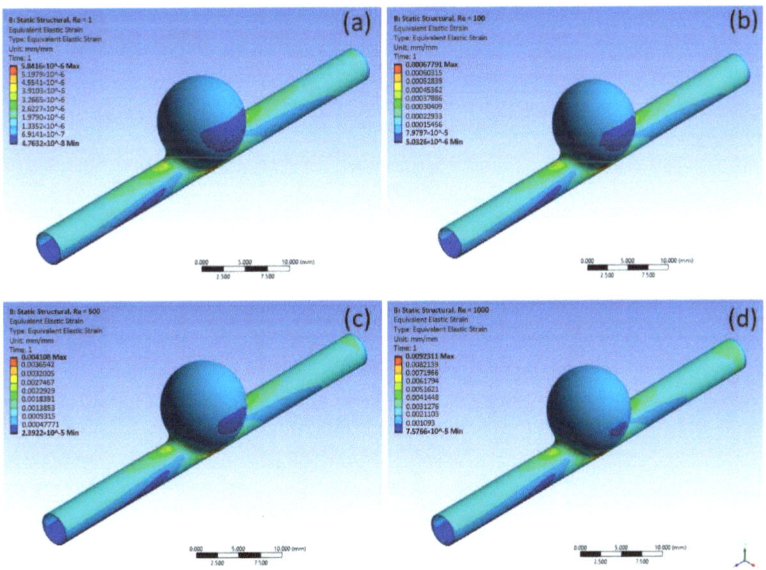

Figure 14. Equivalent Elastic Strain: (**a**) Re = 1; (**b**) Re = 100; (**c**) Re = 500; (**d**) Re = 1000.

In addition to displacement and strain, normal stress and shear stress values were also computed. Next, the images of shear stress (Figure 15) and normal stress (Figure 16) for the highest Re (Re = 1000) and the graph of normal stress for all analyzed Re (Figure 17) are presented.

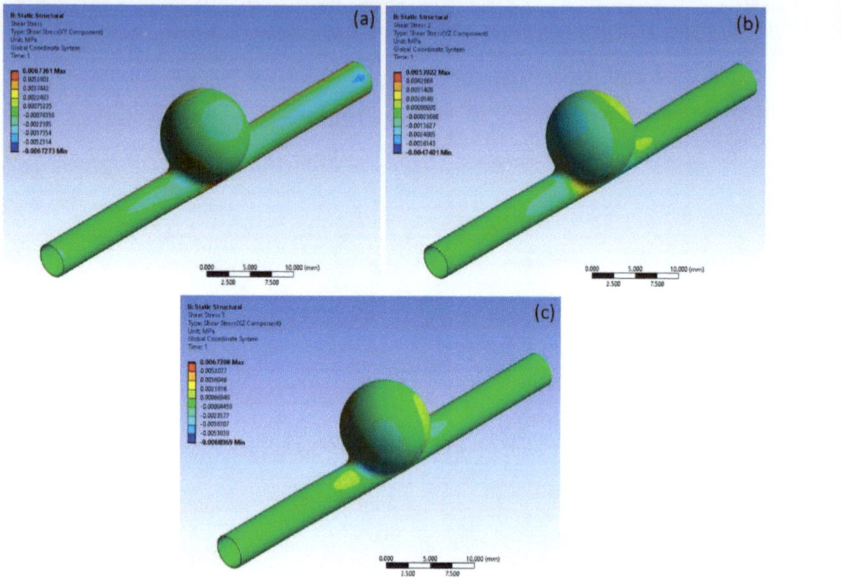

Figure 15. Shear stress to Re = 1000. (**a**) XY, (**b**) YZ and (**c**) XZ component.

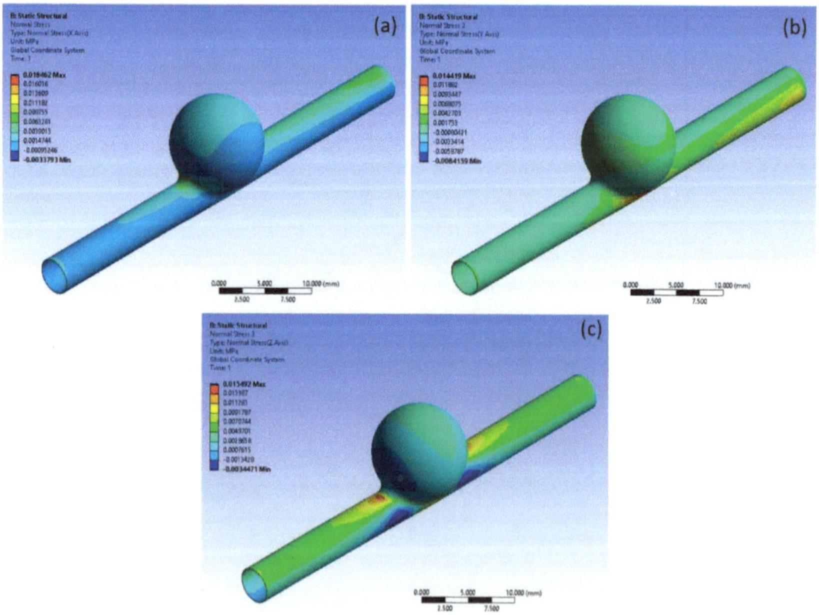

Figure 16. Normal Stress to Re = 1000. (**a**) X, (**b**) Y and (**c**) Z axis.

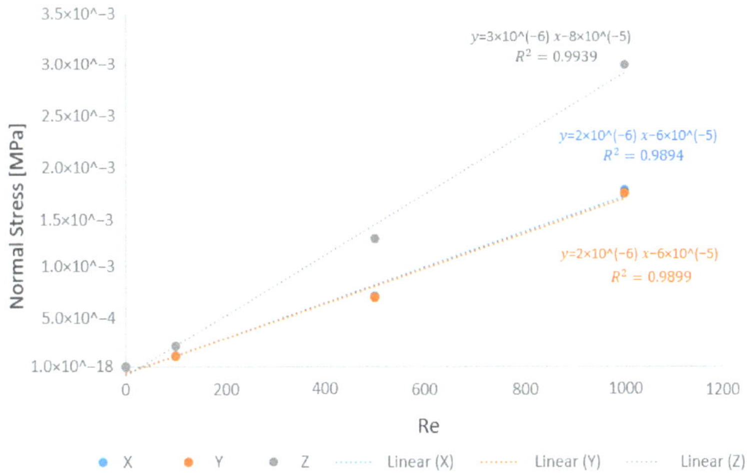

Figure 17. Normal stress as a function of Re.

The behavior of the shear stress changes according to the coordinates, with the change close to the aneurysm being more noticeable (Figure 15). For shear stresses analyzed in component XY, the highest stress values are in the channel zone below the aneurysm dome. In the YZ component, the highest shear stress values are found in the channel along the curvature of the aneurysm outlet. Additionally, in component XZ, the highest values are near the neck of the aneurysm. By analyzing the mean of the values in each direction, we have the highest mean shear stress in the YZ component, which is the flow direction within the channel.

Figure 16 shows the results for normal stress. As can be seen, the normal stress is more accentuated in the Z direction. The higher stress values are close to the aneurysm entrance and exit. When we analyze the graph in Figure 17, with the mean stress values, it is clear that the normal stress in Z, for all analyzed Re, has a greater relevance (higher values) when compared to the other directions (X and Y).

4. Conclusions

The present work aimed to observe the flow behavior in an ideal intracranial aneurysm using a Newtonian blood viscosity model and a non-Newtonian model (Carreau model). In addition, the biomechanical behavior of the wall was analyzed.

By analyzing the distribution of velocities in the two rheological models, it was observed that there was no significant difference in the magnitude of velocities. Regarding the pressure distribution, despite the values being higher in the blood rheology described by the Carreau model, this also proved to be almost independent of the non-Newtonian properties of the blood. Hence, this result indicates that assuming the characteristics of blood, being Newtonian fluid, is acceptable.

In order to obtain more insights into the hemodynamic behavior, different Re was compared, from Re = 1 (without physiological relevance) to Re = 1000, in order to observe the existence or not of recirculations inside the IA. With the obtained results, it was evident that the increase in the number of Re makes the phenomenon of blood recirculation more noticeable. There is only the formation of a vortex in the center of the aneurysm, which becomes increasingly accentuated closer to the wall. Nevertheless, there is no formation of additional vortex regions as the Re number increases. For Re = 1, the phenomenon of blood recirculation does not appear due to the extremely low flow velocities at the channel inlet.

Regarding the structural analysis, it was observed that the displacement was maximum on the sides of the aneurysm and in the passage from the channel to the aneurysm and from the aneurysm to the channel, which are the areas where there is an abrupt change in geometry. Likewise, an increase in the Re leads to an increment of both normal stress and shear stress values. However, for Re = 1, the stress was extremely low, with values close to zero.

The normal stresses have shown more accentuated values at the Z direction, and the shear stresses presented the highest values at the XY plane. Hence, the normal stresses were higher at the transition region from the vessel to the aneurysm. Additionally, the shear stresses were higher at the sides of the aneurysm. Note that the representation of the type of distribution of the normal and shear stresses was similar to the tested Reynolds numbers.

Although in the present work, a simplified vessel model was used, this does not withdraw the validity of results since the form and geometry of the blood vessel vary from patient to patient. The use of a more realistic vessel, for example, constructed from 2D medical images obtained by computed tomography, gives more precise information regarding this patient. In the near future, it is intended to perform this comparison.

Author Contributions: Conceptualization, M.S.S., A.S., J.R. and C.S.F.; methodology, M.S.S.; software M.S.S.; validation, J.R., A.S. and C.S.F.; formal analysis, A.S.; investigation, A.S.; resources, J.R., C.S.F.; data curation, M.S.S.; writing—original draft preparation, M.S.S., A.S. and V.C.; writing—review and editing, J.R., S.T. and R.L.; visualization A.S. and V.C.; supervision, J.R. and R.L.; project administration, J.R.; funding acquisition J.R. and R.L. All authors have read and agreed to the published version of the manuscript.

Funding: The authors acknowledge the financial support from the project EXPL/EME-EME/0732/2021, funded by the NORTE 2020 Portugal Regional Operational Programme, under the PORTUGAL 2020 Partnership Agreement, through the European Regional Development Fund (FEDER) and by Fundação para a Ciência e Tecnologia (FCT). This work was also supported by Fundação para a Ciência e a Tecnologia (FCT) under the strategic grants UIDB/04077/2020, UIDB/04436/2020, UIDB/00319/2020 and UIDB/00532/2020. Andrews Souza and Violeta Carvalho also acknowledge the financial support by FCT through the individual research grants 2021.07961.BD and UI/BD/151028/2021, respectively.

Institutional Review Board Statement: Not applicable.

Informed Consent Statement: Not applicable.

Data Availability Statement: Not applicable.

Conflicts of Interest: The authors declare no conflict of interest.

References

1. Johnston, S.C.; Higashida, R.T.; Barrow, D.L.; Caplan, L.R.; Dion, J.E.; Hademenos, G.; Hopkins, L.N.; Molyneux, A.; Rosenwasser, R.H.; Vinuela, F.; et al. Recommendations for the endovascular treatment of intracranial aneurysms: A statement for healthcare professionals from the Committee on Cerebrovascular Imaging of the American Heart Association Council on Cardiovascular Radiology. *Stroke* **2002**, *33*, 2536–2544. [CrossRef] [PubMed]
2. Schievink, W.I. Intracranial Aneurysms. *N. Engl. J. Med.* **1997**, *336*, 28–40. [CrossRef] [PubMed]
3. Thompson, B.G.; Brown, R.D., Jr.; Amin-Hanjani, S.; Broderick, J.P.; Crokroft, K.M.; Connolly, E.S., Jr.; Duckwiler, G.R.; Harris, C.C.; Howard, V.J.; Johnston, S.C.C.; et al. Guidelines for the Management of Patients With Unruptured Intracranial Aneurysms: A Guideline for Healthcare Professionals From the American Heart Association/American Stroke Association. *Stroke* **2015**, *46*, 2368–2400. [CrossRef] [PubMed]
4. Amenta, P.S.; Yadla, S.; Campbell, P.G.; Maltenfort, M.G.; Dey, S.; Ghosh, S.; Ali, M.S.; Jallo, J.I.; Tjoumakaris, S.I.; Gonzalez, L.F.; et al. Analysis of nonmodifiable risk factors for intracranial aneurysm rupture in a large, retrospective cohort. *Neurosurgery* **2012**, *70*, 693–701. [CrossRef] [PubMed]
5. Lasheras, J.C. The Biomechanics of Arterial Aneurysms. *Annu. Rev. Fluid Mech.* **2007**, *39*, 293–319. [CrossRef]
6. Tromp, G.; Weinsheimer, S.; Ronkainen, A.; Kuivaniemi, H. Molecular basis and genetic predisposition to intracranial aneurysm. *Ann. Med.* **2014**, *46*, 597–606. [CrossRef]
7. Baratchi, S.; Khoshmanesh, K.; Woodman, O.L.; Potocnik, S.; Peter, K.; McIntyre, P. Molecular Sensors of Blood Flow in Endothelial Cells. *Trends Mol. Med.* **2017**, *23*, 850–868. [CrossRef]
8. Hoskins, P.R.; Lawford, P.V.; Doyle, B.J. *Cardiovascular Biomechanics*; Springer: Cham, Switzerland, 2017.
9. Cebral, J.R.; Raschi, M. Suggested connections between risk factors of intracranial aneurysms: A review. *Ann. Biomed. Eng.* **2013**, *41*, 1366–1383. [CrossRef]

10. Gasser, T.C.; Ogden, R.W.; Holzapfel, G.A. Hyperelastic modelling of arterial layers with distributed collagen fibre orientations. *J. R. Soc. Interface* **2006**, *3*, 15–35. [CrossRef]
11. Victor, A.; Ribeiro, J.; Araújo, F.F. Study of PDMS characterization and its applications in biomedicine: A review. *J. Mech. Eng. Biomech.* **2019**, *4*, 1–9. [CrossRef]
12. Ariati, R.; Sales, F.; Souza, A.; Lima, R.A.; Ribeiro, J. Polydimethylsiloxane Composites Characterization and Its Applications: A Review. *Polymers* **2021**, *13*, 4258. [CrossRef] [PubMed]
13. Roloff, C.; Stucht, D.; Beuing, O.; Berg, P. Comparison of intracranial aneurysm flow quantification techniques: Standard PIV vs. stereoscopic PIV vs. tomographic PIV vs. phase-contrast MRI vs. CFD. *J. Neurointerv. Surg.* **2019**, *11*, 275–282. [CrossRef] [PubMed]
14. Doutel, E.; Viriato, N.; Carneiro, J.; Campos, J.B.L.M.; Miranda, J.M. Geometrical effects in the hemodynamics of stenotic and non-stenotic left coronary arteries—numerical and in vitro approaches. *Int. J. Numer. Method. Biomed. Eng.* **2019**, *35*, e3207. [CrossRef] [PubMed]
15. Souza, A.; Souza, M.S.; Pinho, D.; Agujetas, R.; Ferrera, C.; Lima, R.; Puga, H.; Ribeiro, J. 3D manufacturing of intracranial aneurysm biomodels for flow visualizations: Low cost fabrication processes. *Mech. Res. Commun.* **2020**, *107*, 103535. [CrossRef]
16. Querzoli, G.; Fortini, S.; Espa, S.; Costantini, M.; Sorgini, F. Fluid dynamics of aortic root dilation in Marfan syndrome. *J. Biomech.* **2014**, *47*, 3120–3128. [CrossRef] [PubMed]
17. Querzoli, G.; Fortini, S.; Espa, S.; Melchionna, S. A laboratory model of the aortic root flow including the coronary arteries. *Exp. Fluids* **2016**, *57*, 1–9. [CrossRef]
18. Tupin, S.; Saqr, K.M.; Ohta, M. Effects of wall compliance on multiharmonic pulsatile flow in idealized cerebral aneurysm models: Comparative PIV experiments. *Exp. Fluids* **2020**, *61*, 1–11. [CrossRef]
19. Rodrigues, R.O.; Pinho, D.; Bento, D.; Lima, R.; Ribeiro, J. Wall expansion assessment of an intracranial aneurysm model by a 3D Digital Image Correlation System. *Measurement* **2016**, *88*, 262–270. [CrossRef]
20. Carvalho, V.; Maia, I.; Souza, A.; Ribeiro, J.; Costa, P.; Puga, H.; Teixeira, S.; Lima, R.A. In vitro stenotic arteries to perform blood analogues flow visualizations and measurements: A Review. *Open Biomed. Eng. J.* **2020**, *14*, 87–102. [CrossRef]
21. Carvalho, V.; Gonçalves, I.; Lage, T.; Rodrigues, R.O.; Minas, G.; Teixeira, S.F.C.F.; Moita, A.S.; Hori, T.; Kaji, H.; Lima, R.A. 3D printing techniques and their applications to organ-on-a-chip platforms: A systematic review. *Sensors* **2021**, *21*, 3304. [CrossRef]
22. Souza, R.R.; Gonçalves, I.M.; Rodrigues, R.O.; Minas, G.; Miranda, J.M.; Moreira, A.L.N.; Lima, R.; Coutinho, G.; Pereira, J.E.; Moita, A.S. Recent advances on the thermal properties and applications of nanofluids: From nanomedicine to renewable energies. *Appl. Therm. Eng.* **2022**, *201*, 117725. [CrossRef]
23. Sadek, S.H.; Rubio, M.; Lima, R.; Vega, E.J. Blood particulate analogue fluids: A review. *Materials* **2021**, *14*, 2451. [CrossRef] [PubMed]
24. Pinho, D.; Carvalho, V.; Gonçalves, I.M.; Teixeira, S.; Lima, R. Visualization and measurements of blood cells flowing in microfluidic systems and blood rheology: A personalized medicine perspective. *J. Pers. Med.* **2020**, *10*, 249. [CrossRef] [PubMed]
25. Faustino, V.; Rodrigues, R.O.; Pinho, D.; Costa, E.; Santos-Silva, A.; Miranda, V.; Amaral, J.S.; Lima, R. A microfluidic deformability assessment of pathological red blood cells flowing in a hyperbolic converging microchannel. *Micromachines* **2019**, *10*, 645. [CrossRef]
26. Pinho, D.; Muñoz-Sánchez, B.N.; Anes, C.F.; Vega, E.J.; Lima, R. Flexible PDMS microparticles to mimic RBCs in blood particulate analogue fluids. *Mech. Res. Commun.* **2019**, *100*, 18–20. [CrossRef]
27. Rodrigues, R.O.; Lima, R.; Gomes, H.T.; Silva, A.M.T. Polymer microfluidic devices: An overview of fabrication methods. *U. Porto J. Eng.* **2015**, *1*, 67–79. [CrossRef]
28. Miranda, I.; Souza, A.; Sousa, P.; Ribeiro, J.; Castanheira, E.M.S.; Lima, R.; Minas, G. Properties and Applications of PDMS for Biomedical Engineering: A Review. *J. Funct. Biomater.* **2022**, *13*, 2. [CrossRef]
29. Saalfeld, S.; Voß, S.; Beuing, O.; Preim, B.; Berg, P. Flow-splitting-based computation of outlet boundary conditions for improved cerebrovascular simulation in multiple intracranial aneurysms. *Int. J. Comput. Assist. Radiol. Surg.* **2019**, *14*, 1805–1813. [CrossRef]
30. Hoi, Y.; Meng, H.; Woodward, S.H.; Bendok, B.R.; Hanel, R.A.; Guterman, L.R.; Hopkins, L.N. Effects of arterial geometry on aneurysm growth: Three-dimensional computational fluid dynamics study. *J. Neurosurg.* **2004**, *101*, 676–681. [CrossRef]
31. Carvalho, V.; Pinho, D.; Lima, R.A.; Teixeira, J.C.; Teixeira, S. Blood Flow Modeling in Coronary Arteries: A Review. *Fluids* **2021**, *6*, 53. [CrossRef]
32. Carvalho, V.; Arcipreste, B.; Soares, D.; Ribas, L.; Rodrigues, N.; Teixeira, S.F.C.F.; Teixeira, J.C. Numerical modeling of the wave soldering process and experimental validation. *J. Electron. Packag.—ASME* **2021**, *144*, 011011. [CrossRef]
33. Carvalho, V.; Carneiro, F.; Ferreira, A.C.; Gama, V.; Teixeira, J.C.; Teixeira, S. Numerical study of the unsteady flow in simplified and realistic iliac bifurcation models. *Fluids* **2021**, *6*, 284. [CrossRef]
34. Carvalho, V.; Rodrigues, N.; Lima, R.A.; Teixeira, S.F.C.F. Modeling blood pulsatile turbulent flow in stenotic coronary arteries. *Int. J. Biol. Biomed. Eng.* **2020**, *14*, 1998–4510. [CrossRef]
35. Shimogonya, Y.; Ishikawa, T.; Imai, Y.; Matsuki, N.; Yamaguchi, T. Can temporal fluctuation in spatial wall shear stress gradient initiate a cerebral aneurysm? A proposed novel hemodynamic index, the gradient oscillatory number (GON). *J. Biomech.* **2009**, *42*, 550–554. [CrossRef]

36. Rashad, S.; Sugiyama, S.; Niizuma, K.; Sato, K.; Endo, H.; Omodaka, S.; Matsumoto, Y.; Fujimura, M.; Tominaga, T. Impact of bifurcation angle and inflow coefficient on the rupture risk of bifurcation type basilar artery tip aneurysms. *J. Neurosurg.* **2018**, *128*, 723–730. [CrossRef]
37. Saqr, K.M.; Rashad, S.; Tupin, S.; Niizuma, K.; Hassan, T.; Tominaga, T.; Ohta, M. What does computational fluid dynamics tell us about intracranial aneurysms? A meta-analysis and critical review. *J. Cereb. Blood Flow Metab.* **2019**, *40*, 1021–1039. [CrossRef]
38. Hassan, T.; Ezura, M.; Timofeev, E.V.; Tominaga, T.; Saito, T. Computational Simulation of Therapeutic Parent Artery Occlusion to Treat Giant Vertebrobasilar Aneurysm. *AJNR Am. J. Neuroradiol.* **2004**, *25*, 63–68.
39. Usmani, A.Y.; Patel, S. Hemodynamics of a cerebral aneurysm under rest and exercise conditions. *Int. J. Energy Clean Environ.* **2018**, *19*, 119–136. [CrossRef]
40. Li, Y.; Verrelli, D.I.; Yang, W.; Qian, Y.; Chong, W. A pilot validation of CFD model results against PIV observations of haemodynamics in intracranial aneurysms treated with flow-diverting stents. *J. Biomech.* **2019**, *100*, 109590. [CrossRef]
41. Janiga, G.; Daróczy, L.; Berg, P.; Thévenin, D.; Skalej, M.; Beuing, O. An automatic CFD-based flow diverter optimization principle for patient-specific intracranial aneurysms. *J. Biomech.* **2015**, *48*, 3846–3852. [CrossRef]
42. Carvalho, V.; Rodrigues, N.; Lima, R.A.; Teixeira, S. Numerical simulation of blood pulsatile flow in stenotic coronary arteries: The effect of turbulence modeling and non-Newtonian assumptions. In Proceedings of the International Conference on Applied Mathematics & Computer Science, Chania, Greece, 19–22 July 2020; pp. 112–116.
43. Appanaboyina, S.; Mut, F.; Löhner, R.; Putman, C.M.; Cebral, J.R. Computational fluid dynamics of stented intracranial aneurysms using adaptive embedded unstructured grids. *Int. J. Numer. Methods Fluids* **2008**, *57*, 475–493. [CrossRef]
44. Xiang, J.; Natarajan, S.K.; Tremmel, M.; Ma, D.; Mocco, J.; Hopkins, L.N.; Siddiqui, A.H.; Levy, E.I.; Meng, H. Hemodynamic-morphologic discriminants for intracranial aneurysm rupture. *Stroke* **2011**, *42*, 144–152. [CrossRef]
45. Dolan, J.M.; Kolega, J.; Meng, H. High wall shear stress and spatial gradients in vascular pathology: A review. *Ann. Biomed. Eng.* **2013**, *41*, 1411–1427. [CrossRef] [PubMed]
46. Li, Y.S.J.; Haga, J.H.; Chien, S. Molecular basis of the effects of shear stress on vascular endothelial cells. *J. Biomech.* **2005**, *38*, 1949–1971. [CrossRef]
47. Rashad, S.; Han, X.; Saqr, K.; Tupin, S.; Ohta, M.; Niizuma, K.; Tominaga, T. Epigenetic response of endothelial cells to different wall shear stress magnitudes: A report of new mechano-miRNAs. *J. Cell. Physiol.* **2020**, *235*, 7827–7839. [CrossRef] [PubMed]
48. Xu, L.; Sugawara, M.; Tanaka, G.; Ohta, M.; Liu, H.; Yamaguchi, R. Effect of elasticity on wall shear stress inside cerebral aneurysm at anterior cerebral artery. *Technol. Heal. Care.* **2016**, *24*, 349–357. [CrossRef] [PubMed]
49. Mooney, M. A theory of large elastic deformation. *J. Appl. Phys.* **1940**, *11*, 582–592. [CrossRef]
50. Lawrence, K.L. *ANSYS Workbench Tutorial Release 14*; SDC Publications: Mission, KS, USA, 2012.
51. Chen, X.; Liu, Y. *Finite Element Modeling and Simulation with ANSYS Workbench*; CRC Press: Boca Raton, FL, USA, 2014.
52. Parlea, L.; Fahrig, R.; Holdsworth, D.W.; Lownie, S.P. An Analysis of the Geometry of Saccular Intracranial Aneurysms. *AJNR Am. J. Neuroradiol* **1999**, *20*, 1079–1089.
53. Bird, R.B.; Armstrong, R.C.; Hassager, O. *Dynamic of Polymeric Liquids*; John Willey & Sons: New York, NY, USA, 1987; Volume 1.
54. Çengel, Y.A.; Ghajar, A.J. *Heat and Mass Transfer*, 4th ed.; Mc. Graw Hill: New York, NY, USA, 2012.
55. Cardoso, C.; Fernandes, C.S.; Lima, R.; Ribeiro, J. Biomechanical analysis of PDMS channels using different hyperelastic numerical constitutive models. *Mech. Res. Commun.* **2018**, *90*, 26–33. [CrossRef]

fluids

MDPI

Article

CFD to Quantify Idealized Intra-Aneurysmal Blood Flow in Response to Regular and Flow Diverter Stent Treatment

Augusto Fava Sanches [1,*], Suprosanna Shit [2], Yigit Özpeynirci [1] and Thomas Liebig [1]

1 Institute of Neuroradiology, University Hospital LMU Munich, 81377 Munich, Germany; yigit.oezpeynirci@med.uni-muenchen.de (Y.Ö.); thomas.liebig@med.uni-muenchen.de (T.L.)
2 Department of Informatics, Technical University of Munich, 85748 Garching, Germany; suprosanna.shit@tum.de
* Correspondence: augusto.sanches@med.uni-muenchen.de

Abstract: Cerebral aneurysms are pathological dilatations of the vessels supplying the brain. They carry a certain risk of rupture, which in turn, results in a high risk of mortality and morbidity. Flow diverters (FDs) are high-density meshed stents which are implanted in the vessel segment harboring an intracranial aneurysm to cover the entrance of the aneurysm, thus reducing the blood flow into the aneurysm, promoting thrombosis formation and stable occlusion, which prevents rupture or growth of the aneurysm. In the present study, the blood flow in an idealized aneurysm, treated with an FD stent and a regular stent (RS), were modeled and analyzed considering their design, surface area porosity, and flow reduction to investigate the quantitative and qualitative effect of the stent on intra-aneurysmal hemodynamics. CFD simulations were conducted before and after treatment. Significant reductions were observed for most hemodynamic variables with the use of stents, during both the peak systolic and late diastolic cardiac cycles. FD reduces the intra-aneurysmal wall shear stress (WSS), inflow, and aneurysmal flow velocity, and increases the turnover time when compared to the RS; therefore, the possibility of aneurysm thrombotic occlusion is likely to increase, reducing the risk of rupture in cerebral aneurysms.

Keywords: cerebral aneurysms; CFD; flow diverter stent; hemodynamics

Citation: Sanches, A.F.; Shit, S.; Özpeynirci, Y.; Liebig, T. CFD to Quantify Idealized Intra-Aneurysmal Blood Flow in Response to Regular and Flow Diverter Stent Treatment. *Fluids* **2022**, *7*, 254. https://doi.org/10.3390/fluids7080254

Academic Editor: Huidan (Whitney) Yu

Received: 15 June 2022
Accepted: 26 July 2022
Published: 28 July 2022

1. Introduction

Cerebral aneurysms present a disease characterized by the local dilatation of arterial walls in the intracranial vasculature that generally occur on arterial curves and bifurcations in the circle of Willis [1]. The aneurysm may rupture and cause subarachnoid hemorrhage, which is associated with high mortality and morbidity [2]. A clinical study [3] reports that about 2% to 5% of the population is carrying such intracranial aneurysms.

In clinical practice, cerebral aneurysms are occasionally being discovered more frequently because of enhanced and widely utilized imaging technologies. Understanding the hemodynamic mechanisms involved is crucial for reducing the risk of rupture and hemorrhage in cerebral aneurysms and for identifying effective treatment options. Numerical methods may also offer good support for the medical treatment of brain aneurysms.

There are two approaches to effectively treating brain aneurysms. The first one is via clipping the aneurysmal neck and the other is via endovascular intervention. The advantage of the endovascular treatment is the fact that there is no need to do a craniotomy, exposing the surface of the brain vessel. Large or giant aneurysms, defined as wide-necked, dissecting, and fusiform aneurysms having a diameter $\geq$25 mm, are considered more challenging and less tractable to the traditional endovascular coiling [4]. While stent-assisted coiling and balloon-assisted coiling are alternative techniques developed to deal with such complex aneurysms, they offer less than desired efficacy, given their high rate of recanalization [5,6], and the flow diverter stent represents a paradigm change, with the intervention carried out in the parent artery [7,8]. The FD stent is a well-established method of endovascular reconstruction and aneurysm occlusion for large and complex intracranial aneurysms, with an overall porosity metallic mesh set in the parent artery to reduce the blood flow in the intracranial aneurysm to the point of stagnation and continuous aneurysmal thrombosis [9]. This phenomenon is affected by the metal surface area coverage provided by the stent. Rather than porosity, the pore density of the flow diverters seems to be a critical factor modulating device capability [10].

Computational techniques offer new capabilities in healthcare provision for cerebral aneurysms. The availability of a simulation tool for the flow diverter is extremely useful to support the decisions of treatment options by medical experts and to develop and optimize new implant designs.

Among several hemodynamic parameters that are discussed as key factors in the initiation, development, or rupture of intracranial aneurysms, one of the most studied parameters is the WSS. High or low local values of the WSS and non-uniform distribution of instability are negative conditions for the development of an aneurysm. Low WSS may lead to the spatial disorganization of endothelial cells and a dysregulation of antioxidant and anti-inflammatory mediators, resulting in arterial wall remodeling [11]. Consensually, high WSS may lead to the initiation of aneurysm formation, but its influence on growth and rupture is largely unknown. Tremmel et al. [12] suggest that both high WSS and low WSS could lead to rupture and growth of the aneurysm. Jou et al. [13] found that ruptured aneurysms had low WSS, whereas Shogima et al. [14] and Cebral et al. [15] suggested that high WSS was associated with ruptured aneurysms.

Several cerebral aneurysms have been effectively treated with flow diverter devices [16–21], albeit there have been reports of problems connected to late rupture [22–25]. These issues show that the aneurysms are not immediately protected following the procedure. Therefore, obtaining stable aneurysm occlusion quickly is necessary to improve the success of these treatments.

Goubergrits et al. [26] numerically studied the hemodynamic changes in the flow diverter device compared with a non-flow diversion device, rather than another stent; therefore, there is no information regarding the blood flow viscosity assumption. They conclude that the stenting does not affect the pressure in the cerebral aneurysm post-treatment, but significantly alters intra-aneurysmal hemodynamics through flow reduction and a change in flow pattern. Jou et al. [27] analyzed the flow behavior in a giant aneurysm using a CFD simulation, and they show that flow impingement is pointed as a significant factor for aneurysm initiation, growth, and burst. An impingement index is utilized to evaluate the size and effectiveness of flow impingement. However, the analysis was performed on pre-stent giant aneurysms. Chien et al. [28] investigated the hemodynamics of small unruptured and ruptured aneurysms at the same anatomical location, using simulation tools, and they concluded that the wall shear stress is an important parameter related to the development and rupture mechanism in brain aneurysms. This study shows the relevance of WSS, but considered only small aneurysms. Bouillot et al. [29] compared the hemodynamic characteristics in only medium-sized untreated cerebral aneurysm before and after the treatment with stents with different porosities. CFD simulations and particle imaging velocimetry (PIV) showed quantitative and qualitative evidence of the pressure and shear rate mechanisms driving the flow for both pre- and post-stent treated aneurysms, which is consistent which the findings in the present study.

Previous studies [30] show that the risk of rupture in large and giant aneurysms is higher than in small aneurysms. However, there are no studies regarding several hemodynamic changes in giant cerebral aneurysms treated with stents of different porosities and the non-Newtonian blood flow model.

The present numerical study investigates, in detail, the effect of stent porosity on hemodynamics in an idealized giant sidewall, wide-neck aneurysm. It might lead to a better understanding of failed aneurysm occlusion with flow-diverter stents.

2. Materials and Methods

2.1. Vascular Modeling and Stent Geometry

An idealized giant saccular brain aneurysm, cf. Figure 1, is designed based on a 3D rotational angiography. A spherical aneurysm with a diameter of 33 mm is located at 2.0 mm above a straight cylindrical artery of diameter 4.5 mm. The distance of the artery inlet to the aneurysm proximal is 52 mm, and the length of the aneurysm distal to the artery outlet is also 52 mm. Thus, the inlet and outflow conditions were imposed far away from the location of the aneurysm, and the flow characteristics in the aneurysm were not affected.

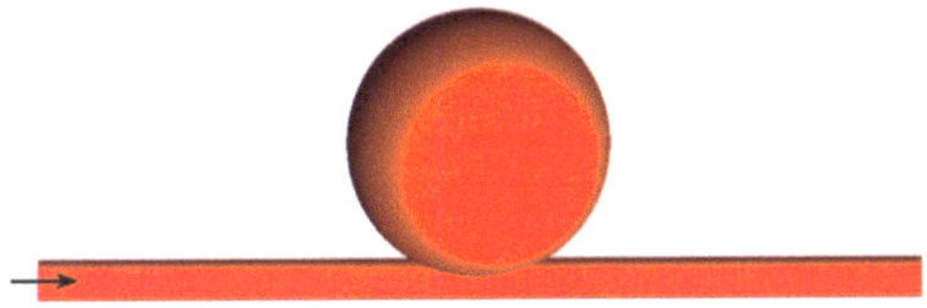

Figure 1. Idealized model of an idealized saccular giant brain aneurysm.

Two stents with meshes made of cylindrical metal wires are considered which fit the shape of the parent artery. Their geometrical configuration is summarized in Figure 2 using a stent unit cell [31]. The stent with a low metal coverage proportion is called a regular stent (RS), and that with a high metal coverage proportion is known as a flow diverter stent (FD). The porosity ε, cf. Figure 2, of a stent is given by

$$\varepsilon = \frac{S_{wm}}{S_s} \tag{1}$$

where S_{wm} is the surface area of the stent without the material, S_s the surface area of the stent, and ε is the porosity.

Dimension	RS	FD
l [mm]	0.96	0.30
m [mm]	0.94	0.30
c [mm]	60	42
β [°]	49	55
ε [-]	0.88	0.73

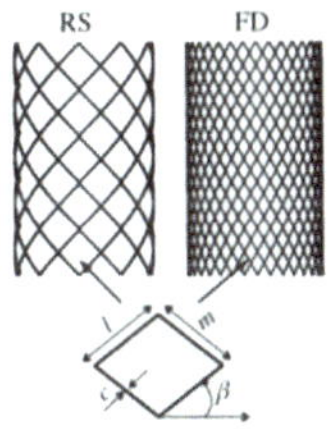

Figure 2. Geometrical configuration [31] of the RS and the FD.

2.2. Numerical Grid and Mathematical Model

The computational meshing is created using the octree technique by the mesh generator software ICEM CFD v.19.2 (ANSYS Inc, Canonsburg, PA, USA). Due to the rapid variations of the hemodynamic parameters near the arterial wall, the numerical meshing consists of five prismatic layer elements near the wall surface, which are combined with tetrahedron grid cells into the flow region. Concerning the parent artery segment, covered with the stent devices, only tetrahedron grid cells are created. The distance of the first layer to the vessel surface is fixed to 0.01 mm with an average nodal space, increasing by a ratio of 1.2.

For the stent, the numerical grid is refined until the computed flow field is independent of the number of grid nodes. For the configuration with the regular stent, 11.4 million, and for the flow diverter, 11.9 million cells are used. The maximum velocities values obtained by the meshes were observed, and a difference of less than 2% was found in the results. For the untreated aneurysms, a similar mesh density (6.2 million) was computed to maintain the consistency between the huge number of elements. The stents have no deformation, due the flow interactions.

The blood flow is assumed as incompressible, and the unsteady Navier–Stokes equations are solved using the continuity and momentum equations [32,33], which are as follows

$$\nabla \cdot \mathbf{u} = 0 \tag{2}$$

$$\rho\left(\frac{\partial \mathbf{u}}{\partial t} + \mathbf{u} \cdot \nabla \mathbf{u}\right) = -\nabla p + \nabla \cdot \boldsymbol{\tau} \tag{3}$$

where $\mathbf{u}$ is the fluid velocity field, ρ is the fluid density, and p is the static pressure. $\boldsymbol{\tau}$ is the deviatoric stress tensor

$$\boldsymbol{\tau} = \mu(\dot{\gamma})\left(\nabla \mathbf{u} + \nabla \mathbf{u}^{\mathrm{T}}\right), \tag{4}$$

where the superscript T denotes the transposed tensor and μ is the shear-dependent dynamic viscosity, cf. Equation (5).

The discretized governing equations are solved using the finite volume-based software platform OpenFOAM V3.1 (OpenCFDLtd, London, England), where a second-order upwind scheme for the convective terms is used, and a semi-implicit method for pressure-linked equations coupled with a solution scheme based on the algebraic multi-grid method is activated.

To close the system of equations, a constitutive law must be given to calculate the local fluid dynamic viscosity. The blood rheology literature provides a robust indication that the non-Newtonian behavior of the blood flow cannot be ignored [34]. Mainly, for predicting the risk of rupture in brain aneurysms, the accurate understanding of quantities such as the WSS and pressure distribution are crucial. Here, we assume the power-law Carreau–Yasuda viscosity model [35], a non-Newtonian viscosity model, to simulate shear-thinning blood, given by

$$\mu = \mu_\infty + [\mu_0 - \mu_\infty]\left[1 + \left((\lambda\dot{\gamma})^2\right)\right]^{\left(\frac{a-1}{2}\right)}, \tag{5}$$

where μ_0 = 0.0456 Pa·s and μ_∞ = 0.0032 Pa·s are the asymptotic viscosities at zero and for infinite shear rate, respectively. The shear rate, which is a scalar measure of the strain rate tensor, is represented by $\dot{\gamma}$. The relation time constant λ equals 10.03 s, and the power law index a is 0.344 [32].

In the present calculations, the inlet flow rate has been found [36] to be equal to the average values of the flow velocity in the internal carotid artery. A zero-pressure condition was used at the outlet. All the vascular walls are assumed rigid, with a no-slip boundary condition. In contrast to the substantial vessel wall motions in the aortic artery, the radial dilation of the arteries in the circle of Willis does not increase by more than 10% of its diameter [37].

The wall shear stress (WSS), recognized as one of the main risk factors for the aneurysm's initiation, growth, and rupture, is analyzed [38]. The WSS is a viscous force, consisting of the tangential component of the stress tensor applied on the arterial wall. The stress tensor σ is given by [32]

$$\sigma = p\mathbf{I} - \boldsymbol{\tau} \tag{6}$$

where $\mathbf{I}$ is the identity tensor.

Thus, the WSS is represented by

$$\sigma_{\mathrm{n}} - (\sigma_{\mathrm{n}} \cdot \mathbf{n})\mathbf{n} = \boldsymbol{\tau}_{\mathrm{n}} - (\boldsymbol{\tau}_{\mathrm{n}} \cdot \mathbf{n})\mathbf{n} \tag{7}$$

where $\mathbf{n}$ is the normal vector to the arterial wall. σ_{n} and $\boldsymbol{\tau}_{\mathrm{n}}$ are the normal components of the stress and deviatoric tensors, respectively.

The intra-aneurysmal flow activity describes the aneurysm's hemodynamics, which is quantified through the averaged magnitude of flow velocity in the aneurysm and the vorticity contours, which will be analyzed in the Results section. The wall distribution of the static pressure in both the untreated and treated aneurysm will also be studied. To quantify the stasis of the flow inside the aneurysm, the turnover time is given by

$$t_t = \frac{V_{aneurysm}}{\vartheta_{inflow}} \quad (8)$$

which is defined as the volume of the aneurysm divided by the aneurysmal volumetric inflow rate at the neck. The magnitude of the average flow velocity is calculated to indicate the flow activity inside the aneurysm.

3. Results and Discussion

Simulations are performed for the untreated aneurysm (UA), the aneurysm with a regular stent (RS), and with a flow diverter stent (FD), cf. Figures 1 and 2, to calculate the change in the intra-aneurysmal hemodynamics.

Speaking of the saccular intracranial aneurysms, measurement at the neck is important to identify the characteristics of the blood flow entering and exiting the aneurysm sac, providing more information about the growth of the aneurysm sac in time, without changing its shape. The dome is the region at which most ruptures occur. To investigate the rupture risk, knowing the hemodynamic properties in this area is necessary. Additionally, a change in the hemodynamics from the neck to the dome (i.e., measurements performed in the middle segment of the sac) could clarify the growth pattern, secondary bleb formation, and rupture risk of the aneurysm.

First, the maximum value of the WSS magnitude on the aneurysmal wall will be analyzed at the different cross-sections, as shown in Figure 3, i.e., at the neck of the aneurysm (A-A′), in the center (B-B′), and in the dome (C-C′). Tables 1 and 2 show the maximum value of the WSS magnitude on the aneurysmal wall for the UA, RS, and FD for the systolic (0.17 s) and diastolic (0.78 s) cardiac flow, respectively. A strong reduction in the WSS for the different cross sections, for both the regular stent and the flow diverter, is observed. However, the WSS reduction of the flow diverter during the peak systole is more significant (up to 99.2% in the dome).

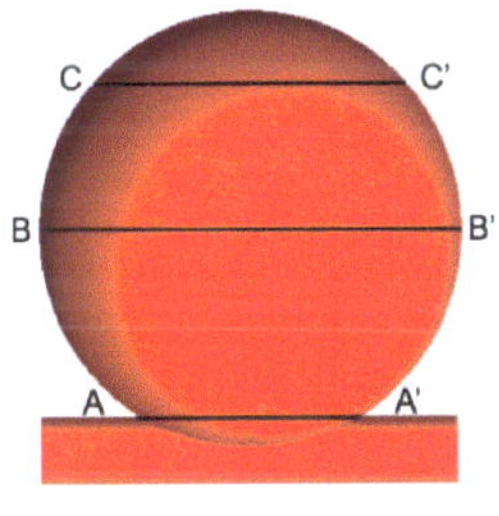

Figure 3. Giant cerebral aneurysm divided into three different cross sections.

Table 1. WSS for the systolic cardiac flow.

	UA	RS	FD
A-A′	13.2000 Pa	2.6000 Pa (−80.3%)	1.3000 Pa (−90.1%)
B-B′	0.4500 Pa	0.0660 Pa (−85.3%)	0.0400 Pa (−91.1%)
C-C′	0.1800 Pa	0.0310 Pa (−82.7%)	0.0015 Pa (−99.2%)

Table 2. WSS for the diastolic cardiac flow.

	UA	RS	FD
A-A′	1.620 Pa	0.8800 Pa (−45%)	0.2400 Pa (−85%)
B-B′	0.0380 Pa	0.0043 Pa (−88.7%)	0.0043 Pa (−88.7%)
C-C′	0.0130 Pa	0.0018 Pa (−86.1%)	0.0001 Pa (−99.2%)

The huge variation in the reduction in the WSS between the RS and the FD occurs during diastolic cardiac flow in the aneurysm's neck, where the RS shows a decrease in WSS of 45% and of 85% for the FD, respectively, compared to the untreated aneurysm. However, during the diastole, the WSS reductions in the center of the aneurysm are similar for the RS and the FD (88.7%), and present a significant discrepancy in the aneurysm neck, with 45% for the RS and 85% for FD.

Figures 4 and 5 show the velocity contour and streamlines, respectively, of the systole, and Figures 6 and 7 display the corresponding plots for the diastole. The contour plots show the cut at the mid-plane (B-B'), cf. top parts of Figures 4 and 6, as well as a zoom of the neck region in the lower parts of those figures. The velocity streamlines refer to the 3D situation. Each figure shows the flow in the untreated aneurysm (left), the aneurysm treated with a regular stent (center), and treated with the flow diverter (right). Note that Figures 5 and 7 show the placement of the RS (center) and the FD (right), which are omitted in Figures 4 and 6 for a better visibility of the flow pattern in the aneurysm neck.

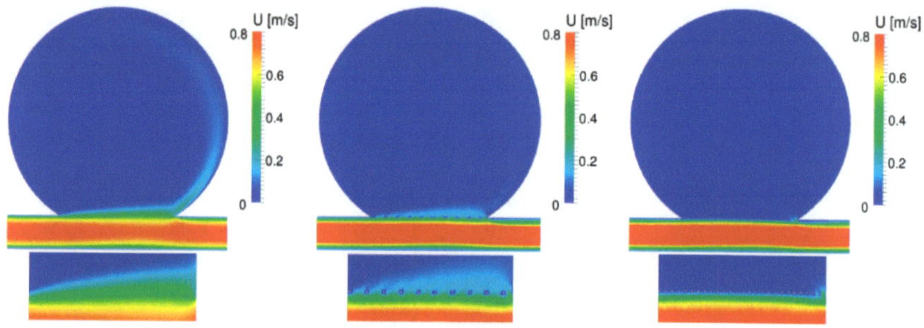

Figure 4. Velocity contour at the mid plane (**top**) and the zoom view of the velocity contour (**bottom**) of the UA (**left**), RS (**center**), and FD (**right**) at peak systole.

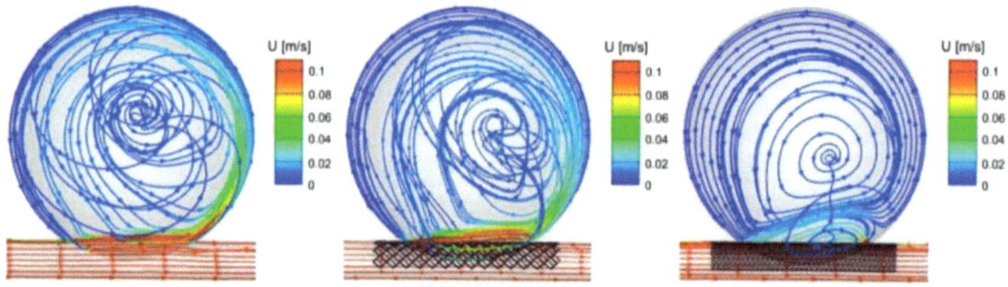

Figure 5. Velocity streamlines of the (**left**) UA, (**center**) RS, and (**right**) FD at peak systole.

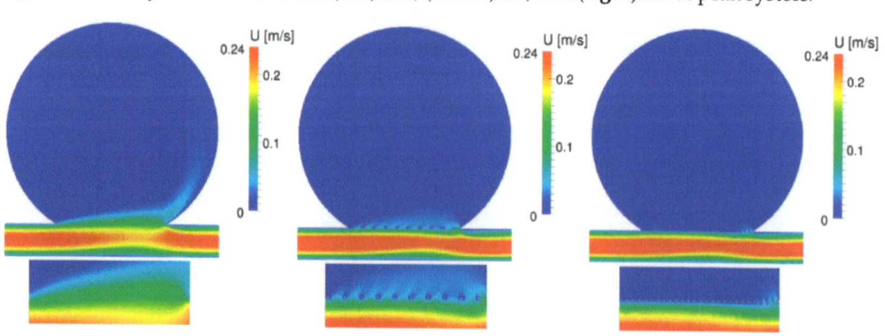

Figure 6. Velocity contour at the mid plane (**top**) and the zoom view of the velocity contour (**bottom**) of the UA (**left**), RS (**center**), and FD (**right**) at late diastole.

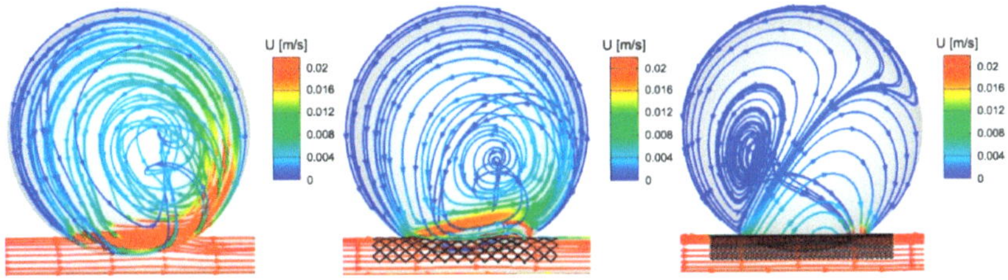

Figure 7. Velocity streamlines of the (**left**) UA, (**center**) RS, and (**right**) FD at late diastole.

During the systole, the flow within the untreated aneurysm is characterized by a regular vortex located at the center of the aneurysm, cf. Figure 5, with inflow at the distal and outflow at the proximal edge of the neck. After the treatment with the regular stent, the flow pattern has similar characteristics as seen in the UA, with a slight displacement of the vortex towards the outflow of the cerebral aneurysm. After the FD implant, the flow field shows a separation of the vortex, where the main vortex circulates inwards towards the proximal side of the neck, as seen in the UA and RS situations, and a separated vortex rotates outwards from the proximal towards the distal side of the neck. The zoom view of the velocity contour plots in Figure 4 shows the strong impact of both the RS and the FD on the flow field in the neck of the aneurysm, which is accompanied by a strong decrease in blood flow into the cerebral aneurysm: the flow is shifted from the neck of the aneurysm towards the parent artery, and the performance of the stent devices is obvious. Note how the struts of the stent affect the velocity flow field. Notice that in the outflow region of the aneurysmal neck, the velocity profiles in both aneurysms treated with the RS and the FD, the flow widens somewhat upward, showing the critical region where a risk of rupture may occur in aneurysms treated with stent devices.

The flow diverter's pore size, albeit sufficiently small to achieve flow re-channeling, is large enough to provide a scaffolding for the growing of endothelial and neointimal tissue across the aneurysmal neck [39]. Regarding the flow direction, the intensity of this effect is proportional to the stent porosity.

For the diastole shown in Figures 6 and 7, the flow field in the UA is characterized by a regular vortex, which is slightly displaced towards the outflow of the aneurysm, with inflow at the distal and outflow at the proximal side of the neck. The velocity streamlines after the implant of the RS reveal some flow division close to the aneurysm's neck, where the fluid circulates counterclockwise towards the distal inflow and clockwise towards the proximal outflow. This flow alteration is accompanied by a strong reduction in the absolute blood flow velocity in the aneurysmal neck and a shift of the peak values towards the parent artery. Both effects are strengthened by the implant of the FD, which shows its superior performance over the RS in this region. Thus, the numerical results clearly show that the flow impingement on the aneurysmal wall in the distal neck region reduces with decreasing stent porosity.

With the decrease in stent porosity, the magnitude of the vortical flow is decreased. Figures 8 and 9 show the vorticity contours at the different cross sections marked in Figure 3. The magnitude of the vorticity perpendicular to the plane is displayed, where blue and red colors indicate clockwise and counterclockwise rotation, respectively. For all cross-sections, the magnitude of the vorticity is significantly reduced through the deployment of the RS and the FD; note the different scales used at the different cross-sections.

During the systole, inside the aneurysm, the rotation occurs from the proximal to distal for the UA, RS, and FD cases, to the neck, middle, and dome cross sections, as shown by Figure 8. During the diastole, as shown in Figure 9, the rotation pattern is the same, except for FD middle and dome of the aneurysm, which shows the direction from the distal to the proximal within the aneurysm.

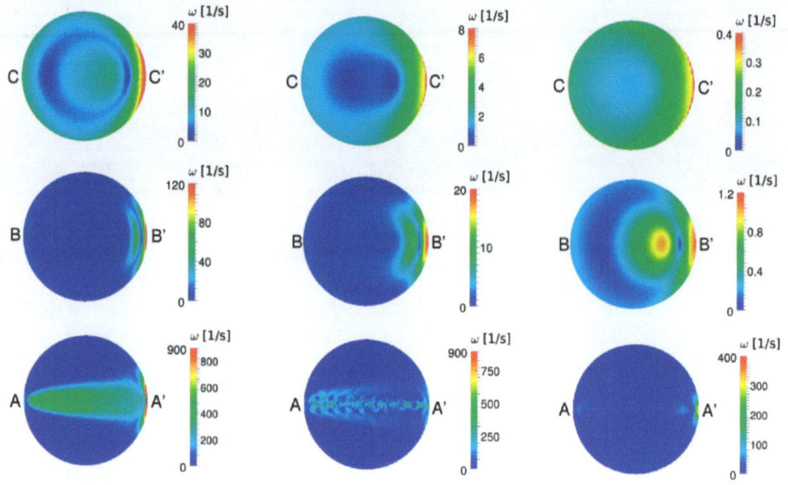

Figure 8. Vorticity contour at different cross-sections of the (**left**) UA, (**center**) RS, and (**right**) FD at peak systole.

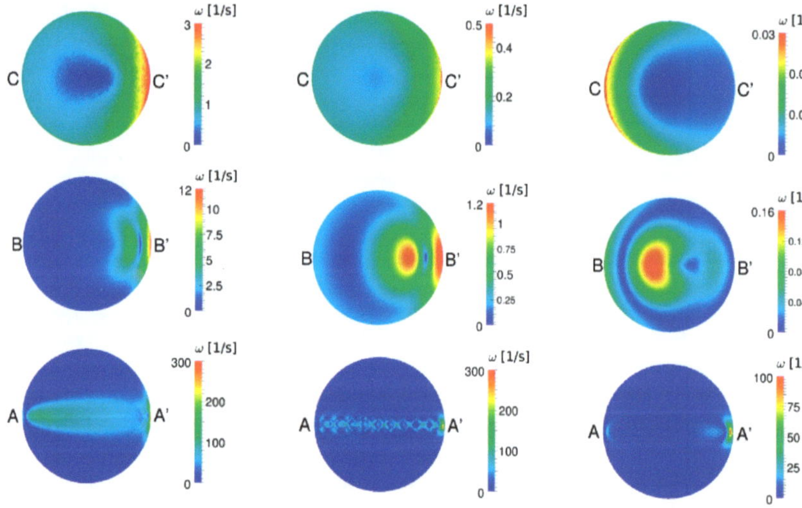

Figure 9. Vorticity contour at different cross-sections of the (**left**) UA, (**center**) RS, and (**right**) FD at late diastole.

The static pressure on the distal neck was decreased after both the RS and the FD implant, for both systole and diastole, as shown in Figures 10 and 11. The mean static pressure in the middle and in the dome of the aneurysmal is not significantly affected by the implant treatment. The primary aim of the stent intervention is the prompt and extensive aneurysmal occlusion by improving the flow diversion, consequently creating favorable conditions to generate aneurysmal thrombotic occlusion, as well as the reconstruction of the parent vessel. An increase in the turnover time, will increase the intra-aneurysmal thrombotic activity, consequently providing a better chance of a successful treatment [40]. The turnover time increases with decreasing stent porosity, as shown in Figure 12. For the UA, a turnover time of 0.63 s is obtained, and it reaches a maximum of 1.4 s after the RS implant and 2.1 s after the FD treatment, respectively, for the systolic cardiac flow. For the diastole, the RS had a turnover time of 3.8 s and after the FD, 6.8 s, starting from 2.1 s for the UA. Thus, the FD again shows its superior performance compared to the RS.

Figure 10. Pressure contour at the mid plane of the UA (**left**), the RS (**center**), and FD (**right**) at peak systole.

Figure 11. Pressure contour at the mid plane of the UA (**left**), RS (**center**), and FD (**right**) at late diastole.

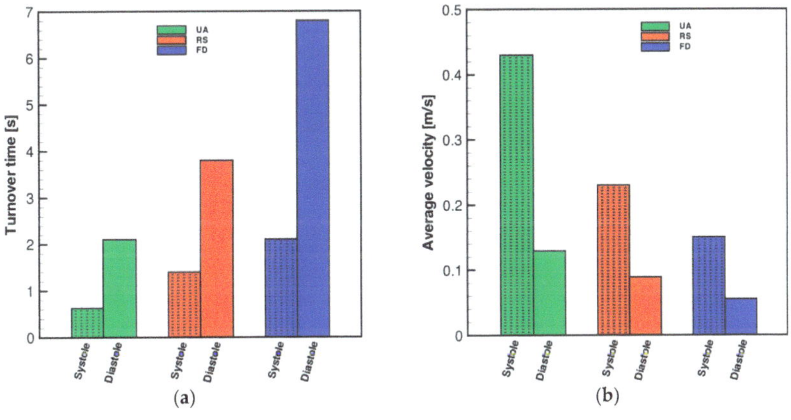

Figure 12. (**a**) Turnover time of the UA, RS, and FD at peak systole and late diastole and (**b**) average intra-aneurysmal velocity magnitude of the UA, RS, and FD at peak systole and late diastole.

The turnover time increases with decreasing stent porosity, as shown in Figure 12a. For the UA, a turnover time of 0.63 s is obtained, and it reaches a maximum of 1.4 s after the RS implant and 2.1 s after the FD treatment, respectively, for the systolic cardiac flow. For the diastole, the RS had a turnover time of 3.8 s and after the FD, 6.8 s, starting from 2.1 s for the UA. Thus, the FD again shows its superior performance compared to the RS. The average intra-aneurysmal flow, as shown in Figure 12b, decreases with an increased stent porosity. This reduced aneurysmal inflow can accelerate the blood clotting condition and the thrombotic occlusion in the aneurysm [41].

For the systole, the average of the flow velocity within the UA is 0.43 m/s, for the RS it is 0.23 m/s, and for the FD, 0.15 m/s is obtained. Considering the diastolic cardiac cycle, the average of the intra-aneurysmal flow for the UA is 0.13 m/s, for the RS it is 0.09 m/s, and 0.05 m/s for the FD. The WSS is higher on the distal neck compared to the proximal neck region of the aneurysm. With the RS implant, the WSS exhibits a strong reduction at the neck, middle, and dome areas of the aneurysm, for both the systole and diastole cardiac flows. However, with the FD implant, the reduction in the WSS is even more significant, avoiding the impact of the WSS on the distal aneurysmal wall.

In the untreated aneurysm, for both the systole and the diastole, the blood inflow into the aneurysm occurs with strong impingement on the distal aneurysmal wall, in a counterclockwise intra-aneurysmal flow. This flow entering the aneurysm is both pressure- and viscous shear-driven, as described by Meng et al. [42]. For the regular stent, a reduction in the velocity gradient occurs, and thereafter, a decrease, but not a complete extinction, in the inflow jet on the distal aneurysmal wall is observed. The implantation of the flow diverter obstructs the flow at the aneurysm neck, promoting a lower shear stress transmission, eliminating the inflow jet at the distal wall, and causing a rise of the pressure gradient along the parent artery. The pressure drives the circulating fluid inward and outward from the brain aneurysm at the distal and proximal sides, respectively.

The pattern of the velocity streamlines is consistent with the PIV experiments performed by Bouillot et al. [43], for the UA, RS, and FD. An exception is noted in the streamlines after the FD for the systole, which for that PIV experiments, showed complete dissolution of the single vortex, which does not happen in the present situation, where the single vortex resides with magnitude; however, the single vortex strongly reduces at all cross sections. These differences may be due the different sizes of the aneurysms studied, with a medium size for the experiment and a giant size in the present study. In the medium-sized aneurysm, the intracranial aneurysmal pressure is higher than the pressure at the proximal neck and lower than the pressure at the distal neck, causing a clockwise flow in the aneurysm. In the present giant aneurysm, a partial clockwise vortex is created beneath the center of the aneurysm, because the intra-aneurysm pressure is only slightly higher than that at the aneurysm proximal neck, which is not sufficient to create a complete clockwise flow within the aneurysm. Thus, in the giant cerebral aneurysm, there is an occurrence of both clockwise and counterclockwise flows in the intracranial aneurysm. The FD implantation is effective in reducing the velocity magnitude within the intracranial aneurysm, but it does not cause extinction of the single vortex in the giant aneurysm.

In agreement with the studies presented by Kerl et al. [44] and by Larrabide et al. [45], the intra-aneurysmal mean static pressure was not affected by the stents devices deployed, except for a small pressure change at the distal neck region. This change in pressure is due to stagnation of impingement flow in that region.

The difference in the reduction in the aneurysm inflow and the flow activity after the implantation of the regular stent and flow diverter stent is a key factor for thrombus formation inside the aneurysm. As the average intra-aneurysmal flow velocity decreases, the turnover time flow increases in an inverse proportion, and therefore, the chance of aneurysm thrombotic occlusion is likely to increase [12]. A delay at the complete aneurysm occlusion exposes the patients to prolonged use of blood-thinners, which in the meantime, could increase the risk of bleeding [46].

4. Limitations and Outlook

There are a few limitations in the present study, such as the assumption of the absence of fluid-structure interaction between the flow and the aneurysm [47,48], and the stent and the vessel. Furthermore, we are considering the flow as a single phase, neglecting the thrombus formation and growing effects and interaction with the stent. Further studies are planned to involve in-vitro and in-vivo 4D flow magnetic resonance imaging patient-specific image models to validate the simulations.

5. Conclusions

The flow diverter effectively reduces the wall shear stress and the blood flow velocity, while also providing a structure that supports the endothelization and reconstruction of the parent vessel. This device also decreases the vorticity magnitude and relocates the center of the vorticity in the intracranial aneurysm. The implantation of the FD does not affect the static cerebral aneurysm pressure. It was observed that due its low porosity, the use of the FD greatly improves the performance compared to the RS, promoting a more efficient reduction in the aneurysm inflow, in addition to an increase in the higher turnover time, and consequently, a gradual thrombosis formation. The hemodynamic alterations in the cerebral aneurysm blood flow dynamics by the deployment of the stent have the potential to induce intra-aneurysmal thrombosis, which is the objective of this treatment paradigm. Further studies are necessary to correlate hemodynamics with intra-aneurysmal thrombosis and to determine the optimum stent design for cerebral aneurysm applications.

Author Contributions: Conceptualization, A.F.S., Y.Ö. and T.L.; methodology, A.F.S.; software, A.F.S. and S.S.; validation, A.F.S.; formal analysis, A.F.S.; investigation, A.F.S.; resources, T.L.; data curation, A.F.S.; writing—original draft preparation, A.F.S.; writing—review and editing, A.F.S., Y.Ö. and T.L.; visualization, A.F.S.; supervision, T.L.; project administration, A.F.S. and T.L.; funding acquisition, T.L. All authors have read and agreed to the published version of the manuscript.

Funding: This research received no external funding.

Institutional Review Board Statement: Not applicable.

Informed Consent Statement: Not applicable.

Data Availability Statement: Not applicable.

Acknowledgments: The authors gratefully acknowledge the Gauss Centre for Supercomputing e.V. (www.gauss-centre.eu (accessed on 1 May 2020) for funding this project by providing computing time on the GCS Supercomputer SuperMUC at Leibniz Supercomputing Centre (www.lrz.de (accessed on 1 May 2020)).

Conflicts of Interest: The authors declare no conflict of interest.

References

1. Willis, T. *The Anatomy of the Brain and the Nerves, Tercentenary ed.*; Feindel, W., Ed.; McGill University Press: Montreal, QB, Canada, 1965.
2. Dennis, K.D.; Rossman, T.L.; Kallmes, D.F.; Dragomir-Daescu, D. Intra-aneurysmal flow rates are reduced by two flow diverters: An experiment using tomographic particle image velocimetry in an aneurysm model. *J. NeuroIntervent. Surg.* **2015**, *7*, 937–942. [CrossRef] [PubMed]
3. Augsburger, L.; Reymond, P.; Rüfenacht, D.A.; Stergiopulos, N. Intracranial stents being modeled as a porous medium: Flow simulation in stented cerebral aneurysms. *Ann. Biomed. Eng.* **2010**, *39*, 850–863. [CrossRef] [PubMed]
4. Möhlenbruch, M.; Herweh, C.; Jestaedt, L.; Stampfl, S.; Schönenberger, S.; Ringleb, P.; Bendszus, M.; Pham, M. The FRED flow-diverter stent for intracranial aneurysms: Clinical study to assess safety and efficacy. *Am. J. Neuroradiol.* **2015**, *36*, 1155–1161. [CrossRef] [PubMed]
5. Shapiro, M.; Becske, T.; Sahlein, D.; Babb, J.; Nelson, P. Stent-supported aneurysm coiling: A literature survey of treatment and follow-up. *Am. J. Neuroradiol.* **2012**, *33*, 159–163. [CrossRef] [PubMed]
6. Coley, S.; Sneade, M.; Clarke, A.; Mehta, Z.; Kallmes, D.; Cekirge, S.; Saatci, I.; Roy, D.; Molyneux, A. Cerecyte coil trial: Procedural safety and clinical outcomes in patients with ruptured and unruptured intracranial aneurysms. *Am. J. Neuroradiol* **2012**, *33*, 474–480. [CrossRef] [PubMed]
7. Alderazi, Y.J.; Shastri, D.; Kass-Hout, T.; Prestigiacomo, C.J.; Gandhi, C.D. Flow diverters for intracranial aneurysms. *Stroke Res. Treat.* **2014**, *2014*, 1–12. [CrossRef]
8. Becske, T.; Kallmes, D.F.; Saatci, I.; McDougall, C.G.; Szikora, I.; Lanzino, G.; Moran, C.J.; Woo, H.H.; Lopes, D.K.; Berez, A.L.; et al. Pipeline for uncoilable or failed aneurysms: Results from a multicenter clinical trial. *Radiology* **2013**, *267*, 858–868. [CrossRef]
9. Girdhar, G.; Li, J.; Kostousov, L.; Wainwright, J.; Chandler, W.L. In-vitro thrombogenicity assessment of flow diversion and aneurysm bridging devices. *J. Thromb. Thromboly* **2015**, *40*, 437–443. [CrossRef] [PubMed]
10. Sadasivan, C.; Cesar, L.; Seong, J.; Rakian, A.; Hao, Q.; Tio, F.O.; Wakhloo, A.K.; Lieber, B.B. An original flow diversion device for the treatment of intracranial aneurysms: Evaluation in the rabbit elastase-induced model. *Stroke* **2009**, *40*, 952–958. [CrossRef]
11. Li, H.; Peng, T.; Wu, J.; Huang, C.; Jiang, Y.; Chen, L. Outflow vessel in the plane of main vortex of large cerebral aneurysms: A study of hemodynamic analyses. *Neurosc. Med.* **2015**, *6*, 65–70. [CrossRef]
12. Tremmel, M.; Xiang, J.; Natarajan, S.K.; Hopkins, L.N.; Siddiqui, A.H.; Levy, E.I.; Meng, H. Alteration of intra-aneurysmal hemody-namics for flow diversion using enterprise and vision stents. *World Neurosurg.* **2010**, *74*, 306–315. [CrossRef] [PubMed]
13. Jou, L.-D.; Lee, D.; Morsi, H.; Mawad, M. Wall shear stress on ruptured and unruptured intracranial aneurysms at the internal carotid artery. *Am. J. Neuroradiol.* **2008**, *29*, 1761–1767. [CrossRef] [PubMed]

14. Shojima, M.; Oshima, M.; Takagi, K.; Torii, R.; Hayakawa, M.; Katada, K.; Morita, A.; Kirino, T. Magnitude and role of wall shear stress on cerebral aneurysm computational fluid dynamic study of 20 middle cerebral artery aneurysms. *Stroke* **2004**, *35*, 2500–2505. [CrossRef] [PubMed]
15. Cebral, J.R.; Castro, M.A.; Burgess, J.E.; Pergolizzi, R.S.; Sheridan, M.J.; Putman, C.M. Characterization of cerebral aneurysms for assessing risk of rupture by using patient-specific computational hemodynamics models. *Am. J. Neuroradiol.* **2005**, *26*, 2550–2559. [PubMed]
16. Wong, G.K.; Kwan, M.C.; Ng, R.Y.; Simon, C.H.; Poon, W.S. Flow diverters for treatment of intracranial aneurysms: Current status and ongoing clinical trials. *J. Clin. Neuroradiol.* **2011**, *18*, 737–740. [CrossRef] [PubMed]
17. Steinman, D.A.; Milner, J.S.; Norley, C.J.; Lownie, S.P.; Holdsworth, D.W. Image-based computational simulation of flow dynamics in a giant intracranial aneurysm. *Am. J. Neuroradiol.* **2003**, *24*, 559–566. [PubMed]
18. Kojima, M.; Irie, K.; Masunaga, K.; Sakai, Y.; Nakajima, M.; Takeuchi, M.; Fukuda, T.; Arai, F.; Negoro, M. Hybrid stent device of flow-diverting effect and stent-assisted coil embolization formed by fractal structure. *Med. Biol. Eng. Comput.* **2016**, *54*, 1–11. [CrossRef]
19. Ley, D.; Mühl-Benninghaus, R.; Yilmaz, U.; Körner, H.; Cattaneo, G.F.M.; Mailänder, W.; Kim, Y.-J.; Scheller, B.; Reith, W.; Simgen, A. The derivo embolization device, a second-generation flow diverter for the treatment of intracranial aneurysms, evaluated in an elastase-induced aneurysm model. *J. Clin. Neurol.* **2017**, *3*, 335–343. [CrossRef]
20. O'kelly, C.J.; Spears, J.; Chow, M.; Wong, J.; Boulton, M.; Weill, A.; Willinsky, R.A.; Kelly, M.; Marotta, T.R. Canadian experience with the pipeline embolization device for repair of unruptured intracranial aneurysms. *Am. J. Neuroradiol.* **2013**, *34*, 381–387. [CrossRef]
21. Ma, J.; You, Z.; Peach, T.; Byrne, J.; Rizkallah, R.R. A new flow diverter stent for direct treatment of intracranial aneurysm. *J. Biomech.* **2015**, *48*, 4206–4213. [CrossRef] [PubMed]
22. Cohen, J.E.; Gomori, J.M.; Moscovici, S.; Leker, R.R.; Itshayek, E. Delayed complications after flow-diverter stenting: Reactive in-stent stenosis and creeping stents. *J. Clin. Neuros.* **2014**, *21*, 1116–1122. [CrossRef]
23. Kuzmik, G.A.; Williamson, T.; Ediriwickrema, A.; Andeejani, A.; Bulsara, K.R. Flow diverters and a tale of two aneurysms. *J. NeuroIntervent Surg.* **2013**, *5*, e23. [CrossRef]
24. Hampton, T.; Walsh, D.; Tolias, C.; Fiorella, D. Mural destabilization after aneurysm treatment with a flow-diverting device: A report of two cases. *J. NeuroIntervent. Surg.* **2011**, *3*, 167–171. [CrossRef]
25. Kulcsár, Z.; Houdart, E.; Bonafé, A.; Parker, G.; Millar, J.; Goddard, A.J.; Renowden, S.; Gál, G.; Turowski, B.; Mitchell, K.; et al. Intra-aneurysmal thrombosis as a possible cause of delayed aneurysm rupture after flow-diversion treatment. *Am. J. Neuroradiol* **2011**, *32*, 20–25. [CrossRef]
26. Goubergrits, L.; Schaller, J.; Kertzscher, U.; Woelken, T.; Ringelstein, M.; Spuler, A. Hemodynamic impact of cerebral aneurysm endovascular treatment devices: Coils and flow diverters. *Expert Rev. Med. Devices* **2014**, *11*, 361–373. [CrossRef]
27. Jou, L.D.; Mawad, M.E. Timing and size of flow impingement in a giant intracranial aneurysm at the internal carotid artery. *Med. Biol. Eng. Comput.* **2011**, *49*, 891–899. [CrossRef]
28. Chien, A.; Tateshima, S.; Castro, M.; Sayre, J.; Cebral, J.; Vinueala, F. Patient-specific flow analysis of brain aneurysms at a single location: Comparison of hemodynamic characteristics in small aneurysms. *Med. Biol. Eng. Comput.* **2008**, *46*, 1113–1120. [CrossRef]
29. Bouillot, P.; Brina, O.; Ouared, R.; Yilmaz, H.; Lovblad, K.-O.; Farhat, M.; Pereira, V.M. Computational fluid dynamics with stents: Quantitative comparison with particle image velocimetry for the three commercials off the shelf intracranial stents. *J. NeuroIntervent. Surg.* **2016**, *8*, 309–315. [CrossRef]
30. Boussel, L.; Rayz, V.; McCulloch, C.; Martin, A.; Acevedo-Bolton, G.; Lawton, M.; Higashida, R.; Smith, W.S.; Young, W.L.; Saloner, D. Aneurysm growth occurs at region of low wall shear stress patient- specific correlation of hemodynamics and growth in a longitudinal study. *Stroke* **2008**, *39*, 2997–3002. [CrossRef] [PubMed]
31. Bouillot, P.; Brina, O.; Ouared, R.; Lovblad, K.-O.; Farhat, M.; Pereira, V.M. Hemodynamic transition driven by stent porosity in sidewall aneurysms. *J. Biomech.* **2015**, *48*, 1300–1309. [CrossRef]
32. Gambaruto, A.; Janela, J.; Moura, A.; Sequeira, A. Sensitivity of hemodynamics in a patient specific cerebral aneurysm to vascular geometry and blood rheology. *Math. Biosc. Eng.* **2011**, *8*, 409–423.
33. Cebral, J.; Castro, M.; Appanaboyina, S.; Putman, C.; Millan, D.; Frangi, A. Efficient pipeline for image-based patient-specific analysis of cerebral aneurysm hemodynamics: Technique and sensitivity. *IEEE Trans. Med. Imaging* **2005**, *24*, 457–467. [CrossRef] [PubMed]
34. Xiang, J.; Tremmel, M.; Kolega, J.; Levy, E.I.; Natarajan, S.K.; Meng, H. Newtonian viscosity model could overestimate wall shear stress in intracranial aneurysm domes and underestimate rupture risk. *J. NeuroIntervent. Surg.* **2012**, *4*, 351–357. [CrossRef] [PubMed]
35. Gambaruto, A.; Janela, J.; Moura, A.; Sequeira, A. Shear-thinning effects of hemodynamics in patient-specific cerebral aneurysms. *Math. Biosc. Eng.* **2013**, *10*, 649–665.
36. Reymond, P.; Merenda, F.; Perren, F.; Rüfenacht, D.; Stergiopulos, N. Validation of one-dimensional model of the systemic arterial tree. *Am. J. Physiol Heart Circ. Physiol.* **2009**, *297*, H208–H222. [CrossRef]
37. Valencia, A.; Morales, H.; Rivera, R.; Bravo, E.; Galvez, M. Blood flow dynamics in patient-specific cerebral aneurysm models: The relationship between wall shear stress and aneurysm area index. *Med. Eng. Phys.* **2008**, *30*, 329–379. [CrossRef] [PubMed]

38. Meng, H.; Tutino, V.M.; Xiang, J.; Siddiqui, A. High WSS or low WSS? Complex interactions of hemodynamics with intracranial aneurysm initiation, growth, and rupture: Toward a unifying hypothesis. *Am. J. Neuroradiol.* **2014**, *35*, 1254–1262. [CrossRef] [PubMed]
39. Kallmes, D.F.; Ding, Y.H.; Dai, D.; Kadirvel, R.; Lewis, D.A.; Cloft, H.J. A new endoluminal, flow-disrupting device for treatment of saccular aneurysms. *Stroke* **2007**, *38*, 2346–2352. [CrossRef]
40. Lieber, B.B.; Sadasivan, C. Endoluminal scaffolds for vascular reconstruction and exclusion of aneurysms from the cerebral circulation. *Stroke* **2010**, *41*, S21–S25. [CrossRef]
41. Wootton, D.M.; Ku, D.N. Fluid mechanics of vascular systems, diseases, and thrombosis. *Ann. Rev. Biomed. Eng.* **1999**, *1*, 299–329. [CrossRef]
42. Meng, H.; Wang, Z.; Kim, M.; Ecker, R.D.; Hopkins, L.N. Saccular aneurysms on straight and curved vessels are subject to different hemodynamics: Implications of intravascular stenting. *Am. J. Neuroradiol.* **2006**, *27*, 1861–1865. [PubMed]
43. Bouillot, P.; Brina, O.; Ouared, R.; Lovblad, K.-O.; Farhat, M.; Pereira, V.M. Particle imaging velocimetry evaluation of intracranial stents in sidewall aneurysm: Hemodynamic transition related to the stent design. *PLoS ONE* **2014**, *9*, e113762. [CrossRef]
44. Kerl, H.U.; Boll, H.; Fiebig, T.; Figueiredo, G.; Förster, A.; Nölte, I.S.; Nonn, A.; Groden, C.; Brockmann, M.A. Implantation of pipeline flow-diverting stents reduces aneurysm inflow without relevantly affecting static intra-aneurysmal pressure. *Neurosurgery* **2014**, *4*, 321–334. [CrossRef] [PubMed]
45. Larrabide, I.; Aguilar, M.; Morales, H.G.; Geers, A.; Kulcsar, Z.; Rüfenacht, D.; Frangi, A. Intra-aneurysmal pressure and flow changes induced by flow diverters: Relation to aneurysm size and shape. *Am. J. Neuroradiol.* **2013**, *34*, 816–822. [CrossRef]
46. Xiang, J.; Ma, D.; Snyder, K.V.; Levy, E.I.; Siddiqui, A.H.; Meng, H. Increasing flow diversion for cerebral aneurysm treatment using a single flow diverter. *Neurosurgery* **2014**, *75*, 286–294. [CrossRef] [PubMed]
47. Gholampour, S.; Mehrjoo, S. Effect of bifurcation in the hemodynamic changes and rupture risk of small intracranial aneurysm. *Neurosurg. Rev.* **2021**, *44*, 1703–1712. [CrossRef] [PubMed]
48. Hajirayat, K.; Gholampour, S.; Sharif, I.; Bizaria, D. Biomechanical Simulation to Compare the Blood Flow Hemodynamics and Cerebral Anurysm Rupture Risk in Patients with Different Aneurysms Necks. *J. Appl. Mech. Tech. Phys.* **2017**, *58*, 968–974. [CrossRef]

fluids

MDPI

Article

Hemodynamic Investigation of the Flow Diverter Treatment of Intracranial Aneurysm

Maria Antonietta Boniforti *, Roberto Magini and Tania Orosco Salinas

Department of Civil, Building, and Environmental Engineering, Sapienza University, 00184 Rome, Italy
* Correspondence: antonietta.boniforti@uniroma1.it

Abstract: Flow diverter stents (FDS) are increasingly used for the treatment of complex intracranial aneurysms such as fusiform, giant, or wide-neck aneurysms. The primary goal of these devices is to reconstruct the diseased vascular segment by diverting blood flow from the aneurysm. The resulting intra-aneurysmal flow reduction promotes progressive aneurysm thrombosis and healing of the disease. In the present study, a numerical investigation was performed for modeling blood flow inside a patient-specific intracranial aneurysm virtually treated with FDS. The aim of the study is to investigate the effects of FDS placement prior to the actual endovascular treatment and to compare the effectiveness of devices differing in porosity. Numerical simulations were performed under pulsatile flow conditions, taking into account the non-Newtonian behavior of blood. Two possible post-operative conditions with virtual stent deployment were simulated. Hemodynamic parameters were calculated and compared between the pre-operative (no stent placement) and post-operative (virtual stent placement) aneurysm models. FDS placement significantly reduced intra-aneurysmal flow velocity and increased the Relative Residence Time (RRT) on the aneurysm, thus promoting thrombus formation within the dilatation and aneurysm occlusion. The results highlighted an increase in the effectiveness of FDS as its porosity increased. The proposed analysis provides pre-operative knowledge on the impact of FDS on intracranial hemodynamics, allowing the selection of the most effective treatment for the specific patient.

Keywords: hemodynamics; intracranial aneurysm; flow diverter stent; image-based computational fluid dynamics (CFD); patient-specific modelling; wall shear stress (WSS); oscillatory shear index (OSI); endothelial cell activation potential (ECAP); relative residence time (RRT)

Citation: Boniforti, M.A.; Magini, R.; Orosco Salinas, T. Hemodynamic Investigation of the Flow Diverter Treatment of Intracranial Aneurysm. *Fluids* **2023**, *8*, 189. https://doi.org/10.3390/fluids8070189

Academic Editors: D. Andrew S. Rees and Huidan (Whitney) Yu

Received: 12 June 2023
Revised: 20 June 2023
Accepted: 22 June 2023
Published: 24 June 2023

1. Introduction

Intracranial aneurysms (IA) are pathological dilatations of the arterial wall in the intracranial vasculature, most frequently observed in curves and bifurcations in the circle of Willis. They have a 3–5% incidence in the adult population [1].

Based on their shape, they are categorized into saccular and non-saccular types. Most IA remain asymptomatic during the lifetime of the patients. Nevertheless, if they rupture, subarachnoid hemorrhage occurs, which is associated with a high mortality and morbidity rate [2].

The evaluation of the risk of intracranial aneurysm rupture is still difficult and controversial. Several studies have suggested that hemodynamic changes associated with arterial dilatation might play an important role in aneurysm rupture. Many hemodynamic parameters have been correlated with the growth and rupture of the cerebral aneurysm, including low and/or high wall shear stress (WSS) and blood recirculation [3]. During the cardiac cycle, the luminal arterial surface is constantly subjected to the action of shear stress as a result of the blood flow. The temporal and spatial variations in WSS significantly affect the rates at which endothelial cells are remodeled, so that they are hypothesized to be associated with the growth and rupture of the intracranial aneurysm [4–6]. Conventionally, these vascular lesions were treated by surgical clipping or endovascular coiling. Insertion of coils promotes blood coagulation inside the aneurysm, reducing inflow into the arterial dilatation. Consequently, the aneurysm is gradually excluded from the main blood stream, and the risk of rupture decreases [7]. Nowadays, treatments include flow diverter stents, which offer a minimally invasive alternative to traditional methods, especially for giant or fusiform aneurysms and those with a wide neck that might be untreatable by conventional coiling [8]. The primary function of flow diverters is to redirect the flow reaching the aneurysm towards the original stream direction, decreasing inflow into the dilatation. This allows for reducing blood recirculation into the aneurysm, inducing a gradual thrombosis with subsequent healing of the disease [9]. The final result of the endovascular treatment will depend on the efficacy of the selected FDS. Thus, it is relevant to investigate the effect of different FDS on the aneurysmal hemodynamics in order to evaluate their efficacy.

Flow diverter stents can differ in porosity and pore density. The first is defined as the ratio of the metal-free surface area to the total surface area of the stent, the second indicates the number of pores per unit surface area [10]. Recently, several computational investigations analyzed the influence of stent porosity on the effectiveness of the endovascular treatment and the hemodynamic changes deter-mined by FDS placement [11–13].

Despite the interest of recent findings, some limitations recur in the numerical investigations on the FDS efficacy for the treatment of intracranial aneurysms. Many hemodynamic studies on cerebral aneurysms treated with flow diverter stents consider idealized models, disregarding the influence of real patient-specific geometries on WSS and the associated hemodynamic indices [14–18]. Furthermore, most numerical simulations of intracranial aneurysm hemodynamics model blood as a Newtonian fluid. This last hypothesis significantly affects the computational results since the characteristics of blood flow strongly depend on the rheological properties of the blood, and evidence for non-Newtonian behavior of intracranial blood flow was recently confirmed from Doppler ultrasonography measurements conducted in the cerebral arteries of 16 selected patients [19]. Furthermore, several researchers avoid the complex virtual reconstruction of the flow diverter stent and simulate the FDS as a porous medium with porosity and permeability equivalent to those of the actual FDS [11,20–22]. This kind of approach considers the flow through a medium with uniform spatial characteristics, whereas the mesh of flow diverter stents can be coarser in some regions and denser in others, influencing blood flow dynamics.

At last, it should be noted that, although FDS represents an effective alternative to conventional treatments, postoperative complications have been reported [23,24]. Furthermore, no firm conclusion has been reached yet about hemodynamic changes inside the aneurysm once flow diverting devices are placed into the diseased artery. For all these reasons, a better understanding of the effect of FDS placement in the intracranial aneurysm is critical for evaluating the most effective aneurysm treatment and planning surgery.

Taking into account the previous considerations, an accurate image-based computational investigation was performed for transient modeling of the blood stream inside a patient-specific intracranial aneurysm treated with flow diverter stents. The aim of the present study was to investigate the effects of FDS placement prior to the actual endovascular treatment and to compare the effectiveness of devices differing in porosity. In this way, it was possible to provide valuable insight into the most effective way to achieve the desired therapeutic outcome.

The patient-specific aneurysm model was derived from Computed Tomography Angiography (CTA) images. Following the reconstruction of the intracranial aneurysm, two FDS geometric models, differing in porosity, were realized. The flow diverter stents that, in the actual treatment, would be implanted in the patient-specific aneurysm were virtually inserted within the aneurysm, thus simulating real endovascular treatment.

Subsequently, the effect of FDS placement on the aneurysm hemodynamics was analyzed. The effectiveness of the devices was compared based on their ability to redirect blood flow towards the parent flow direction and exclude the aneurysm from blood circulation. For this purpose, numerical simulations were performed under pulsatile flow conditions in the pre-operative model (IA without stent placement) and in the post-operative ones (IA virtually treated with FDS deployment). Blood was modeled as a non-Newtonian fluid using the Carreau rheological model. As time-averaged wall shear stress (TAWSS) was found to play an important role in aneurysm growth, rupture, and thrombosis [2,3,24], hemodynamic parameters including TAWSS, oscillatory shear index (OSI), and endothelial cell activation potential (ECAP) were compared between the pre-operative and post-operative models. In addition, instantaneous velocity streamlines were analyzed. At last, a further hemodynamic parameter, the relative residence time (RRT), was evaluated to quantify the tendency toward thrombus formation and consequently the healing of the disease. This index made it possible to estimate the relative time that blood resided close to the aneurysmal wall. It was found to be a fundamental indicator of altered aneurysm hemodynamics and thrombus deposition [25–27], thus allowing the evaluation of the FDS's effectiveness.

Importantly, the methodology used for the FDS reconstruction and its deployment in a real (non-ideal) patient-specific geometry allowed for evaluating the effect of FDS placement on the hemodynamics of the intracranial aneurysm and identifying the most effective treatment.

2. Materials and Methods

2.1. Intracranial Aneurysm and FDS Reconstruction Process

The first step to simulating the pulsatile blood flow inside patient-specific aneurysms is their accurate geometrical reconstruction. In this study, the intracranial aneurysm model was derived from anonymous patient data received in Digital Imaging and Communications in Medicine (DICOM) format from "Università Cattolica del Sacro Cuore, Policlinico A. Gemelli", Italy. Informed consent was obtained from the patient. Computed Tomography Angiography revealed the presence of a saccular aneurysm in the vertebral artery, with a maximum diameter of about 14 mm and a wide neck of approximately 10 mm. No radiological signs of rupture were found. Endovascular repair with the deployment of a flow diverter stent was indicated for the aneurysm treatment. The procedures for patient-specific aneurysm reconstruction from the CTA images and FDS reconstruction are illustrated below.

First, a rough 3D model was reconstructed from the CTA data using the image analysis software ITK-SNAP (v.3.6.0). Subsequently, smoothing of the arterial wall was performed using the Meshmixer software (v.3.5). At last, further post-processing was performed with the software Vascular Modeling Toolkit (VMTK, v.1.4.0) to obtain the final model of the patient-specific aneurysm. In this last step, cylindrical flow extensions were added at the inlet and outlet of the aneurysm model to reduce boundary effects on numerical results and allow the flow to become fully developed before affecting the aneurysmal dilatation. Once the processing of CTA images was completed, the SpaceClaim software (v.2022 R1, ANSYS Inc., Canonsburg, PA, USA) was used to extract the luminal surface of the aneurysm, which was needed to perform the numerical simulations. The main steps of the aneurysm reconstruction process are shown in Figure 1.

Following the reconstruction of the patient-specific aneurysm, two FDS geometric models differing in porosity were created using DesignModeler software (v.2022 R1, ANSYS Inc., Canonsburg, PA, USA). The flow diverter stents that, in the actual endovascular treatment, would be implanted in the patient-specific aneurysm have been virtually inserted within the vascular model.

The method of realizing and inserting flow diverter stents into the patient-specific IA model required several steps. First, an unperforated reference tube, representing the scaffold of the flow diverter stent, was realized using DesignModeler software, approximately following the centerline of the non-dilated vessel. Then, several circular contours identifying the hypothetical healthy artery were placed inside the aneurysmal dilatation, as shown in Figure 2a. These circles identified the "skeleton" of the reference tube and allowed its reconstruction (Figure 2b).

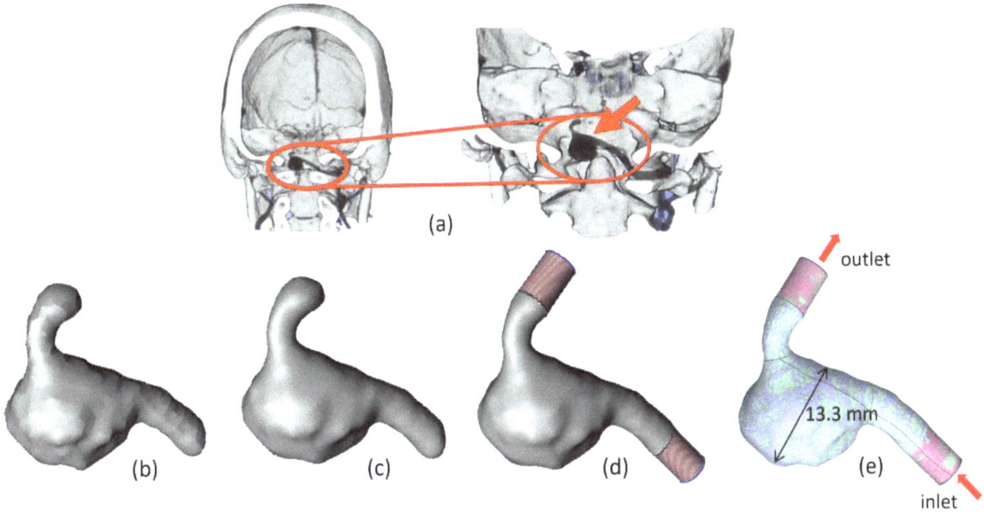

Figure 1. Schematic workflow chart of the intracranial aneurysm reconstruction from CTA images: (**a**) Volume rendering of the patient-specific intracranial aneurysm; (**b**) rough segmentation; (**c**) smoothing of the luminal surface; (**d**) insertion of extensions and global smoothing; (**e**) patient-specific final model. The arrows indicate the flow direction.

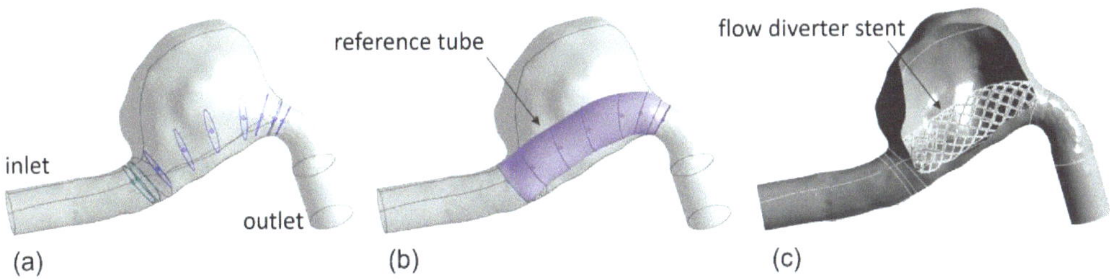

Figure 2. (**a**) Patient-specific aneurysm and circles identifying the "skeleton" of the reference tube; (**b**) patient-specific aneurysm with the reference tube representing the scaffold of the flow diverter stents; (**c**) section view of the patient-specific aneurysm with virtual FDS placement.

The initial and final cross-sections of the reference tube were made equal to those of the healthy artery. The surface of the scaffold in the initial area, to a limited extent, was accurately superimposed on the arterial wall. Checking the adherence between the scaffold surface and the vessel surface in the initial area of the scaffold (before the aneurysm dilatation) limited leakage problems and ensured the correct functionality of the flow diverter. Thus, preliminary tests were performed to determine the correct placement of the scaffold in the dilated artery.

Subsequently, suitable Boolean subtraction operations allowed the holing of the reference pipe to obtain two FDS models, named FD1 and FD2, which had the same pore density but differed in porosity. The FDS porosity was calculated by the ratio of the void surface area to the total surface area of the devices, that is, the surface of the reference tube. The porosity of the reconstructed FD1 and FD2 stents was equal to 76% and 49%, respectively.

Each FDS model was virtually deployed in the previously reconstructed patient-specific aneurysm, thus simulating the clinical endovascular treatment. The deployment of the flow diverter devices inside the intracranial aneurysm was achieved using a further Boolean subtraction operation. In particular, the considered flow diverter stent was subtracted from the aneurysmal geometry, thus obtaining the final computational fluid domain corresponding to the intracranial aneurysm with virtual implantation of the flow diverter stent (Figure 2c).

2.2. Governing Equations and Numerical Setup

The mass and momentum conservation equations for incompressible fluids and negligible gravity forces are [28]:

$$\nabla \cdot \vec{u} = 0 \tag{1}$$

$$\rho\left(\frac{\partial \vec{u}}{\partial t} + \vec{u} \cdot \nabla \vec{u}\right) = -\nabla p + \nabla \cdot \boldsymbol{\tau} \tag{2}$$

where $\vec{u}$ is the velocity vector, p the pressure, and ρ the density of the fluid. The deviatoric stress tensor $\boldsymbol{\tau}$ is a function of the strain rate tensor $\boldsymbol{D}$, $\boldsymbol{D} = \left(\nabla \vec{u} + \nabla \vec{u}^{T}\right)/2$, according to the relation $\boldsymbol{\tau} = 2\mu(\dot{\gamma})\boldsymbol{D}$, where μ is the dynamic viscosity of the fluid and $\dot{\gamma}$ is the shear rate. The system of Equations (1) and (2), with the associated boundary and initial conditions, determines the blood flow inside the aneurysm.

The pulsatile nature of the blood flow was modeled by assigning physiological boundary conditions. A no-slip condition, $\vec{u} = 0$, was imposed on the arterial wall of the patient-specific models; a time-dependent velocity v = v(t) was uniformly assigned at the inlet; and the pressure p = 100 mm Hg was prescribed at the outlet of the aneurysm models, that is, in the cases with and without FDS placement. Since the patient-specific velocity waveform was not available, a typical flow-rate waveform for the vertebral cerebral artery was taken from the literature [29]. Then, the inlet area of the patient-specific aneurysm was used to calculate the time-dependent velocity waveform, which was assigned at the inlet of the models (Figure 3). The period of the velocity waveform was equal to 0.8 s, the maximum velocity occurred at the systolic peak instant t = 0.16 s, and the minimum velocity was observed at the diastolic instant t = 0.432 s. Further significant instants were considered for describing hemodynamics inside the aneurysm, as illustrated in Figure 3. The instants t = 0.096 s, t = 0.112 s, and t = 0.208 s referred to the systolic phase; the instant t = 0.64 s referred to the diastolic phase.

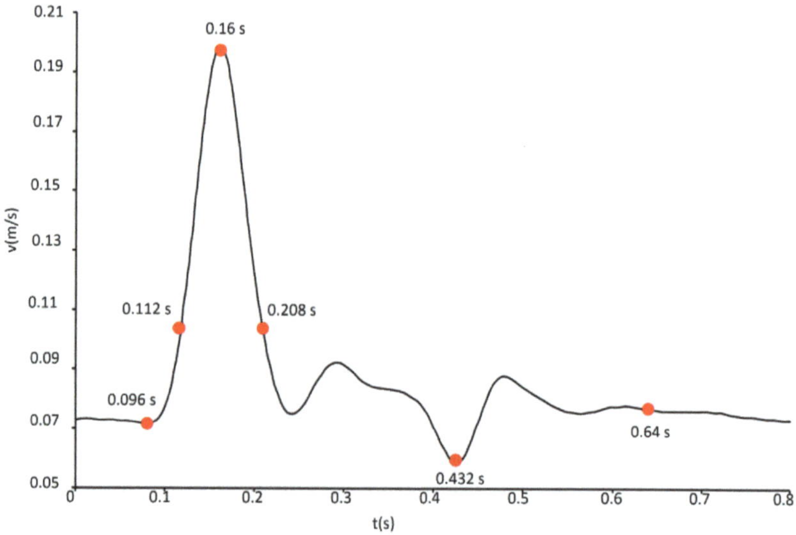

Figure 3. Pulsatile velocity waveform assigned at the inlet of the IA models. Red points indicate the instants of the cardiac cycle selected for the hemodynamic investigation.

The non-Newtonian rheological properties of blood were taken into account. The shear-thinning behavior of blood was modeled by the Carreau model, using the following relationship for the dynamic viscosity μ:

$$\mu(\dot{\gamma}) = \mu_\infty + (\mu_0 - \mu_\infty)\left[1 + (\lambda\dot{\gamma})^2\right]^{\frac{n-1}{2}} \tag{3}$$

where μ_0 = 0.056 kg/(m·s) is the viscosity at zero shear rate $\dot{\gamma}$, μ_∞ = 0.0035 kg/(m·s) is the viscosity for an infinite shear rate, λ = 3.313 is the relaxation time, and n = 0.3568 is the power-law index [30].

A laminar blood flow was assumed, as suggested by the low value of the averaged Reynold number Re = ρUD/μ, which is based on the diameter D of the healthy artery at the model inlet (D = 0.00427 m) and the time-averaged velocity U assigned at the inlet (U = 0.1034 m/s). Assuming the blood density ρ = 1060 kg/m^3 and the dynamic viscosity $\mu = \mu_\infty$ = 0.0035 kg/(m·s), the averaged Reynolds number is about Re $\approx$ 134, and the maximum Reynolds number, corresponding to the systolic peak velocity, is $Re_{max} \approx 257$. Furthermore, the rigid wall assumption was made.

A computational investigation was performed for modeling blood flow in the patient-specific aneurysm under pulsatile flow conditions. The analyzed cases referred to the pre-operative vascular geometry obtained from the image segmentation process and the simulated post-operative ones obtained by virtual placement of the FD1 and FD2 stents inside the aneurysm. The numerical simulations were performed using ANSYS Fluent.v.2022 R1 [31]. The SIMPLE (Semi-Implicit Method for Pressure-Linked Equations) method for the pressure-velocity coupling and a second-order upwind scheme for the spatial discretization of momentum were adopted [32]. For the discretization of the temporal terms, a first-order implicit method was used.

The cardiac cycle was divided into 100 time-steps of 0.008 s, and 200 iterations were performed for each time step. The convergence criteria were set such that the residuals of velocity components and continuity were below 10^{-5} at each time-step. To minimize the effect of initial numerical transients at the beginning of the computation, three cardiac cycles were simulated, and only the third cycle was considered for the hemodynamic analysis.

The computational domains associated with the pre-operative condition (IA with no FDS placement) and the post-operative ones (IA with FD1 placement and IA with FD2 placement) were discretized into a large number of tetrahedral computational cells. The size of the mesh elements in the lumen was the same in all cases. However, in the cases of FD1 and FD2, the mesh was refined in the proximity of the devices.

2.3. Mesh Convergence Analysis

The size of the mesh elements can greatly affect the accuracy of numerical solutions. Therefore, a mesh sensitivity analysis was performed for each IA model to ensure that the numerical results were not sensitive to the mesh size. For the mesh independence study, TAWSS values were monitored for different mesh sizes along a curve located in the middle of the aneurysmal surface. Figure 4 shows the comparison of the TAWSS values along the selected curve for the last three mesh sizes considered. The values refer to the analyzed cases, that is, the pre-operative model and the two post-operative ones. As the number of mesh elements increased, a greater agreement between the TAWSS profiles was observed, and a mesh independence condition could be assumed (Figure 4). As a further validation of the mesh independence, Table 1 shows the influence of the mesh size on the instantaneous WSS value averaged on the artery wall for the considered models. The values refer to the systolic peak instant. In all cases, only a minimal variation of less than 1% was found between the WSS values associated with the two last meshes, i.e., the meshes with the largest number of elements.

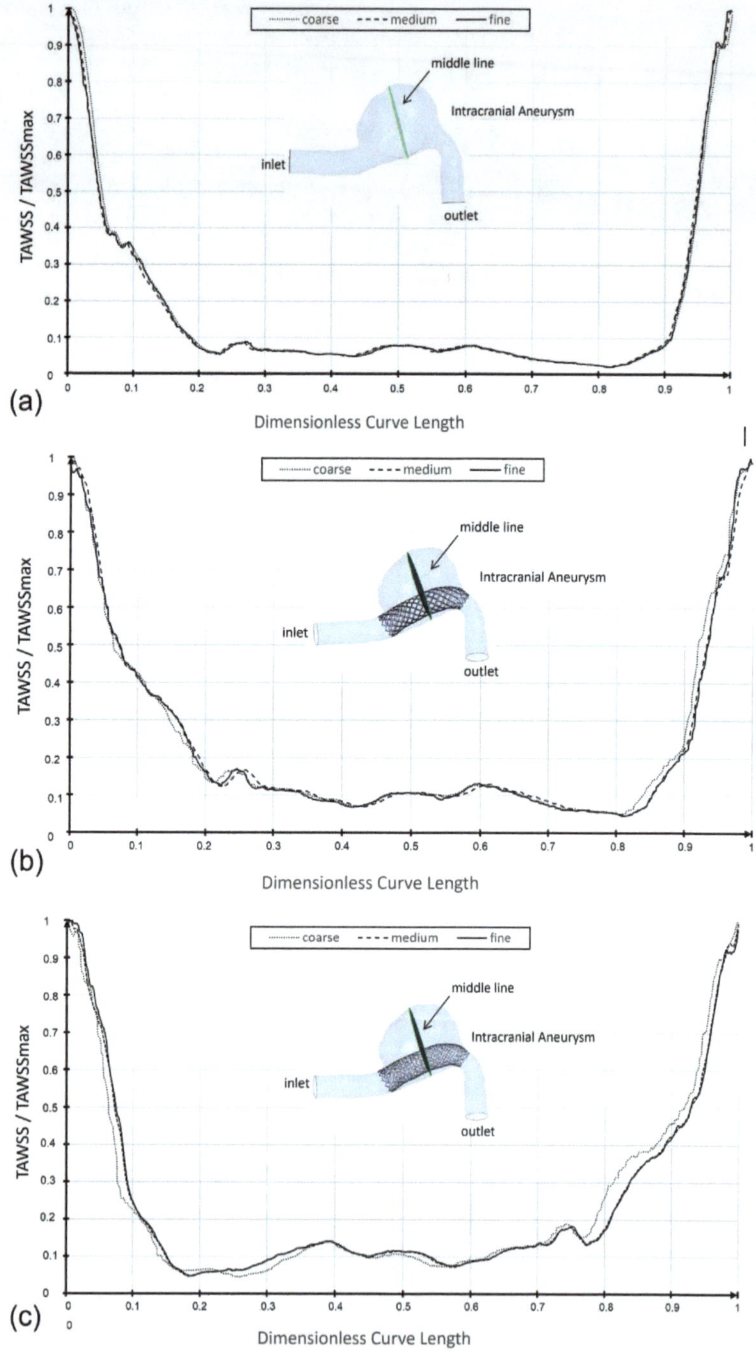

Figure 4. Dimensionless TAWSS along the middle curve in the patient-specific aneurysm for three different mesh sizes (coarse, medium, and fine): (**a**) pre-operative model with no FDS placement, number of mesh elements approximately equal to 1×10^6, 2×10^6, and 3×10^6; (**b**,**c**) post-operative models with FD1 and FD2 placement respectively, number of mesh elements approximately equal to 2×10^6, 3×10^6, and 4×10^6 in both cases.

Table 1. WSS at the systolic peak instant (t = 0.16 s), averaged on the artery wall, for different numbers of mesh elements in the pre-operative IA model and in the two post-operative ones.

No FDS Placement		**FD1 Placement**		**FD2 Placement**	
Elements (approximate number)	**WSS** (Pa)	**Elements** (approximate number)	**WSS** (Pa)	**Elements** (approximate number)	**WSS** (Pa)
1×10^6	1.24014	2×10^6	1.2718	2×10^6	1.22263
2×10^6	1.23136	3×10^6	1.23366	3×10^6	1.21016
3×10^6	1.24095	4×10^6	1.2331	4×10^6	1.2181

The results shown in Table 1 and the good correspondence in the TAWSS profiles for the highest numbers of mesh elements suggested the choice of the optimal mesh. In particular, the final mesh in the absence of flow diversion devices reached approximately 3×10^6 elements, as opposed to approximately 4.1×10^6 and 4.3×10^6 elements in the presence of FD1 and FD2 devices. These values were obtained using an element size equal to 0.12 mm in the lumen for all cases and 0.05 mm in the vicinity of the flow diverter devices for FD1 and FD2 cases.

Figure 5 shows the computational mesh generated in the case of virtual FDS placement (post-operative condition) and the refinement near the flow-diverting stent. The finer mesh cells locate the FDS position.

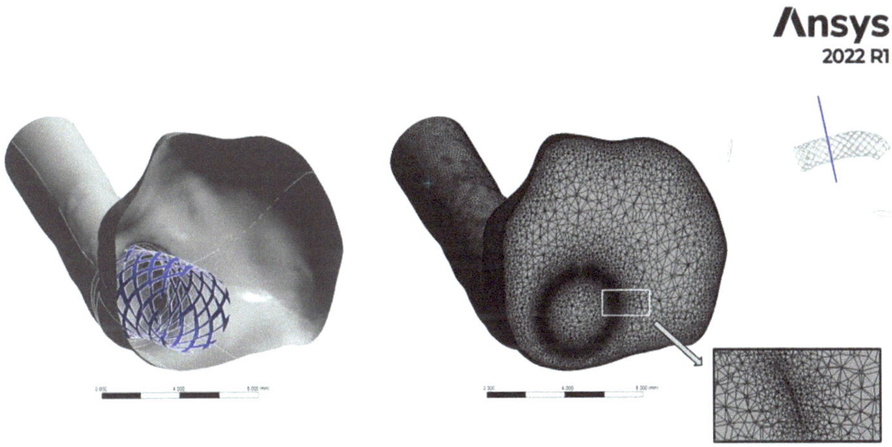

Figure 5. Mesh view of the cross section identified by the blue line in the intracranial aneurysm in the case of FDS placement. The finer mesh cells locate the position of the flow diverter stent.

2.4. Hemodynamic Parameters

To analyze the effect of FDS placement in the patient-specific aneurysm, the hemodynamic indicators TAWSS, OSI, ECAP, and RRT were calculated. These time-averaged parameters are able to quantify the unsteady nature of blood flow and were found to be essential for describing the altered hemodynamics in intracranial aneurysms after FDS implantation and evaluating the performance of the devices.

The time-averaged wall shear stress TAWSS furnishes the average over the cardiac cycle of the magnitude of the WSS vector, i.e.,

$$\text{TAWSS} = \frac{1}{T}\int_0^T \left|\overrightarrow{WSS}\right| dt$$

where T is the period of the cardiac cycle and $\overrightarrow{WSS}$ indicates the instantaneous wall shear stress vector.

The oscillatory shear index OSI is a non-dimensional parameter that takes into account directional changes of the WSS vector during the cardiac cycle with respect to the dominant direction of the flow [33].

$$\mathrm{OSI} = \frac{1}{2}\left(1 - \frac{\left|\int_0^T \overrightarrow{WSS}\,dt\right|}{\int_0^T \left|\overrightarrow{WSS}\right| dt}\right)$$

OSI values range from 0 to 0.5. Unidirectional shear stress, corresponding to a zero OSI value, is related to a healthy condition, while a high OSI value was recognized as inducing an inflammatory response of the artery wall [34].

Another important hemodynamic parameter is the endothelial cell activation potential (ECAP), which correlates the TAWSS values with the OSI values:

$$\mathrm{ECAP} = \mathrm{OSI}/\mathrm{TAWSS}$$

This parameter is generally used to characterize the "thrombogenic susceptibility" of the aneurysm wall. The knowledge of the ECAP distribution allows us to locate areas with high ECAP values. These regions are exposed to both high OSI and low TAWSS at the same time, which identifies conditions of endothelial susceptibility [35,36].

Furthermore, according to Himburg and co-authors [25], the residence time of particles near the vessel wall can be evaluated using the RRT index:

$$\mathrm{RRT} = \frac{1}{(1 - 2\cdot\mathrm{OSI})\cdot\mathrm{TAWSS}} = \frac{1}{1/T\left|\int_0^T \overrightarrow{WSS}\,dt\right|} \tag{4}$$

This last parameter allows us to quantify the reduction of the velocity of blood flow in the aneurysmal dilatation, thus identifying the possibility of thrombus formation and subsequent aneurysm healing. For this reason, it is particularly useful for determining the effectiveness of flow-diverting devices.

3. Results and Discussion

The primary function of flow diverters is to optimally alter the hemodynamics within the aneurysm sac, reducing flow recirculation and allowing for thrombus formation within the aneurysm. The results of the computational investigation highlighted that in the absence of FDS, the flow pattern in the aneurysmal sac was characterized by a single large, organized vortex at each instant of the cardiac cycle. The blood recirculating region was characterized by a counterclockwise vortex, as depicted in Figure 6, at a selected instant of the cardiac cycle.

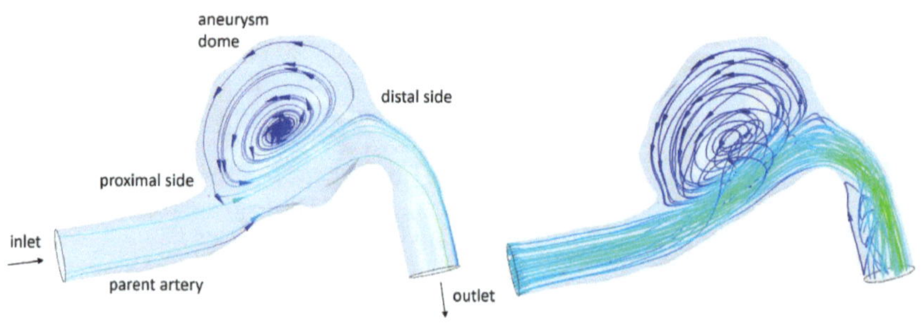

Figure 6. The 2D and 3D streamlines in the patient-specific aneurysm without stent placement at the instant t = 0.208 s of the cardiac cycle. The arrows highlight the direction of blood flow, coming from the parent artery, through the aneurysm dome, and towards the outlet.

3.1. FDS Effects on Blood Flow

The deployment of the flow-diverting devices in the patient-specific geometry significantly altered the aneurysm hemodynamics. Figure 7 shows the time-averaged velocity contours on a longitudinal cross-section of the intracranial aneurysm in the analyzed conditions: the pre-operative condition, i.e., aneurysm with no FDS placement, and the simulated post-operative conditions, i.e., virtual placement of the FD1 and FD2 stents, respectively. The presence of flow diverter stents redirected blood flow within them, thus reducing the enlargement of the fluid vein in the dilatation, dampening the region of high velocity in the inflow zone, and restoring the original healthy stream direction, as shown in Figure 7. The results of the numerical investigation highlighted a significant decrease in the time-averaged velocity inside the aneurysm after FDS implantation and a greater blockage of blood supply to the aneurysm for the lower-porosity FD2 stent.

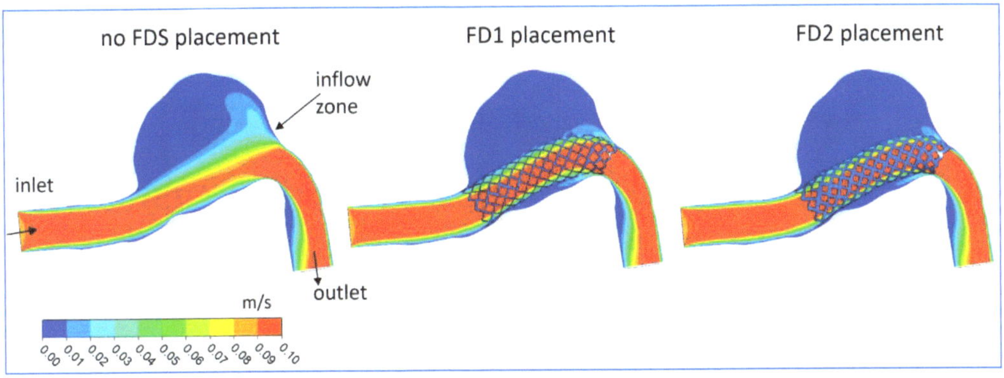

Figure 7. Time-averaged velocity contours on a longitudinal cut plane in the patient-specific intracranial aneurysm in the cases of aneurysms with no stent placement, with FD1 stent placement, and with FD2 stent placement. The inflow zone is the area of the aneurysm neck where the blood flow enters the aneurysm.

The instantaneous velocity contours on the longitudinal section of the aneurysm confirmed the diversion of blood flow from the aneurysmal sac in the majority of the cardiac cycle (Figure 8). It is interesting to note the FDS efficacy in reducing velocity inside the aneurysmal dome in almost all phases of the cardiac cycle, that is, in the systolic acceleration phase, the diastolic minimum, and the late diastole phase. On the contrary, in the absence of FDS placement, extensive regions of high velocity values were observed in the aneurysm dome, as depicted in the first column of Figure 8. At the systolic peak instant (t = 0.16 s), the FDS efficacy in decreasing the high-velocity regions in the dome was less than in the other instants, as the flow diverting devices were not able to completely contain inside them the incoming fast flow. However, unlike the other phases of the cardiac cycle, this instant was not associated with a large recirculation involving the entire aneurysm dome, as shown in Figure 9.

The last consideration suggested that the effectiveness of the devices for the treatment of intracranial aneurysms could be better evaluated by considering additional hemodynamic indicators able to take into account the complexity of the blood flow. With this aim, the distribution on the aneurysm surface of the time-averaged parameters TAWSS, OSI, ECAP, and RRT was evaluated in order to quantify the tendency toward thrombus formation and the consequent evaluation of the FDS efficacy. In addition, streamline evolution during the cardiac cycle was analyzed. The knowledge of the flow evolution during the cardiac cycle was significant because recirculating flow associated with low TAWSS has been considered responsible for the deterioration of the arterial wall [37,38]. In particular, the dynamics of the recirculation regions give rise to non-physiological WSS, which in turn appears to be negatively correlated with thrombus deposition [39].

The 2D streamline evolution during the cardiac cycle is illustrated in Figure 9. In addition, Figure 10 shows the 3D streamlines in the considered models at the instants of systolic peak (t = 0.16 s) and diastolic minimum (t = 0.64 s).

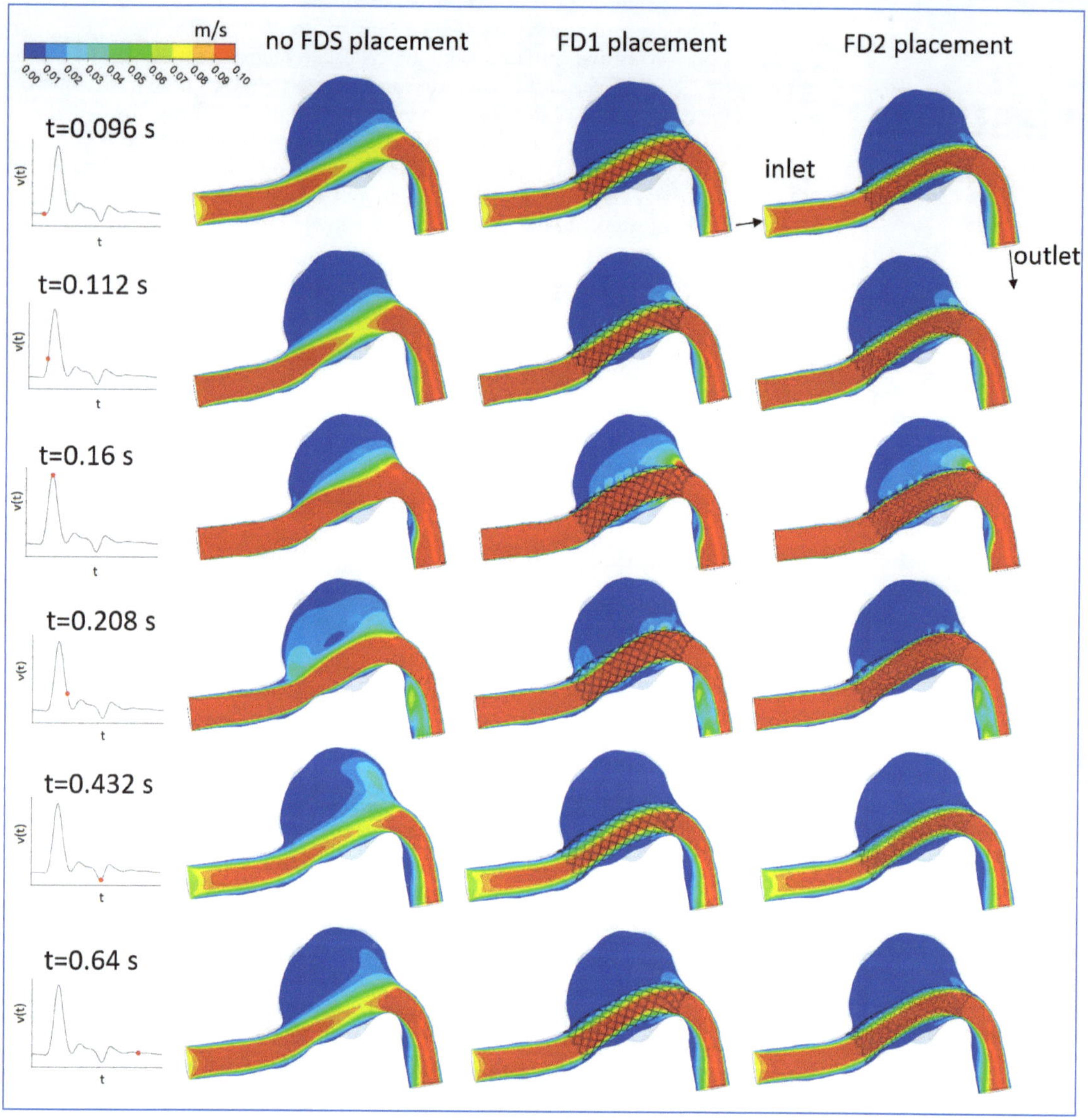

Figure 8. Velocity magnitude contours on the longitudinal section of the intracranial aneurysm at the selected instants of the cardiac cycle. The first column refers to the pre-operative condition (absence of FDS placement); the second and third columns refer to the simulated post-operative conditions, that is, FD1 and FD2 placement, respectively.

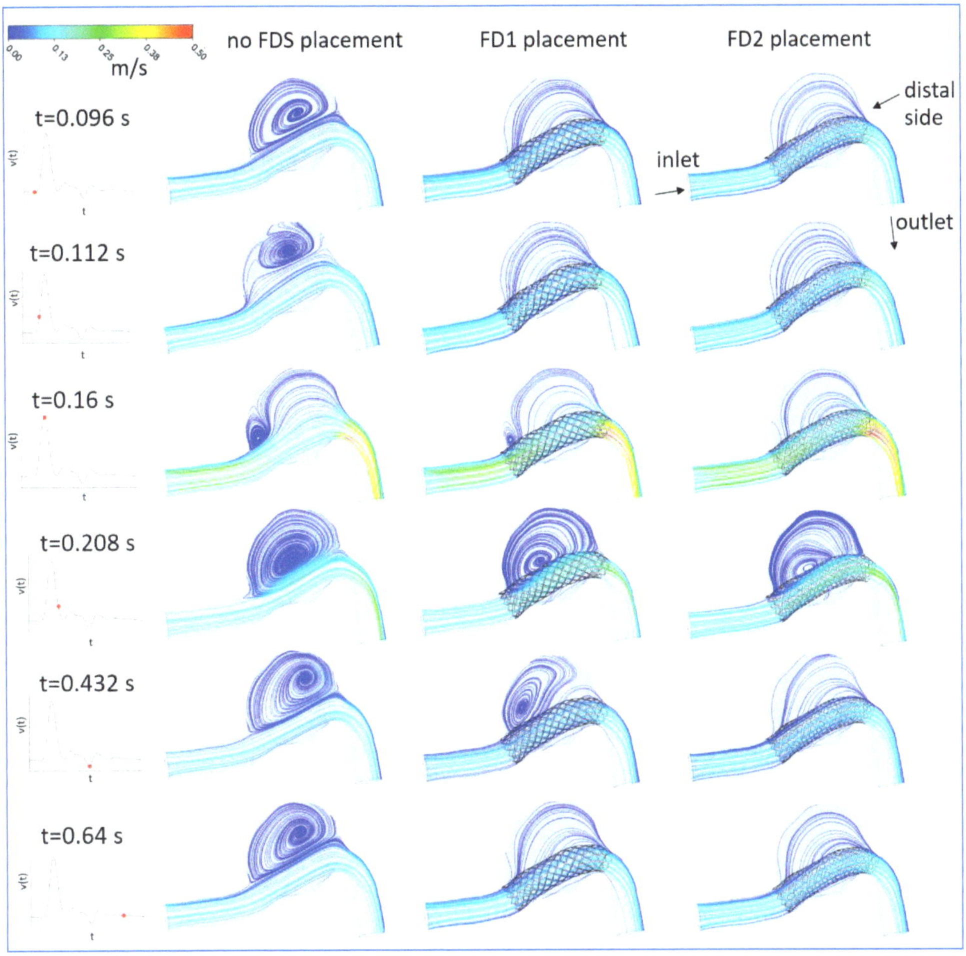

Figure 9. The 2D streamlines evolution during the cardiac cycle in the patient-specific aneurysm without a flow diverter stent (first column), with the FD1 stent placement (second column), and the FD2 stent placement (third columns). The streamlines were colored by the local velocity magnitude. The same number of streamlines was considered in all three cases.

At the selected instants of the cardiac cycle, different flow patterns were observed. The flow distribution in the longitudinal plane revealed a large recirculation area in the aneurysm that was found at all instants of the cardiac cycle in the absence of FDS treatment (Figure 9, first column). Referring to this pre-operative condition, it can be observed as follows. A counterclockwise recirculation area involved the entire aneurysm dome at all instants of the cardiac cycle except the instant of the systolic peak. In the presence of this large vortex, the flow pattern was divided into two separate regions: a recirculating region confined to the aneurysm dome and an underlying fluid vein advancing in the original stream direction. In the phase of systolic acceleration, the pre-existing vortical region emerging from the previous cycle was reduced due to the action of the blood flow entering the aneurysm that compressed the vortex region, confining it to the top of the dome (Figure 9, t = 0.112 s). At the instant of the systolic peak, t = 0.16 s, only a small recirculating region caused by flow detachment from the wall was observed. This vortex was confined to the proximal area of the aneurysm by the fast blood flow advancing in the mainstream direction that widened into the dilatation. As the velocity assigned at the inlet of the IA model decreased (during the cardiac cycle), the vortex gradually widened to the entire aneurysm dome (Figure 9, t = 0.208 s). In the last phase of the cardiac cycle, the low velocity inlet values characterizing the late diastole allowed for the maintenance of a large recirculating region at all instants of the diastolic phase, as shown in Figure 9 for t = 0.432 s and t = 0.64 s. In this phase, a single large vortex was observed, caused by the slow flow entering the dilatation distally and flowing back along the aneurysm wall.

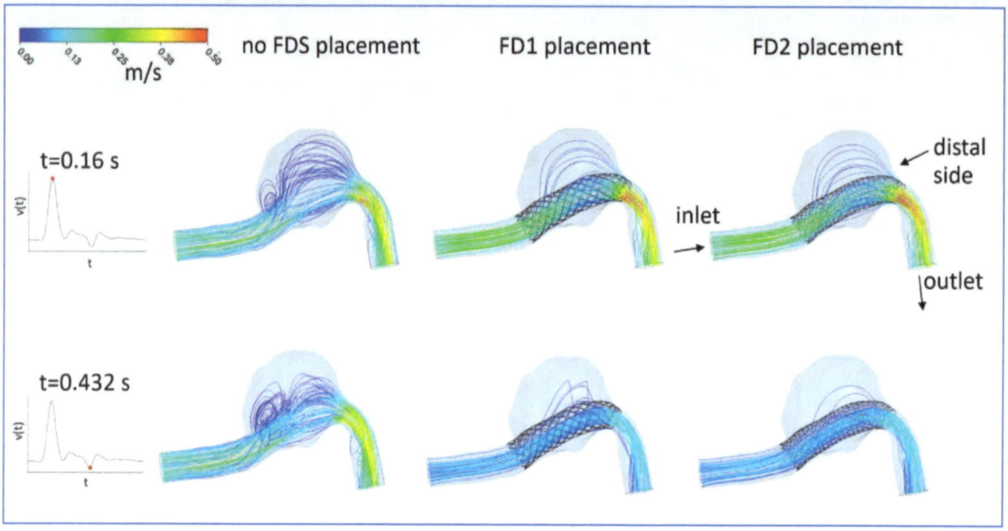

Figure 10. The 3D streamlines in the patient-specific aneurysm without a flow diverter stent (first column), with the FD1 stent placement (second column), and the FD2 stent placement (third columns) at the systolic peak instant t = 0.16 s and the diastolic minimum instant t = 0.64 s. The streamlines were colored by the local velocity magnitude. The same number of streamlines was considered in all three cases.

Figure 10 shows the 3D streamlines at the systolic peak instant t = 0.16 s and the diastolic minimum instant t = 0.64 s in the considered pre- and post-operative models of the intracranial aneurysm.

The presence of flow diverter stents significantly altered the described blood flow dynamics. The flow redirection effect, due to the FD1 or FD2 placement, determined a substantial reduction in flow recirculation and velocity magnitude inside the aneurysm, as depicted in the second and third columns of Figures 8–10. The lower porosity of the FD2 stent enhanced this effect, determining a further decrease in the velocity values inside the aneurysm and preventing or weakening the formation of the vortex region.

3.2. *Hemodynamic Parameters*

The hemodynamic parameters averaged over the cardiac cycle were significantly modified by the virtual deployment of the flow diverter stents. The distributions of TAWSS, OSI, ECAP, and RRT for the pre-operative model of the aneurysm and the post-operative ones are shown in Figure 11. After FDS placement, significant increases inside the aneurysm were observed for all hemodynamic indices except TAWSS, which, on the contrary, decreased on the surface of the aneurysm (Figure 11). In fact, the flow redirection effect due to the stent placement translated into a reduction of TAWSS values on the aneurysmal dome, determining an advantageous effect on the aneurysmal flow pattern.

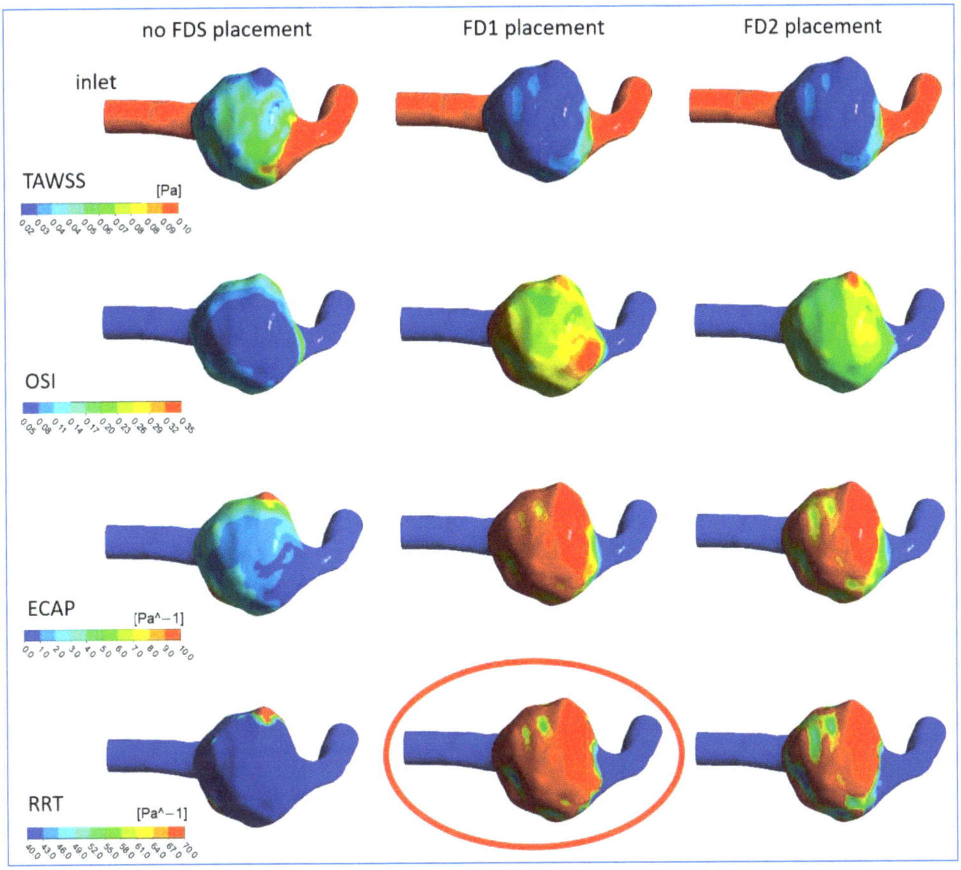

Figure 11. TAWSS, OSI, ECAP, and RRT contours on the luminal surfaces of the patient-specific aneurysm in the pre-operative condition (absence of FDS placement, first column) and in the considered post-operative conditions (FD1 and FD2 flow diverter placement, second and third columns, respectively). The red circle highlights the high RRT values due to the FD1 placement.

Furthermore, the alteration of the flow due to the FDS treatment produced an elevation of the OSI and modified its distribution on the aneurysm wall. The higher OSI values observed in the dome indicated remarkable fluctuations of the WSS vector during the cardiac cycle in the aneurysm.

Numerical results highlighted that the high OSI region increased with the increase in the FDS porosity, that is, in the case of the FD1 placement. However, high OSI alone is not sufficient for identifying critical regions for thrombus formation, while the regions where both high OSI and low TAWSS were found, have a high probability of thrombus deposition [39]. Thus, the ECAP parameter was analyzed since it combines the effects of OSI and TAWSS, taking into account the level of the shear and its oscillatory character [36]. As well documented in the literature, this parameter is an appropriate hemodynamic indicator for identifying the endothelial susceptibility of the aneurysm wall [35]. The deployment of FD1 and FD2 stents in the parent vessel harboring the aneurysm resulted in a substantial increase in the ECAP values, especially at the top of the aneurysmal dome. Again, the higher porosity of the FD1 stent, which was less effective in reducing intra-aneurysm velocity and recirculation, determined a greater increase in the high ECAP region than in the FD2 case.

At last, a substantial increase in RRT on the aneurysm wall compared to the preoperative condition was found in the presence of FDS placement, as shown in the last row of Figure 11 and in the different aneurysm views illustrated in Figure 12.

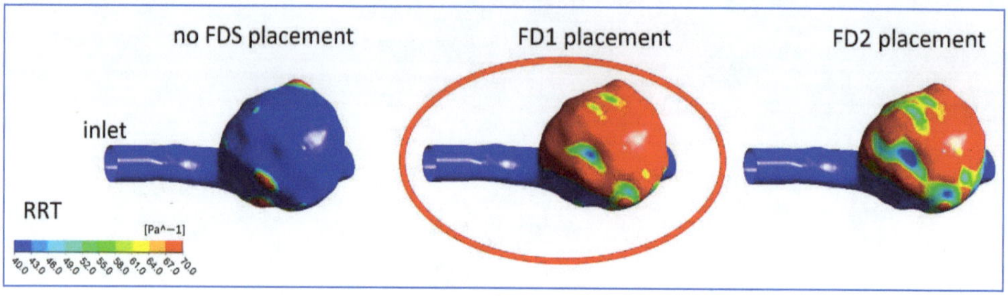

Figure 12. RRT contours on the luminal surfaces of the patient-specific aneurysm in the considered cases: absence of FDS placement, FD1 flow diverter placement, and FD2 flow diverter placement. The red circle highlights the high RRT values due to the FD1 placement.

The extensive regions with high RRT observed in the presence of FDS treatments indicated blood flow stagnation in the aneurysmal dome and suggested the formation of thrombogenic conditions advantageous for aneurysm occlusion, confirming the efficacy of the FDS treatments. The use of the FD1 stent, characterized by the higher porosity, enhanced this beneficial effect, further increasing the residence time of blood (RRT) in the neighborhood of the vascular endothelium, as depicted in the last row of Figure 11 and in Figure 12. Although the higher porosity of this stent partially allowed blood inflow into the aneurysm, the more extensive regions of high RRT suggested the FD1's greater efficacy in producing blood thrombosis and aneurysm occlusion.

3.3. Study Limitations

The main assumptions adopted in the current study are analyzed in the following.

Because the patient-specific velocity profile was not available, a pulsatile velocity waveform typical of the vertebral artery was taken from the literature and used as inlet boundary condition for the computational models. Despite the fact that patient-specific inflow is desirable for flow simulations in patient-specific analysis, the use of a literature waveform does not limit the validity of the numerical investigation since patient-specific inlet waveforms minimally affect the hemodynamics of patient-specific intracranial aneurysms [40]. Confirming this, a recent study conducted on 156 intracranial aneurysms indicated that patient-specific inflow boundary conditions may not be a must-have condition in CFD simulations, and hemodynamic parameters associated with rupture are the same using patient-specific and generalized inflow boundary conditions [41].

Another assumption that may affect the results of this study is the hypothesis of a rigid wall of the aneurysm, although it is commonly adopted in most computational studies on the hemodynamics of cerebral aneurysms. This assumption is justified by the loss of elastic lamina and substantial variability in collagen architecture found in cerebral aneurysms, which results in reduced wall compliance [42,43]. On the other hand, fluid–structure interaction investigations, which account for wall compliance, require information on patient-specific material properties of the arterial wall, including aneurysm wall stiffness, thrombus properties, and the local thickness of the aneurysm wall. These quantities are difficult to evaluate, and the lack of this information can result in marked differences in the distributions of the fundamental hemodynamic parameters. As an example, accounting for non-uniform wall thickness distribution in fluid-structure simulations, strong differences were found in the wall stress distribution, and much higher stress values were found in the rupture site of an intracranial aneurysm compared to a configuration with constant wall thickness [44]. Future studies will address these limitations. Furthermore, it could be interesting to investigate the influence of pore density on the efficacy of the FDS endovascular treatment.

4. Conclusions

In recent years, flow diverting devices have become an effective alternative to conventional treatments of intracranial aneurysms. Understanding FDS's effectiveness in inducing aneurysm occlusion may be fundamental for treatment planning. To achieve this aim, a computational investigation of the hemodynamics of a patient-specific intracranial aneurysm reconstructed from medical images and virtually treated with FDS was performed. The FDS porosity was taken into account as it plays a crucial role in the effectiveness of the device. This parameter affects the stent's ability to redirect blood flow away from the aneurysm into the parent artery. A low porosity is desirable, but if the porosity is too low, the stent might become too rigid for deployment. Two possible post-operative conditions were simulated, with the deployment of flow diverter stents differing in porosity. A comparison of the devices in modifying cerebral aneurysm hemodynamics was analyzed. The flow diverter stents were virtually implanted in the parent vessel harboring the aneurysm, with the devices covering the entire neck area. The placement of the devices determined a substantial reduction in flow recirculation and velocity magnitude inside the aneurysm. A reduction of the TAWSS values and an increase in the OSI and ECAP values were found. In particular, the relative residence time RRT on the aneurysmal wall markedly increased, indicating blood flow stagnation in the dome and thrombogenic conditions for the aneurysm occlusion.

Interestingly, the different impact of the considered flow diverter stents on the aneurysmal hemodynamics. An increase in the region occupied by higher RRT was observed for an increase in the porosity of the flow diverter device, that is, for virtual implantation of the FD1 stent. Although this higher porous stent was less effective in decreasing flow recirculation and velocity within the aneurysm dome, it determined more marked increases in the ECAP and RRT values on the aneurysm wall, thus resulting more effective in thrombus formation and aneurysm occlusion.

At last, the current study highlighted how a computational investigation of the effects of FDS placement performed before the actual surgical procedure can provide valuable information on the most effective treatment to achieve the desired therapeutic result.

Author Contributions: Conceptualization, M.A.B.; methodology, M.A.B. and T.O.S.; software, M.A.B. and T.O.S.; validation, M.A.B., R.M. and T.O.S.; formal analysis, M.A.B.; investigation, M.A.B., R.M. and T.O.S.; resources, M.A.B.; writing—original draft preparation, M.A.B. and T.O.S.; writing—review and editing, M.A.B., R.M. and T.O.S.; supervision, M.A.B.; project administration, M.A.B.; funding acquisition, M.A.B. All authors have read and agreed to the published version of the manuscript.

Funding: This research was funded by "Sapienza University of Rome" Progetti di Ricerca (Piccoli, Medi)—Progetti Piccoli, 2021. Protocol number: RP12117A5D8437A8.

Data Availability Statement: Not applicable.

Acknowledgments: The authors would like to acknowledge Enrico Marchese for providing the anonymous patient-specific medical images and useful comments about them.

Conflicts of Interest: The authors declare no conflict of interest.

References

1. Etminan, N.; Rinkel, G.J. Unruptured intracranial aneurysms: Development, rupture and preventive management. *Nat. Rev. Neurol.* **2016**, *12*, 699–713. [CrossRef] [PubMed]
2. Texakalidis, P.; Sweid, A.; Mouchtouris, N.; Peterson, E.C.; Sioka, C.; Rangel-Castilla, L.; Reavey-Cantwell, J.; Jabbour, P. Aneurysm Formation, Growth, and Rupture: The Biology and Physics of Cerebral Aneurysms. *World Neurosurg.* **2019**, *130*, 277–284. [CrossRef] [PubMed]
3. Meng, H.; Tutino, V.M.; Xiang, J.; Siddiqui, A. High WSS or low WSS? Complex interactions of hemodynamics with intracranial aneurysm initiation, growth, and rupture: Toward a unifying hypothesis. *AJNR Am. J. Neuroradiol.* **2014**, *35*, 1254–1262. [CrossRef] [PubMed]

4. Qiu, T.; Jin, G.; Xing, H.; Lu, H. Association between hemodynamics, morphology, and rupture risk of intracranial aneurysms: A computational fluid modeling study. *Neurol. Sci.* **2017**, *38*, 1009–1018. [CrossRef]
5. Hsiai, T.K.; Cho, S.K.; Honda, H.M.; Hama, S.; Navab, M.; Demer, L.L.; Ho, C.M. Endothelial cell dynamics under pulsating flows: Significance of high versus low shear stress (d(tau)/dt). *Ann. Biomed. Eng.* **2002**, *30*, 646–656. [CrossRef] [PubMed]
6. Murayama, Y.; Fujimura, S.; Suzuki, T.; Takao, H. Computational fluid dynamics as a risk assessment tool for aneurysm rupture. *Neurosurg. Focus* **2019**, *47*, E12. [CrossRef]
7. Jin, Z.H.; Gerdroodbary, M.B.; Valipour, P.; Faraji, M.; Abu-Hamdeh, N.H. CFD investigations of the blood hemodynamic inside internal cerebral aneurysm (ICA) in the existence of coiling embolism. *Alex. Eng. J.* **2023**, *66*, 797–809. [CrossRef]
8. Byrne, J.V.; Beltechi, R.; Yarnold, J.A.; Birks, J.; Kamran, M. Early experience in the treatment of intra-cranial aneurysms by endovascular flow diversion: A multicentre prospective study. *PLoS ONE* **2010**, *5*, e12492. [CrossRef]
9. Venguru, S.; Nguyen, K.; Davidson, J. Intracranial Aneurysms: Evaluation of Braid Pore Configurations on Flow Disruption with Flow Diverter Devices. *Biomed. Sci. Technol.* **2019**, *1*, 18–26.
10. Sadasivan, C.; Cesar, L.; Seong, J.; Rakian, A.; Hao, Q.; Tio, F.O.; Wakhloo, A.K.; Lieber, B.B. An original flow diversion device for the treatment of in-tracranial aneurysms: Evaluation in the rabbit elastase-induced model. *Stroke* **2009**, *40*, 952–958. [CrossRef] [PubMed]
11. Mutlu, O.; Olcay, A.B.; Bilgin, C.; Hakyemez, B. Evaluating the Effectiveness of 2 Different Flow Diverter Stents Based on the Stagnation Region Formation in an Aneurysm Sac Using Lagrangian Coherent Structure. *World Neurosurg.* **2019**, *127*, e727–e737. [CrossRef]
12. Zhang, Y.; Wang, Y.; Kao, E.; Flórez-Valencia, L.; Courbebaisse, G. Towards optimal flow diverter porosity for the treatment of intracranial aneurysm. *J. Biomech.* **2019**, *82*, 20–27. [CrossRef]
13. Dholakia, R.; Sadasivan, C.; Fiorella, D.J.; Woo, H.H.; Lieber, B.B. Hemodynamics of flow diverters. *J. Biomech. Eng.* **2017**, *139*, 021002. [CrossRef]
14. Suzuki, T.; Takao, H.; Fujimura, S.; Dahmani, C.; Ishibashi, T.; Mamori, H.; Fukushima, N.; Yamamoto, M.; Murayama, Y. Selection of helical braided flow diverter stents based on hemodynamic performance and mechanical properties. *J. Neurointerv. Surg.* **2017**, *9*, 999–1005. [CrossRef] [PubMed]
15. Kim, S.; Yang, H.; Hong, I.; Oh, J.H.; Kim, Y.B. Computational Study of Hemodynamic Changes Induced by Overlapping and Compacting of Stents and Flow Diverter in Cerebral Aneurysms. *Front Neurol.* **2021**, *12*, 705841. [CrossRef] [PubMed]
16. Nada, A.; Hassan, M.A.; Fakhr, M.A.; El-Wakad, M.T.I. Studying the effect of stent thickness and porosity on post-stent implantation hemodynamics. *J. Med. Eng. Technol.* **2021**, *45*, 408–416. [CrossRef] [PubMed]
17. Catalán-Echeverría, B.; Kelly, M.E.; Peeling, L.; Bergstrom, D.; Chen, X.; Malvè, M. CFD-Based Comparison Study of a New Flow Diverting Stent and Commercially-Available Ones for the Treatment of Cerebral Aneurysms. *Appl. Sci.* **2019**, *9*, 1341. [CrossRef]
18. Bouillot, P.; Brina, O.; Ouared, R.; Yilmaz, H.; Lovblad, K.O.; Farhat, M.; Pereira, V.M. Computational fluid dynamics with stents: Quantitative comparison with particle image velocimetry for three commercial off the shelf intracranial stents. *J. Neurointerv. Surg.* **2016**, *8*, 309–315. [CrossRef]
19. Saqr, K.M.; Mansour, O.; Tupin, S.; Hassan, T.; Ohta, M. Evidence for non-Newtonian behavior of intracranial blood flow from Doppler ultrasonography measurements. *Med. Biol. Eng. Comput.* **2019**, *57*, 1029–1036. [CrossRef]
20. Abdehkakha, A.; Hammond, A.L.; Patel, T.R.; Siddiqui, A.H.; Dargush, G.F.; Meng, H. Cerebral aneurysm flow diverter modeled as a thin inhomogeneous porous medium in hemodynamic simulations. *Comput. Biol. Med.* **2021**, *139*, 104988.20. [CrossRef]
21. Ngoepe, M.N.; Ventikos, Y. Computational modelling of clot development in pa-tient-specific cerebral aneurysm cases. *J. Thromb Haemost.* **2016**, *14*, 262–272. [CrossRef] [PubMed]
22. Xu, J.; Karmonik, C.; Yu, Y.; Lv, N.; Shi, Z.; Liu, J.M.; Huang, Q. Modeling flow diverters using a porous medium approach: A fast alternative to virtual flow diverter deployment. *World Neurosurg.* **2022**, *164*, e501–e508. [CrossRef] [PubMed]
23. Rouchaud, A.; Brinjikji, W.; Lanzino, G.; Cloft, H.J.; Kadirvel, R.; Kallmes, D.F. Delayed Hemorrhagic Complications after Flow Diversion for Intracranial Aneurysms: A Literature Overview. *Neuroradiology* **2016**, *58*, 171–177. [CrossRef]
24. Cebral, J.R.; Mut, F.; Raschi, M.; Scrivano, E.; Ceratto, R.; Lylyk, P.; Putman, C.M. Aneurysm rupture following treatment with flow diverting stents: Computational hemodynamics analysis of treatment. *AJNR Am. J. Neuroradiol.* **2011**, *32*, 27–33. [CrossRef] [PubMed]
25. Himburg, H.A.; Grzybowski, D.M.; Hazel, A.L.; LaMack, J.A.; Li, X.M.; Friedman, M.H. Spatial comparison between wall shear stress measures and porcine arterial endothelial permeability. *Am. J. Physiol. Heart Circ. Physiol.* **2004**, *286*, H1916–H1922. [CrossRef]
26. Rayz, V.; Boussel, L.; Ge, L.; Leach, J.; Martin, V.; Lawton, V.; McCulloch, V.; Saloner, D. Flow Residence Time and Regions of Intraluminal Thrombus Deposition in Intracranial Aneurysms. *Ann. Biomed. Eng.* **2010**, *38*, 3058–3069. [CrossRef] [PubMed]
27. Lee, S.W.; Antiga, L.; Steinman, D.A. Correlation among indicators of disturbed flow at the normal carotid bifurcation. *J. Biomech. Ing.* **2009**, *131*, 061013. [CrossRef]
28. Schlichting, H. *Boundary Layer Theory*, 7th ed.; McGraw-Hill: New York, NY, USA, 1979.
29. Reymond, P.; Merenda, F.; Perren, F.; Rüfenacht, D.; Stergiopulos, N. Validation of a one-dimensional model of the systemic arterial tree. *Am. J. Physiol. Heart Circ. Physiol.* **2009**, *297*, H208–H222. [CrossRef]
30. Shibeshi, S.S.; Collins, W.E. The rheology of blood flow in a branched arterial system. *Appl. Rheol.* **2005**, *15*, 398–405. [CrossRef]
31. ANSYS. *Workbench User's Guide, Release 2022 R1, January 2022*; ANSYS, Inc.: Canonsburg, PA, USA, 2022.

32. Ferziger, J.H.; Peric, M. *Computational Methods for Fluid Dynamics*; Springer: Berlin/Heidelberg, Germany, 2001; ISBN 978-3-540-42074-3.
33. He, X.; Ku, D.N. Pulsatile flow in the human left coronary artery bifurcation: Average conditions. *J. Biomech. Eng.* **1996**, *118*, 74–82. [CrossRef]
34. Sorescu, G.P.; Song, H.N.; Tressel, S.L.; Hwang, J.; Dikalov, S.; Smith, D.A.; Boyd, N.L.; Platt, M.O.; Lassègue, B.; Griendling, K.K.; et al. Bone morphogenic protein 4 produced in endothelial cells by oscillatory shear stress induces monocyte adhesion by stimulating reactive oxygen species production from a nox1-based NADPH oxidase. *Circ. Res.* **2004**, *95*, 773–779. [CrossRef]
35. Di Achille, P.; Tellides, G.; Figueroa, C.A.; Humphrey, J.D. A haemodynamic predictor of intraluminal thrombus formation in abdominal aortic aneurysms. *Proc. R. Soc. Lond. Math. Phys. Eng. Sci.* **2014**, *470*, 20140163. [CrossRef]
36. Boniforti, M.A.; Cesaroni, M.C.; Magini, R.; Pasqui, E.; de Donato, G. Image-Based Numerical Investigation in an Impending Abdominal Aneurysm Rupture. *Fluids* **2022**, *7*, 269. [CrossRef]
37. Cecchi, E.; Giglioli, C.; Valente, S.; Lazzeri, C.; Gensini, G.F.; Abbate, R.; Mannini, L. Role of hemodynamic shear stress in cardiovascular disease. *Atherosclerosis.* **2011**, *214*, 249–256. [CrossRef] [PubMed]
38. Tarbell, J.M.; Shi, Z.D.; Dunn, J.; Jo, H. Fluid mechanics, arterial disease, and gene expression. *Annu. Rev. Fluid Mech.* **2014**, *46*, 591–614. [CrossRef]
39. Boniforti, M.A.; Di Bella, L.; Magini, R. On the role of hemodynamics in predicting rupture of the abdominal aortic aneurysm. *J. Zhejiang Univ. Sci.* **2021**, *A 22*, 957–978. [CrossRef]
40. Xiang, J.; Siddiqui, A.H.; Meng, H. The effect of inlet waveforms on computational hemodynamics of patient-specific intracranial aneurysms. *J. Biomech.* **2014**, *47*, 3882–3890. [CrossRef] [PubMed]
41. Li, W.; Wang, S.; Tian, Z.; Zhu, W.; Zhang, Y.; Zhang, Y.; Wang, Y.; Wang, K.; Yang, X.; Liu, J. Discrimination of intracranial aneurysm rupture status: Patient-specific inflow boundary may not be a must-have condition in hemodynamic simulations. *Neuroradiology* **2020**, *62*, 1485–1495. [CrossRef] [PubMed]
42. Robertson, A.M.; Duan, X.; Aziz, K.M.; Hill, M.R.; Watkins, S.C.; Cebral, J.R. Diversity in the strength and structure of unruptured cerebral aneurysms. *Ann. Biomed. Eng.* **2015**, *43*, 1502–1515. [CrossRef] [PubMed]
43. Rayz, V.L.; Cohen-Gadol, A.A. Hemodynamics of cerebral aneurysms: Connecting medical imaging and biomechanical analysis. *Annu. Rev. Biomed. Eng.* **2020**, *22*, 231–256. [CrossRef]
44. Voß, S.; Glaßer, S.; Hoffmann, T.; Beuing, O.; Weigand, S.; Jachau, K.; Preim, B.; Thévenin, D.; Janiga, G. Berg, Fluid-structure simulations of a ruptured intracranial aneurysm: Constant versus patient-specific wall thickness. *Comput. Math. Methods Med.* **2016**, *2016*, 9854539. [CrossRef]

fluids

MDPI

Article

Image-Based Numerical Investigation in an Impending Abdominal Aneurysm Rupture

Maria Antonietta Boniforti [1,*], Maria Chiara Cesaroni [1], Roberto Magini [1], Edoardo Pasqui [2] and Gianmarco de Donato [2]

[1] Department of Civil, Building, and Environmental Engineering, Sapienza University, 00184 Rome, Italy
[2] Department of Medicine, Surgery and Neuroscience, University of Siena, 53100 Siena, Italy
* Correspondence: antonietta.boniforti@uniroma1.it

Abstract: Blood flow dynamics plays a crucial role in the growth and rupture of abdominal aortic aneurysms. The aim of this study was to analyze the possibility of predicting aneurysmal rupture by numerical investigations based on diagnostic images. The blood flow dynamics was analyzed in a patient-specific abdominal aortic aneurysm, reconstructed from CT images of an aneurysm while it was rupturing. The patient-specific geometry was virtually repaired in order to obtain a non-ruptured model representative of the geometry immediately preceding the rupture. To reproduce physiological conditions, numerical simulations were performed under pulsatile flow conditions, and blood was modelled as a non-Newtonian fluid, using the Carreau rheological model. Hemodynamic parameters that influence the rupture of the aneurysm were investigated, and their possible association with vascular disease was discussed. The results of the numerical simulations indicated regions of slow recirculation and low values of Time Averaged Wall Shear Stress (TAWSS) in the region of rupture. Unlike literature results, a high Oscillatory Shear Index (OSI) was not clearly found in this region. Nevertheless, just in the region where the rupture will occur, high values of Endothelial Cell Activation Potential index (ECAP) were found. This index is therefore extremely significant for assessing the vulnerability of the aortic wall and locating the critical rupture region.

Keywords: abdominal aneurysm; computational fluid dynamics; wall shear stress; oscillatory shear index; patient-specific modelling

Citation: Boniforti, M.A.; Cesaroni, M.C.; Magini, R.; Pasqui, E.; de Donato, G. Image-Based Numerical Investigation in an Impending Abdominal Aneurysm Rupture. *Fluids* **2022**, *7*, 269. https://doi.org/10.3390/fluids7080269

Academic Editors: Huidan (Whitney) Yu and Mehrdad Massoudi

Received: 7 July 2022
Accepted: 2 August 2022
Published: 5 August 2022

1. Introduction

Abdominal aortic aneurysms (AAAs) are pathological dilatations of the abdominal aorta, which represent a life-threatening condition, as their rupture is often lethal. They involve an increase in diameter above 50% with respect to the diameter of the healthy aorta, which is about 2.0 cm in the abdomen [1,2].

Surgery is generally recommended if the maximum aortic diameter exceeds 5.0 cm in women and 5.5 cm in men, or if the maximum diameter shows a growth rate greater than 0.5–1.0 cm in one year [3,4].

Nevertheless, rupture of small and medium abdominal aortic aneurysms is found as well as perfect integrity of large AAAs over long periods [2,5–10]. Consequently, maximum AAA diameter cannot be the only indicator to predict the AAA's rupture and plan the surgery.

A rupture potential index (RPI), defined as the ratio of the locally acting wall stress to the wall strength, was firstly proposed by Vande Geest et al., which found for ruptured aneurysm a mean RPI value equal to 0.48 [11]. This index was found to be high in ruptured aortic aneurysms. The evaluation of this index allowed a significant differentiation of the risk of rupture for diameters in the range of 55–75 mm, for which the criterion of the maximum diameter was not always useful [12]. Although this index and some others can help to predict AAA rupture risk, their validation is often difficult. Furthermore, they are often based only on mechanical quantities without considering the effect of the pulsatile flow that interacts with the arterial wall.

Actually, intraluminal thrombus (ILT) accumulation and progression of the AAA pathology determine variations in the physiological geometry of the artery, which result in a disturbed blood flow. The perturbed flow plays in turn a crucial role in the growth of the abdominal aortic aneurysm (AAAs) and contributes to the progression of the AAA pathology. In fact, it indirectly determines damage to the endothelium and increases the degeneration of the arterial wall.

As matter of fact, the biological relevance of the wall shear stress (WSS) and its relationship with the endothelium was largely emphasized [13]. Changes in magnitude and direction of the WSS vector act on blood vessels owing to the presence of the endothelial cells, which form the inner lining of the vessel walls. WSS sensed by endothelial cells plays an important role in the development, remodeling and maintenance of the vascular system [13]. Non-physiological shear stress associated with the perturbed flow in the aneurysm correlates with atherosclerotic lesion, as well documented in literature [14,15]. This non-physiologic shear stress was also indicated to promote the growth and possible rupture of the aneurysm [16].

Deterioration of the arterial wall was also associated with low Time Averaged Wall Shear Stresses (TAWSS) and recirculating blood regions [16–19]. While unidirectional wall shear stress characterizes healthy regions of the arterial wall, regions exposed to disturbed flow conditions are associated with Oscillatory Shear Stress (OSI), which induces monocyte adhesion in endothelial cells and inflammatory response [20]. In abdominal aortic aneurysms, compared to healthy aorta values, lower WSS and higher OSI have been found [18]. However, some contradictory results were found, since ILT growth was observed in regions of low OSI [21,22].

To help clinicians in the difficult decision on the need for surgery, this study highlights how some hemodynamic parameters can provide useful indications for the prediction of aneurysm rupture.

In particular, numerical simulations of the blood flow have been employed to investigate the relationship between hemodynamics and aneurysm rupture. A patient-specific model of an abdominal aortic aneurysm was reconstructed from a CT angiography of a ruptured aneurysm, obtained immediately after the rupture in the Hospital of the University of Siena (Italy). The geometry of the aneurysm immediately preceding the rupture was assumed to be similar to the ruptured geometry since it does not change significantly during the rupture. Based on this hypothesis, the patient-specific model reconstructed from the CT diagnostic images was virtually repaired, and a non-ruptured model, which was representative of the geometry immediately preceding the rupture, was used to investigate the blood flow dynamics.

Numerical simulations in physiological conditions of pulsating motion were carried out, and hemodynamic parameters that can predict possible rupture of the aneurysm were analyzed. In particular, 2D streamlines on suitable longitudinal cross-sections were studied throughout the cardiac cycle. Moreover, hemodynamic quantities that can help in identifying artery regions that could be subject to rupture, i.e., time-averaged velocity, TAWSS, OSI, and Endothelial Cell Activation Potential (ECAP), were evaluated.

This work is organized as follows. Firstly, in paragraph 2, the methods and setting up of the study are presented. In particular, the procedure for the 3D reconstruction of the pre-rupture patient-specific aneurysms from the CT images is described, and the governing equations and hypotheses for the numerical fluid dynamics model are furnished. Special attention is also paid to the fluid dynamics parameters used to describe the blood flow in the aneurysm, providing a physics-based interpretation of their role. In paragraph 3, the main results obtained from the computational investigation are presented, with the aim of a better understanding of the blood flow in the AAA's pre-rupture condition. Finally, some concluding remarks are provided, taking into account the link between the hemodynamic fields and the presence and position of rupture regions. In particular, the important role of the ECAP index was highlighted.

It is important to underline the innovative methodology used to obtain the pre-rupture model of the aneurysm, which allows a hemodynamic investigation in an unruptured aneurysm in impending rupture, with the advantage of knowing the real future position of the rupture region.

The results of this work highlight the importance of a patient-specific fluid dynamics investigation to predict a possible rupture of the aneurysm and to aid clinicians in the difficult decision regarding the need for surgery.

2. Materials and Methods

2.1. Pre-Rupture Patient-Specific Model Reconstruction

To simulate the pulsatile blood flow in diseased arteries, accurate reconstruction of the patient-specific model is a fundamental step, as it furnishes the computational domain to be used in the numerical investigation. Incorrect segmentation leads to incorrect hemodynamic analysis which results in wrong conclusions in the assessment of the progression of the disease. The method of reconstruction of the patient-specific geometry from the CT images used in this work is illustrated below.

The AAA model was derived from patient data selected from the clinical case database of the Vascular Surgery Unit, Department of Medicine, Surgery and Neuroscience, University of Siena, Italy. The clinical history was the following. An 81-year-old woman was admitted to the emergency department with abdominal pain. A thoracoabdominal computed tomography angiography was performed, and the exam highlighted the presence of an infrarenal abdominal aortic aneurysm with no radiological signs of rupture or impending rupture. The patient was hospitalized to perform in the following days an endovascular repair of the AAA with the implantation of a bifurcated aortic endograft. [23,24]. In the next hours, the patient started to be hemodynamically unstable with significant hemoglobin loss. In this light, an emergent CT was repeated that revealed a clear rupture of the AAA. The subject gave informed consent to use data for research purposes at the moment of hospital admission. The ethical committee of the hospital was informed of the no-experimental, retrospective analysis of the case, and no objections were raised.

The Angio-CT of the ruptured aneurysm was used for the 3D model reconstruction. As illustrated, the Angio-CT was acquired immediately after the rupture of the aneurysm. Since it is reasonable that the geometry of the aneurysm does not change significantly immediately before and after the rupture, the CT scans of the ruptured aneurysm were considered representative of the conditions immediately preceding the rupture. Based on this assumption, the ITK-SNAP software (v.3.8.0) was used for the segmentation process, in order to reconstruct the pre-rupture geometry, i.e., the geometry that immediately precedes the rupture. Therefore, a first coarse model of the luminal surface of the rupturing patient-specific aneurysm was obtained. The region of rupture, which corresponds to the leakage of blood, was artificially closed and smoothed, by the Meshmixer 3.5 software (http://www.meshmixer.com/ (accessed on 7 October 2021) Autodesk, Inc., San Rafael, CA, USA). In this step of model reconstruction, visceral arteries and minor arteries that branch off the aorta were occluded to simplify the model, and the blood flow was confined into the aorta.

A further pre-processing phase of the aneurysm model reconstruction was made using the open-source VMTK software (Vascular Modeling Tool Kit, version 1.4.0) in order to obtain the final geometry of the pre-rupture aneurysm. In this step, a global smooth of the luminal surface was made, and cylindrical flow extensions were added at the inlet and outlets of the patient-specific model. The insertion of the artificial extensions reduces boundary effects on numerical results and allows the flow to become fully developed before affecting the aneurysmal dilatation [25,26].

The final refined model, in STL format, reflects all the patient-specific morphological features of the examined abdominal aneurysm and allows analyzing the effects of hemodynamics on its rupture. Once the preliminary process on medical images was completed, the specific geometry of the aneurysm was imported into SpaceClaim software (ANSYS Inc., 2020 R2), which allowed extracting surfaces from the STL file. The main steps of the aneurysm model reconstructions and the final pre-rupture geometry are shown in Figure 1.

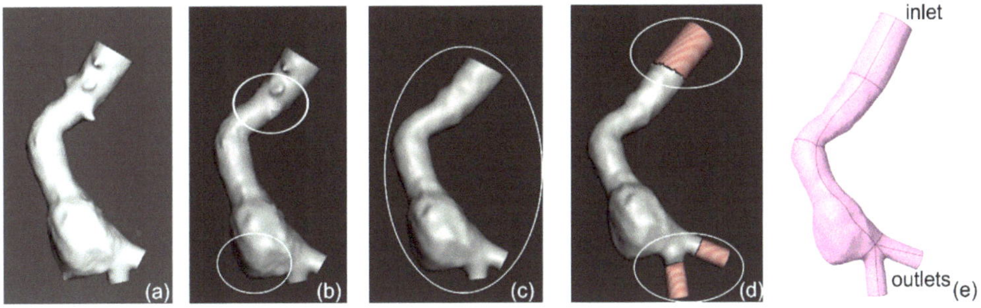

Figure 1. Main steps of the patient-specific aneurysm reconstruction from the Angio-CT medical images of a ruptured aneurysm: (**a**) rough segmentation (ITK_SNAP), (**b**) virtual repair and cut of minor arteries (Meshmixer), (**c**) global smooth (VMTK) (**d**) insertion of extensions (VMTK), (**e**) surface extraction (SpaceClaim).

2.2. Governing Equations and Numerical Setup

To achieve reliable results on the influence of hemodynamics on the growth and rupture of abdominal aortic aneurysms, numerical simulations of blood flow in aneurysms must be carried out considering the complexity of the physical problem. In the abdominal artery, the flow is pulsating, the blood is characterized by non-Newtonian rheological behavior, and the patient-specific aneurysmal geometries are extremely complex and must be accurately reconstructed. The above considerations were taken into account in the present computational investigations.

The governing equations of the fluid dynamics problem are the mass and momentum conservation equations, which for incompressible fluid and negligible gravity force are [25]:

$$\nabla \cdot \vec{u} = 0 \tag{1}$$

$$\rho\left(\frac{\partial \vec{u}}{\partial t} + \vec{u} \cdot \nabla \vec{u}\right) = -\nabla p + \nabla \cdot \tau \tag{2}$$

where $\vec{u}$ is the velocity vector, p the pressure, and ρ the density of the fluid. The deviatoric stress tensor τ is a function of the strain rate tensor D, according to the relation $\tau = 2\,\mu(\dot{\gamma})\,D$, where μ is the dynamic viscosity of the fluid, $\dot{\gamma}$ is the shear rate, and $D = (\nabla\vec{u} + \nabla\vec{u}^{T})/2$.

The assigned boundary conditions are: $\vec{u} = 0$ on the wall of the patient-specific model (no-slip condition), a flat (plug) velocity profile $v = v(t)$ normal to the inlet section, and a time-dependent pressure $p = p(t)$ at the outlet sections (Figure 2).

The system of Equations (1) and (2), with the associate boundary and initial conditions, determines the blood flow. It was numerically solved using the computational fluid dynamics software ANSYS Fluent (v. 2021 R2). A first order implicit method was used for the discretization of the temporal terms, the SIMPLE (Semi-Implicit Method for Pressure Linked Equations) scheme was used for the pressure-velocity coupling, and a second order upwind scheme was selected for the spatial discretization of momentum [19,27–29].

Furthermore, the following assumptions were adopted. Blood was modelled as an incompressible fluid, characterized by a non-Newtonian behavior, due to both its composition of suspended cells and the ability of red blood cells to deform and aggregate.

The shear-thinning behavior of blood was modelled by the Carreau model. Thus, the following relationship holds for the dynamic viscosity μ:

$$\mu(\dot{\gamma}) = \mu_\infty + (\mu_0 - \mu_\infty)[1 + (\lambda\dot{\gamma})^2]^{\frac{n-1}{2}} \quad (3)$$

where μ_0 = 0.056 kg/(m·s) is the viscosity at zero shear rate $\dot{\gamma}$, μ_∞ = 0.0035 kg/(m·s) is the viscosity for an infinite shear rate, λ = 3.313 is the relaxation time, and n = 0.3568 is the power-law index [30].

A laminar flow condition was assumed. This assumption is suggested by the low value of the average Reynolds number $Re = \rho UD/\mu$, where D is the diameter of the healthy aorta at the inlet of the model (D = 0.024 m), and U is the time averaged velocity (U = 0.09 m/s) assigned at the inlet where a velocity distribution constant along the radial direction was assumed. Assuming the blood density ρ = 1060 kg/m^3 and the dynamic viscosity μ = 0.0035 kg/(m·s), the Reynolds number is approximately $Re \approx 680$. The maximum Reynolds number, corresponding to the systolic peak velocity, is equal to 2190. This value is less than 2300 which is the threshold for turbulence of Newtonian steady flow in pipes, when Poiseuille's law holds. In this study, unsteady pulsatile flow conditions of a non-Newtonian fluid are considered. Thus, the laminar flow condition is only a reasonable assumption, frequently used in hemodynamic investigations of the abdominal aortic aneurysm [19,27,31,32].

The unsteady nature of the pulsatile blood flow was modelled by assigning appropriate physiological boundary conditions at the inlet and the outlets of the patient specific model. In particular, a pulsating flow was set at the inlet of the model and a pressure pulse at the iliac outlets. Due to the fact that the patient-specific velocity and pressure waveforms were not available, typical pulses of the abdominal aortic segment, derived from the literature [33], were assigned at the inlet of the model (Figure 2). As shown in Figure 2, peak systolic velocity occurs at t = 0.25 s, and a reverse flow condition is observed at the diastolic minimum instant t = 0.51 s. Other instants were considered for the computational simulations, as they were significant for the evolution of the hemodynamic field. In particular, they refer to the systolic acceleration phase (t = 0.18 s), the deceleration phase (t = 0.31 s and t = 0.43 s), and the diastolic acceleration phase (t = 0.66 s) (Figure 2).

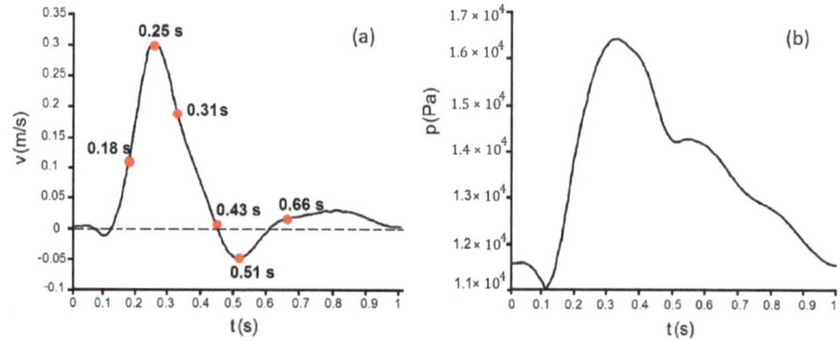

Figure 2. (**a**) Pulsatile velocity waveform assigned at the inlet of the model, (**b**) pressure waveform assigned at the outlets of the model. Red points indicate the instants of the cardiac cycle selected for the numerical investigation.

Furthermore, a no-slip condition was imposed on the walls, and the aneurysm wall was assumed to be rigid, due to the fact that abdominal aortic aneurysms become stiffer with the progression of the disease [5,34]. This last assumption is also supported by degradation of elastin and increase in collagen fibers of the aortic wall that characterize the aneurysm growth [35–37].

In this perspective, numerical simulations were performed under pulsatile flow conditions, using the representative vascular geometry obtained from the segmentation process. A computational mesh was generated with the ANSYS meshing software, with a refinement near the wall in order to capture the high velocity gradients in that area (Figure 3).

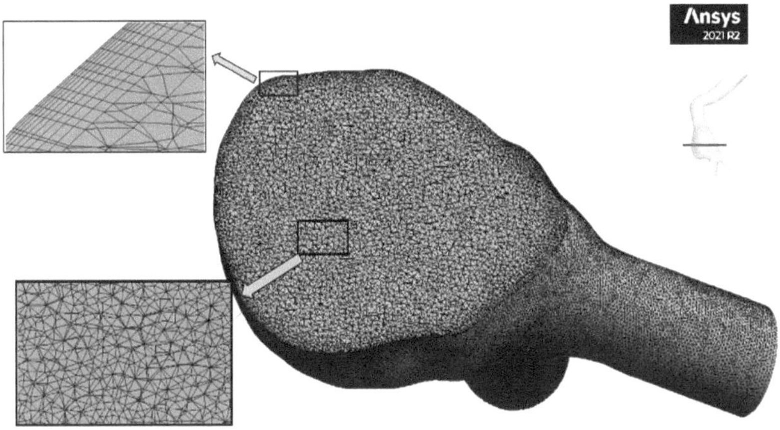

Figure 3. Mesh cross section through the aneurysm in the central region and enlarged view near the wall highlighting the local refinement.

The number of mesh elements can greatly affect the accuracy of the numerical solutions. Therefore, a mesh sensitivity analysis was performed, with element size ranging from 1.3 mm to 0.42 mm, to ensure the numerical results were not sensitive to the mesh size. In particular, the maximum values of the Time Averaged Wall Shear Stress (TAWSS) evaluated at the third numeric cycle was monitored for the different mesh size. Between the last two values, only 1.5% variation was found (Table 1).

Table 1. Maximum TAWSS values at the third numeric cycle for different mesh size.

Mesh Element Size (mm)	TAWSS (Pa)
1.3	2.78073
0.9	2.91181
0.6	2.98067
0.5	3.10283
0.42	3.05512

Moreover, for different sizes of the mesh elements, the velocity profiles, plotted in section A-A passing from the healthy aorta to the aneurysm, were compared at the instant of the systolic peak of the pulsatile motion (Figure 4). As the element size decreased, a minor discordance among the velocity profiles was observed. The good correspondence in the velocity distributions for the lower values of the element size, shown in Figure 3, highlighted the independence of the solutions from the number of computational cells and suggested the choice of the element size of the mesh. Therefore, a maximum size of 0.42 mm was selected for the mesh elements, which corresponds to 3,486,569 elements in the mesh. The element size and the number of elements obtained from the sensitivity analysis are similar to what others have reported for similar cases [31,38–40].

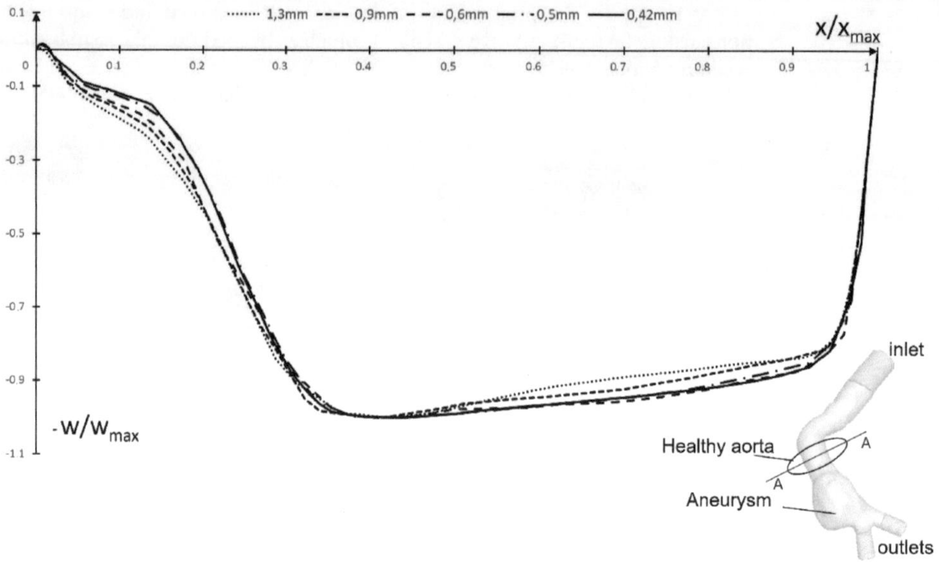

Figure 4. Dimensionless w-velocity at systolic peak instant (t = 0.25 s), plotted in section A-A passing from the healthy aorta to the aneurysm, for different sizes of the mesh element: 1.3 mm, 0.9 mm, 0.6 mm, 0.5 mm, and 0.42 mm.

The time step used for numerical integration was equal to 0.005 s, and 200 time steps were necessary to represent cardiac cycle, which was equal to 1 s. A time-step independence was assumed at the selected time step, in line with similar studies in the literature [32,41,42]

A maximum value of 200 iterations were performed for each time step. The threshold value for residual error convergence of velocity components and continuity was set as 10^{-5}. The numerical investigations were carried out for three cycles in order to minimize the influence of initialization effects, but only the third cycle was analyzed.

2.3. Hemodynamic Parameters

To furnish useful information on the hemodynamic field in the pre-rupture model of the patient-specific aneurysm, 2D and 3D streamline evolutions as well as the instantaneous WSS contours were analyzed at the selected significant instants of the cardiac cycle shown in Figure 2. Moreover, the hemodynamic indicators TAWSS, OSI and ECAP were calculated. These parameters are able to characterize significantly the unsteady nature of the flow in the diseased artery and to identify areas of potential rupture. In particular, the time averaged wall shear stress TAWSS furnishes the average over the cardiac cycle of the magnitude of WSS vector:

$$\mathrm{TAWSS} = \frac{1}{T}\int_0^T \left|\overrightarrow{WSS}\right| dt \tag{4}$$

where T is the period of the cardiac cycle, and $\overrightarrow{WSS}$ indicates the instantaneous WSS vector.

The oscillatory shear index OSI takes into account changes in the direction of the WSS vector with respect to the dominant direction of the flow during the cardiac cycle [15]:

$$\mathrm{OSI} = \frac{1}{2}\left(1 - \frac{\left|\int_0^T \overrightarrow{WSS}\, dt\right|}{\int_0^T \left|\overrightarrow{WSS}\right| dt}\right) \tag{5}$$

Values of OSI range from 0 to 0.5, where 0 indicates unidirectional shear stress. High OSI value has been recognized as inducing an inflammatory response of the artery wall, while unidirectional shear stress is related to a healthy condition [20]. Moreover, a positive correlation between OSI and ILT growth was found [43].

The endothelial cell activation potential ECAP relates TAWSS and OSI through ratio of the two indices and characterizes the degree of 'thrombogenic susceptibility' of the vessel wall [38]:

$$\mathrm{ECAP} = \mathrm{OSI}/\mathrm{TAWSS} \quad (6)$$

This index is generally used to locate areas of the arterial wall that are exposed to both high OSI and low TAWSS at the same time. High ECAP values correspond to large OSI and small TAWSS, that is, conditions of endothelial susceptibility [38].

3. Results and Conclusions

Non-Newtonian pulsatile blood flow in the rupturing patient-specific aneurysm was numerically investigated to evaluate the effect of hemodynamics on the evolution of the disease. Due to the fact that the rupture zone of the aneurysm was artificially repaired, fundamental information was known about the incipient rupture in the analyzed geometry and the location of the rupture zone. Consequently, the link between the flow patterns and the rupture of the aneurysm could be investigated to establish the role of hemodynamic parameters in predicting the rupture of the abdominal aortic aneurysm.

The results of the numerical simulations are shown below. In particular, the temporal evolution of 2D and 3D streamlines in the pre-rupture model of the aneurysm was analyzed to understand how flow patterns and recirculation regions develop in time (Figures 5 and 6). The 2D and 3D streamlines were examined in the same representative phases of the cardiac cycle (Figure 2a). The 2D streamlines refer to a longitudinal section of the aneurysm, which was chosen to capture the maximum area of the dilatation in the direction of the mean flow and to highlight the blood flow around the rupture zone.

The geometry of the considered aneurysm was characterized by high surface complexity and tortuosity, which significantly affected the blood motion. At the selected instants of the cardiac cycle, different flow patterns were observed. In the first instants of the phase of systolic acceleration, the residual vortices emerging from the previous cycle were quickly suppressed by the accelerating flow and were not found in the aneurysm, even if a helical flow persists (Figures 5 and 6, t = 0.18 s).

As the systolic peak was reached, the flow velocity increased significantly, highlighting the maximum values (Figures 5 and 6, t = 0.25 s). Despite this temporal acceleration, the enlargement of the vessel leads to a consistent spatial deceleration in the bulge with respect to the flow condition in the healthy aorta region. The consequent increase in pressure determines the separation of the flow within the dilatation already at the first instants of the deceleration phase (Figures 5 and 6, t = 0.31 s). This is highlighted by the small recirculation area originating in the region of the most significant widening of the aneurysm, marked with A in Figure 5 (t = 0,31 s). At the same time, a consistent vortical flow originated in the healthy aorta upstream of the aneurysm, due to the detachment of the fluid in correspondence with the pronounced angle of the vessel (region B in Figure 5, t = 0,31 s).

In the following phase of the cardiac cycle, the inlet velocity reaches zero values, and the vortices gradually reduce their velocities but widen, affecting the entire bulge of the aneurysm and a large region of the upstream non-dilated aorta (Figures 5 and 6, t = 0.43 s). The vanishing of the inlet velocity at this instant contributes to the onset of a strongly disturbed blood flow in the vessel, with a large number of new vortices in the healthy aorta slowly recirculating.

In the successive diastolic phase, a retrograde flow interacts with the preexisting vortices, determining the weakening of the larger recirculation regions and the disappearance of the smaller vortices close to the wall (Figures 5 and 6, t = 0.51 s and t = 0.66 s). In the later phases of the cardiac cycle, the vortices decrease in number and gradually disappear when a subsequent cycle starts.

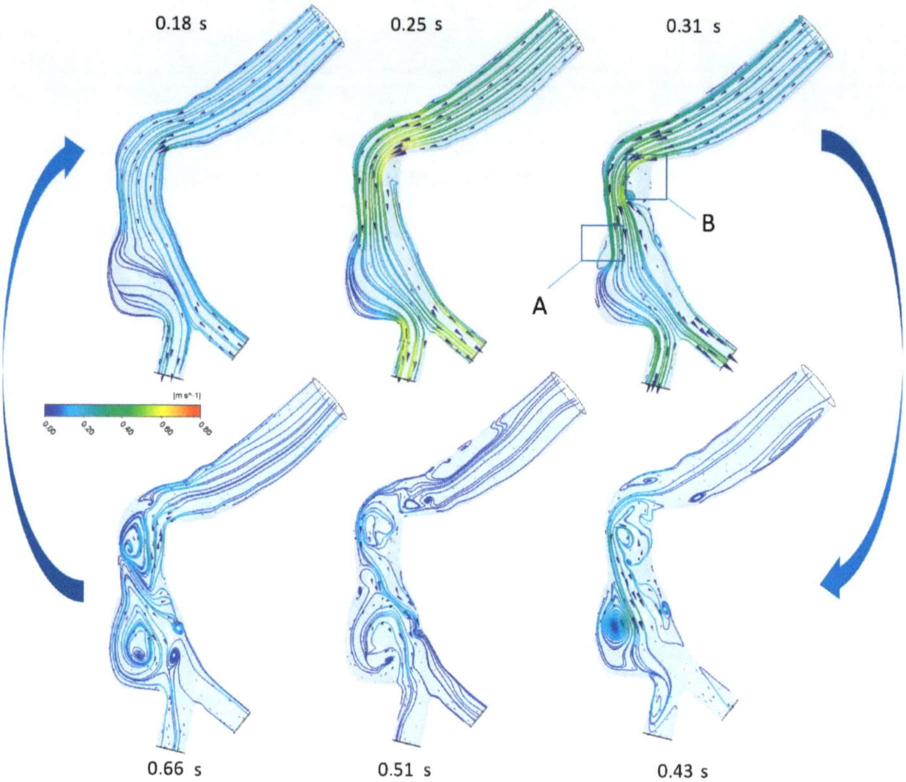

Figure 5. 2D streamlines on the longitudinal cross-section of the patient-specific model of the rupturing aneurysm at the selected instants of the cardiac cycle. In the figure, the recirculation areas in the aneurysm and the healthy aorta are indicated respectively with the letters A and B.

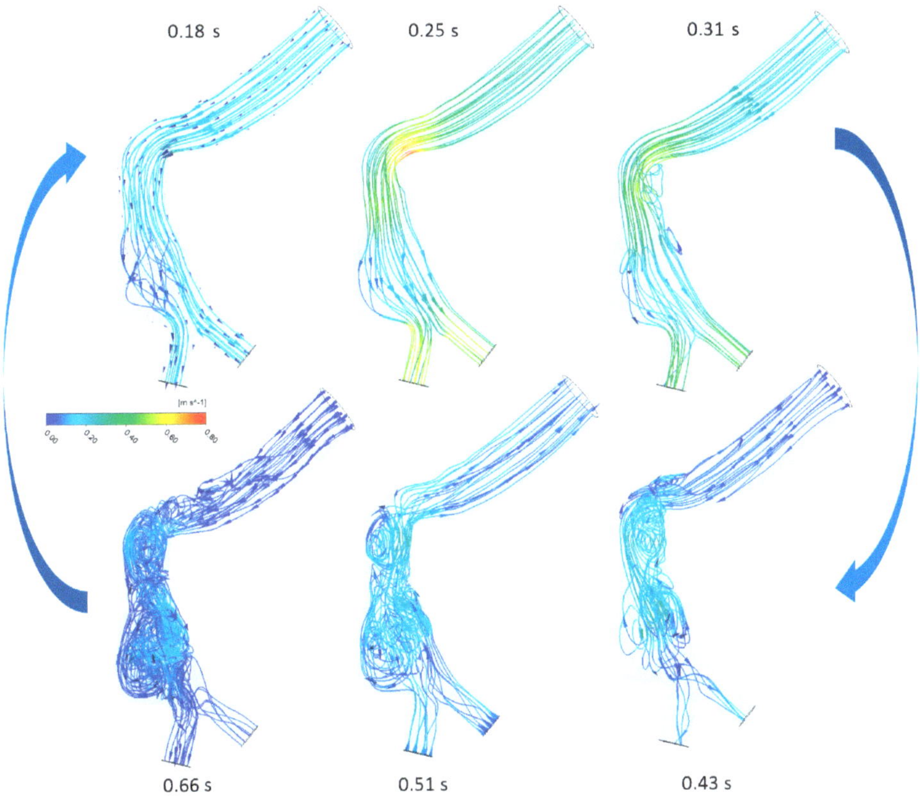

Figure 6. 3D streamlines at the selected instants of the cardiac cycle.

The spatial distributions of WSS were analyzed during the cardiac cycle, since low and oscillating WSS characterizing the pulsatile motion seems to be associated with possible rupture of the aneurysm [27,29,44]. The 3D streamlines' evolution during the cardiac cycle is simulated in Video S1.

The understanding of the flow evolution during the cardiac cycle is significant, because recirculating blood associated with low TAWSS have been considered as responsible for the deterioration of the arterial wall [16–19]. In particular, the dynamics of the recirculation regions give rise to non-physiological WSS that in turn appears to be negatively correlated with ILT accumulation [19,43] and disruption of artery wall integrity [45]. Moreover, regions characterized by low WSS and high OSI are susceptible to thrombus deposition and consequently to a higher risk of rupture [40]. Lastly, low and oscillating WSS characterizing the pulsatile motion was associated with possible rupture of the aneurysm [27,29,44,46].

Taking into account the above consideration, the instantaneous WSS distributions at the selected instants of the cardiac cycle, TAWSS, OSI and ECAP distributions, were determined in the present study.

Figure 7 shows the instantaneous WSS contours obtained from the computational simulations for the image-based reconstructed model of the pre-rupture aneurysm at the selected instants of the cardiac cycle.

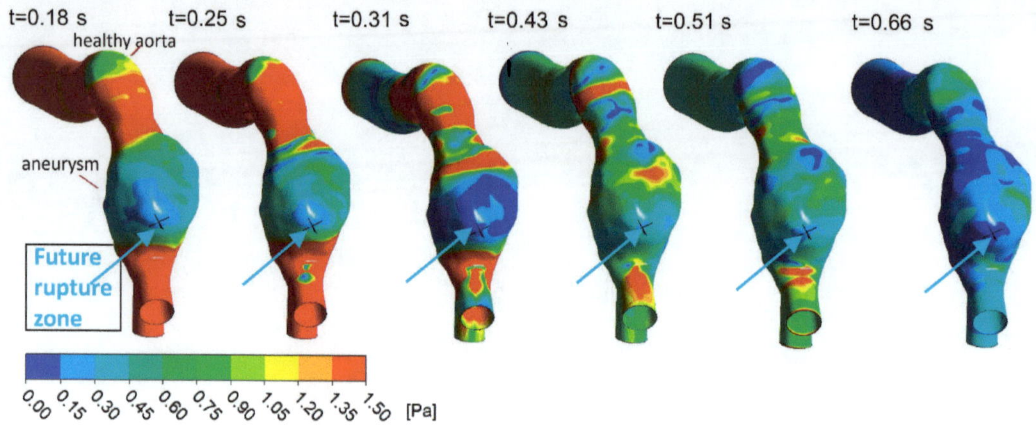

Figure 7. WSS contours distribution in the rupturing aneurysm at the selected instants of the cardiac cycle (t = 0.18 s, t = 0.25 s, t = 0.31 s, t = 0.43 s, t = 0.51 s, and t = 0.66 s). The arrow and the black cross indicate the future rupture point.

The temporal and spatial distribution of WSS inside the rupturing abdominal aortic aneurysm was due to its specific geometry, which caused flow separation and the consequent emergence of recirculation regions. As highlighted in Figure 7, an altered WSS was found in the aneurysm concerning the physiological values that characterize the non-diseased artery. During the entire cycle, lower WSS compared with the values in the healthy aorta were found on the aneurysmal wall, in agreement with literature results [44,46,47]. Relatively high values characterized the proximal and distal regions of the aneurysm. High WSS were also found in the iliac arteries, due to their tortuosity and the narrowing of the vessels, but these areas are not of interest in this study.

It is interesting to note that, at the instant of systolic peak (t = 0.25 s) the healthy arterial wall was associated with an extremely high WSS that decayed in the subsequent instants. Furthermore, at any instant of the cardiac cycle, there were always minimum relative values around the area of the incipient rupture (Figure 5). During the systolic deceleration phase (t = 0.31s), the WSS values in the bulge significantly decreased around the future rupture region but assumed high values in the region of high curvature at the inlet of the dilatation.

At the diastolic minimum, i.e., t = 0.51 s, the retrograde flow, although associated with low velocities, interacted with the pre-existing vortical flow, determining local small increases in WSS. Finally, at the diastolic acceleration phase, t = 0.66s, the WSS values decrease drastically in the bulge.

To describe in more detail the interaction between pulsating blood flow and possible rupture of the aneurysm, the spatial distribution of fundamental hemodynamic quantities, averaged over the entire cardiac cycle, was analyzed, i.e., distribution of time-averaged velocity, TAWSS, OSI and ECAP.

Figure 8 illustrates time-averaged velocity contours on the longitudinal cross-section of the patient-specific model of the rupturing aneurysm. Figures 9–11 show the TAWSS, OSI and ECAP on the luminal surfaces for the patient-specific aneurysm examined in the pre-rupture condition.

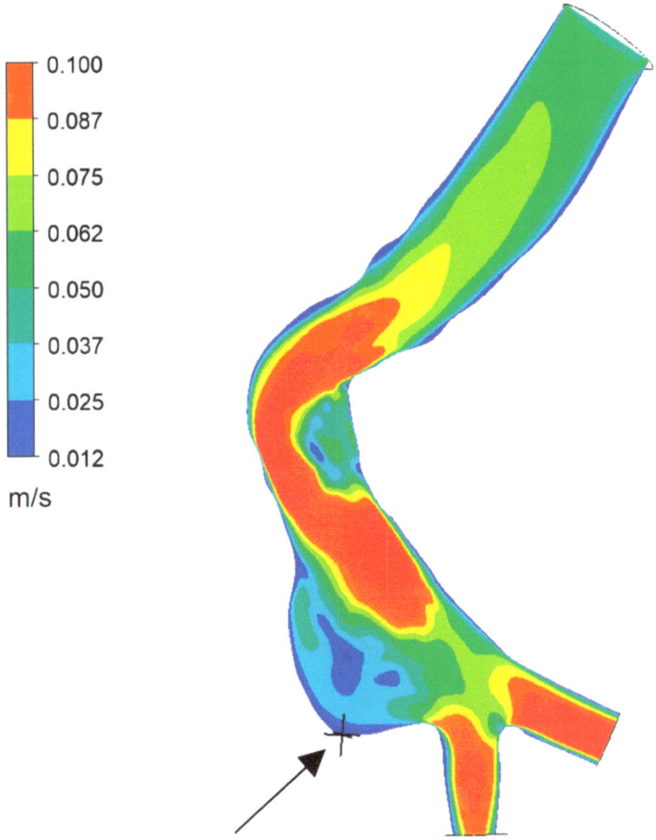

Figure 8. Time-averaged velocity contours on the longitudinal cross-section of the rupturing aneurysm. The arrow and the black cross indicate the region where rupture really occurred (future rupture region).

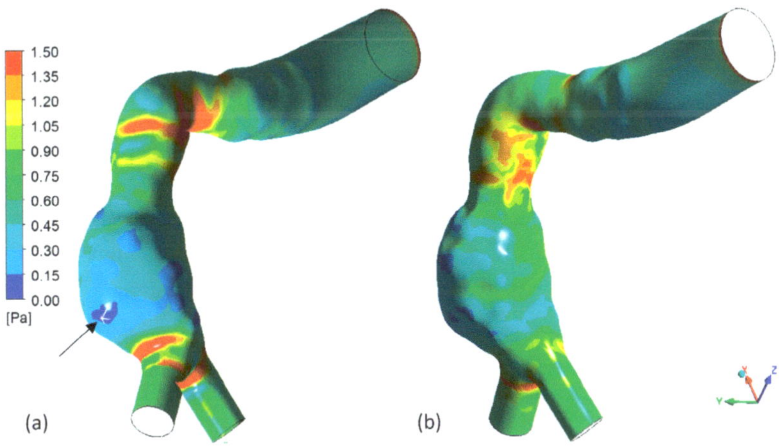

Figure 9. TAWSS contours distribution on the luminal surfaces of the rupturing aneurysm: (**a**) front view, (**b**) back view. The arrow and the white cross indicate the region where rupture really occurred (future rupture region).

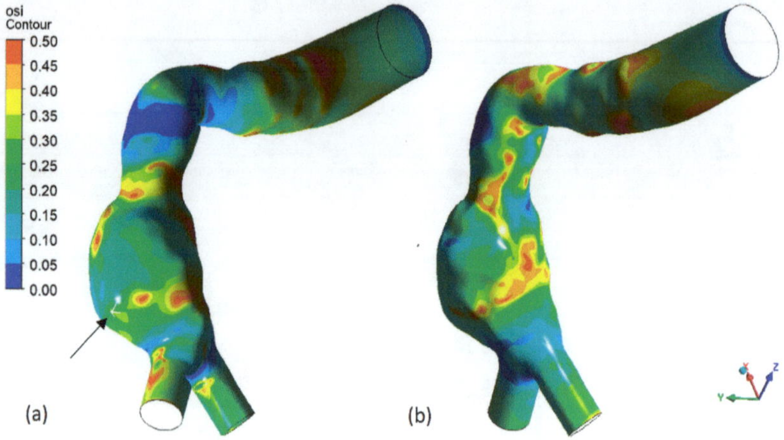

Figure 10. OSI contours distribution on the luminal surfaces of the rupturing aneurysm: (**a**) front view, (**b**) back view. The arrow and the white cross indicate the region where rupture really occurred (future rupture region).

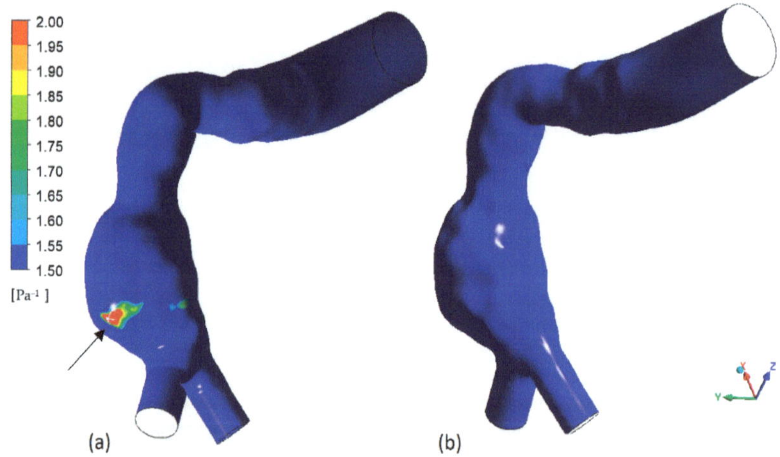

Figure 11. ECAP contours distribution on the luminal surfaces of the rupturing aneurysm: (**a**) front view, (**b**) back view. The arrow and the white cross indicate the region where rupture really occurred (future rupture region).

The results of the numerical simulations indicated regions of slow recirculation flows and low values of Time Averaged Wall Shear Stress (TAWSS) in the region of rupture (Figures 6 and 7). Unlike literature results, high Oscillatory Shear Index (OSI) was not clearly found in this region, and further regions of high OSI values were detected that were not linked to the ruptured zone (Figure 8).

Nevertheless, just in the region where the rupture will occur, the maximum value of the Endothelial Cell Activation Potential index (ECAP) was found (Figure 9), and, unlike what happens when analyzing TAWSS and OSI contour maps, no uncertainty arises regarding further possible areas of rupture of the aneurysm. This index is therefore extremely significant for assessing the vulnerability of the aortic wall and locating the critical rupture region.

In the present study, the wall of the aneurysm was assumed rigid, as widely accepted in the literature [32,39,46–48]. The hypothesis is due to the increase in wall stiffness of the abdominal aortic aneurysm AAAs with the progression of the disease [5,34]. In support of this, degradation of elastin and increase in collagen fibers of the wall have been found in the development of the aortic aneurysm [35–37]. On the other hand, fluid–structure interaction models, which take into account the elasticity of the wall, require patient-specific parameters, e.g., artery wall stiffness or thrombus properties, difficult to evaluate. However, in future research, it might be of interest to consider the elasticity of blood vessels in the analysis of the earlier stages of the disease.

Another limitation in this study is the assumption of non-patient-specific velocity and pressure pulses. As the patient-specific data were not available, only typical pulses of the abdominal aortic segment were used to model blood flow. In this regard, although more accurate results could be achieved with patient-specific boundary conditions, some limitations however could occur, related to the spatial resolution of the current imaging techniques, such as 4D MRI. Moreover, it does not provide information on the pressure field [49].

Despite the above limitations, this study highlights the importance of hemodynamics in assessing the vulnerability of the aortic wall. Moreover, it furnishes a contribution to identifying significant parameters for the prediction of the risk of aneurysm rupture as an alternative to the criterion of maximum diameter, supporting the clinical management of AAA disease.

Supplementary Materials: The following supporting information can be downloaded at: https://www.mdpi.com/article/10.3390/fluids7080269/s1, Video S1: 3D streamlines' evolution during the cardiac cycle.

Author Contributions: Conceptualization, M.A.B. and M.C.C.; methodology, M.A.B. and M.C.C.; software, M.A.B., M.C.C. and R.M.; validation, M.A.B. and M.C.C.; formal analysis, M.A.B. and R.M.; investigation, M.C.C.; resources, G.d.D. and E.P.; writing—original draft preparation, M.A.B. and M.C.C.; writing—review and editing, M.A.B., M.C.C., R.M., E.P. and G.d.D.; supervision, M.A.B. and G.d.D.; project administration, M.A.B.; funding acquisition G.d.D. All authors have read and agreed to the published version of the manuscript.

Funding: This research received no external funding.

Institutional Review Board Statement: Not applicable.

Informed Consent Statement: Informed consent was obtained from all subjects involved in the study.

Data Availability Statement: Not applicable.

Conflicts of Interest: The authors declare no conflict of interest.

References

1. Solomon, C.G.; Kent, K.C. Abdominal aortic aneurysms. *New Engl. J. Med.* **2014**, *371*, 2101–2108. [CrossRef]
2. Gopalakrishnan, S.S.; Pier, B.; Biesheuvel, A. Dynamics of pulsatile flow through model abdominal aortic aneurysms. *J. Fluid Mech.* **2014**, *758*, 150–179. [CrossRef]
3. Chaikof, E.L.; Dalman, R.L.; Eskandari, M.K.; Jackson, B.M.; Lee, W.A.; Mansour, M.A.; Mastracci, T.M.; Mell, M.; Murad, M.H.; Nguyen, L.L.; et al. The Society for Vascular Surgery practice guidelines on the care of patients with an abdominal aortic aneurysm. *J. Vasc. Surg.* **2018**, *67*, 2–77. [CrossRef] [PubMed]
4. Wanhainen, A.; Verzini, F.; Van Herzeele, I.; Allaire, E.; Bown, M.; Cohnert, T.; Dick, F.; van Herwaarden, J.; Karkos, C.; Koelemay, M.; et al. Editor's Choice—European Society for Vascular Surgery (ESVS) 2019 Clinical Practice Guidelines on the Management of Abdominal Aorto-iliac Artery Aneurysms. *Eur. J. Vasc. Endovasc. Surg.* **2019**, *57*, 8–93. [CrossRef] [PubMed]
5. Vorp, D.A. Biomechanics of abdominal aortic aneurysm. *J. Biomech.* **2007**, *40*, 1887–1902. [CrossRef]
6. Huang, Y.; Teng, Z.Z.; Elkhawad, M.; Tarkin, J.M.; Joshi, N.; Boyle, J.R.; Buscombe, J.R.; Fryer, T.D.; Zhang, Y.; Park, A.Y.; et al. High structural stress and presence of intraluminal thrombus predict abdominal aortic aneurysm 18F-FDG uptak. *Circ. Cardiovasc. Imaging* **2016**, *9*, 004656. [CrossRef]
7. Haller, S.J.; Crawford, J.D.; Courchaine, K.M.; Bohannan, C.J.; Landry, G.J.; Moneta, G.L.; Azarbal, A.F.; Rugonyi, S. Intraluminal thrombus is associated with early rupture of abdominal aortic aneurysm. *J. Vasc. Surg.* **2018**, *67*, 1051–1058. [CrossRef]
8. Laine, M.T.; Vänttinen, T.; Kantonen, I.; Halmesmäki, K.; Weselius, E.; Laukontaus, S.; Salenius, J.; Aho, P.; Venermo, M. Rupture of Abdominal Aortic Aneurysms in Patients Under Screening Age and Elective Repair Threshold. *Eur. J. Vasc. Endovasc. Surg.* **2016**, *51*, 511–516. [CrossRef]

9. Nicholls, S.C.; Gardner, J.B.; Meissner, M.H.; Johansen, K.H. Rupture in small abdominal aortic aneurysms. *J. Vasc. Surg.* **1998**, *28*, 884–888. [CrossRef]
10. Darling, R.C.; Messina, C.R.; Brewster, D.C.; Ottinger, L.W. Autopsy study of unoperated abdominal aortic aneurysms. The case for early resection. *Circulation* **1977**, *56* (Suppl. S3), II161-4.
11. Vande Geest, J.P.; Di Martino, E.S.; Bohra, A.; Makaroun, M.S.; Vorp, D.A. A biomechanics-based rupture potential index for abdominal aortic aneurysm risk assessment: Demonstrative application. *Ann. N. Y. Acad. Sci.* **2006**, *1085*, 11–21. [CrossRef] [PubMed]
12. Maier, A.; Gee, M.W.; Reeps, C.; Pongratz, J.; Eckstein, H.-H.; Wall, W.A. A Comparison of Diameter, Wall Stress, and Rupture Potential Index for Abdominal Aortic Aneurysm Rupture Risk Prediction. *Ann. Biomed. Eng.* **2010**, *38*, 3124–3134. [CrossRef] [PubMed]
13. Gijsen, F.; Katagiri, Y.; Barlis, P.; Bourantas, C.; Collet, C.; Coskun, U.; Daemen, J.; Dijkstra, J.; Edelman, E.; Evans, P.; et al. Expert recommendations on the assessment of wall shear stress in human coronary arteries: Existing methodologies, technical considerations, and clinical applications. *Eur. Heart J.* **2019**, *40*, 3421–3433. [CrossRef] [PubMed]
14. Chiu, J.J.; Chien, S. Effects of Disturbed Flow on Vascular Endothelium: Pathophysiological Basis and Clinical Perspectives. *Physiol. Rev.* **2011**, *91*, 327–387. [CrossRef] [PubMed]
15. Ku, D.N.; Giddens, D.P.; Zarins, C.K.; Glagov, S. Pulsatile flow and atherosclerosis in the human carotid bifurcation. Positive correlation between plaque location and low oscillating shear stress. *Arteriosclerosis* **1985**, *5*, 293–302. [CrossRef]
16. Cecchi, E.; Giglioli, C.; Valente, S.; Lazzeri, C.; Gensini, G.F.; Abbate, R.; Mannini, L. Role of hemodynamic shear stress in cardiovascular disease. *Atherosclerosis* **2011**, *214*, 249–256. [CrossRef]
17. Tarbell, J.M.; Shi, Z.D.; Dunn, J.; Jo, H. Fluid mechanics, arterial disease, and gene expression. *Annu. Rev. Fluid Mech.* **2014**, *46*, 591–614. [CrossRef]
18. Takehara, Y.; Isoda, H.; Takahashi, M.; Unno, N.; Shiiya, N.; Ushio, T.; Goshima, S.; Naganawa, S.; Alley, M.; Wakayama, T.; et al. Abnormal flow dynamics result in low wall shear stress and high oscillatory shear index in abdominal aortic dilatation: Initial in vivo assessment with 4D-flow MRI. *Magn. Reson. Med. Sci.* **2020**, *19*, 235–246. [CrossRef]
19. Kelsey, L.J.; Powell, J.T.; Norman, P.E.; Miller, K.; Doyle, B.J. A comparison of hemodynamic metrics and intraluminal thrombus burden in a common iliac artery aneurysm. *Int. J. Numer. Methods Biomed. Eng.* **2017**, *33*, e2821. [CrossRef]
20. Sorescu, G.P.; Song, H.N.; Tressel, S.L.; Hwang, J.; Dikalov, S.; Smith, D.A.; Boyd, N.L.; Platt, M.O.; Lassègue, B.; Griendling, K.; et al. Bone morphogenic protein 4 produced in endothelial cells by oscillatory shear stress induces monocyte adhesion by stimulating reactive oxygen species production from a nox1-based NADPH oxidase. *Circ. Res.* **2004**, *95*, 773–779. [CrossRef]
21. Arzani, A.; Suh, G.-Y.; Dalman, R.L.; Shadden, S.C. A longitudinal comparison of hemodynamics and intraluminal thrombus deposition in abdominal aortic aneurysms. *Am. J. Physiol. Heart Circ. Physiol.* **2014**, *307*, H1786–H1795. [CrossRef] [PubMed]
22. O'Rourke, M.J.; McCullough, J.P.; Kelly, S. An investigation of the relationship between hemodynamics and thrombus deposition within patient-specific models of abdominal aortic aneurysm. *Proc. Inst. Mech. Eng. Part H J. Eng. Med.* **2012**, *226*, 548–564. [CrossRef] [PubMed]
23. Pasqui, E.; de Donato, G.; Giannace, G.; Panzano, C.; Setacci, C.; Palasciano, G. Management of abdominal aortic aneurysm in nonagenarians: A single-centre experience. *Vascular* **2021**, *29*, 27–34. [CrossRef] [PubMed]
24. de Donato, G.; Pasqui, E.; Nano, G.; Lenti, M.; Mangialardi, N.; Speziale, F.; Ferrari, M.; Michelagnoli, S.; Tozzi, M.; Palasciano, G.; et al. Long-term results of treatment of infrarenal aortic aneurysms with low-profile stent grafts in a multicenter registry. *J. Vasc Surg.* **2022**, *75*, 1242–1252.e2. [CrossRef] [PubMed]
25. Schlichting, H. *Boundary Layer Theory*, 7th ed.; McGraw-Hill: New York, NY, USA, 1979.
26. Wood, N.B. Aspects of fluid dynamics applied to the larger arteries. *J. Biol.* **1999**, *199*, 137–161. [CrossRef] [PubMed]
27. Boyd, A.J.; Kuhn, D.C.S.; Lozowy, R.J.; Kulbisky, G.P. Low wall shear stress predominates at sites of abdominal aortic aneurysm 460 rupture. *J. Vasc. Surg.* **2016**, *63*, 1613–1619. [CrossRef]
28. Ferziger, J.H.; Peric, M. *Computational Methods for Fluid Dynamics*; Springer: Berlin/Heidelberg, Germany, 2001; ISBN 978-3-540-42074-3.
29. Forneris, A.; Marotti, F.B.; Satriano, A.; Moore, R.D.; Di Martino, E.S. A novel combined fluid dynamic and strain analysis approach identified abdominal aortic aneurysm rupture. *J. Vasc. Surg. Cases Innov. Tech.* **2020**, *6*, 172–176. [CrossRef]
30. Shibeshi, S.S.; Collins, W.E. The rheology of blood flow in a branched arterial system. *Appl. Rheol.* **2005**, *15*, 398–405. [CrossRef]
31. Zambrano, B.; Gharahi, H.; Lim, C.; Jaberi, F.; Choi, J.; Lee, W.; Baek, S. Association of intraluminal thrombus, hemodynamic forces, and abdominal aortic aneurysm expansion using longitudinal CT images. *Ann. Biomed. Eng.* **2016**, *44*, 1502–1514. [CrossRef]
32. Qiu, Y.; Wang, Y.; Fan, Y.B.; Peng, L.; Liu, R.; Zhao, J.; Yuan, D.; Zheng, T. Role of intraluminal thrombus in abdominal aortic aneurysm ruptures: A hemodynamic point of view. *Med. Phys.* **2019**, *46*, 4263–4275. [CrossRef]
33. Xenos, M.; Rambhia, S.H.; Alemu, Y.; Einav, S.; Labropoulos, N.; Tassiopoulos, A.; Ricotta, J.J.; Bluestein, D. Patient-based abdominal aortic aneurysm rupture risk prediction with fluid structure interaction modeling. *Ann. Biomed. Eng.* **2010**, *38*, 3323–3337. [CrossRef]
34. Vande Geest, J.P.; Sacks, M.S.; Vorp, D.A. The effects of aneurysm on the biaxial mechanical behavior of human abdominal aorta. *J. Biomech.* **2006**, *39*, 1324–1334. [CrossRef] [PubMed]

35. Thubrikar, M.J.; Labrosse, M.; Robicsek, F.; Al-Soudi, J.; Fowler, B. Mechanical properties of abdominal aortic aneurysm wall. *J. Med. Eng. Technol.* **2001**, *25*, 133–142. [CrossRef] [PubMed]
36. Tong, J.; Cohnert, T.; Holzapfel, G.A. Diameter-related variations of geometrical, mechanical, and mass fraction data in the anterior portion of abdominal aortic aneurysms. *Eur. J. Vasc. Endovasc. Surg.* **2015**, *49*, 262–270. [CrossRef] [PubMed]
37. Kolipaka, A.; Illapani, V.S.P.; Kenyhercz, W.; Dowell, J.D.; Go, M.R.; Starr, J.E.; Vaccaro, P.S.; White, R.D. Quantification of abdominal aortic aneurysm stiffness using magnetic resonance elastography and its comparison to aneurysm diameter. *J. Vasc. Surg.* **2016**, *64*, 966–974. [CrossRef] [PubMed]
38. Di Achille, P.; Tellides, G.; Figueroa, C.A.; Humphrey, J.D. A haemodynamic predictor of intraluminal thrombus formation in abdominal aortic aneurysms. *Proc. R. Soc. Lond Math. Phys. Eng. Sci.* **2014**, *470*, 20140163. [CrossRef]
39. Lozowy, R.J.; Kuhn, D.C.S.; Ducas, A.A.; Boyd, A. The relationship between pulsatile flow impingement and intraluminal thrombus deposition in abdominal aortic aneurysms. *Cardiovasc. Eng. Technol.* **2017**, *8*, 57–69. [CrossRef]
40. Les, A.S.; Shadden, S.C.; Figueroa, C.A.; Park, J.M.; Tedesco, M.M.; Herfkens, R.J.; Dalman, R.L.; Taylor, C.A. Quantification of hemodynamics in abdominal aortic aneurysms during rest and exercise using magnetic resonance imaging and computational fluid dynamics. *Ann. Biomed. Eng.* **2010**, *38*, 1288–1313. [CrossRef]
41. Poelma, C.; Watton, P.N.; Ventikos, Y. Transitional flow in aneurysms and the computation of haemodynamic parameters. *J. R. Soc. Interface* **2015**, *12*, 20141394. [CrossRef]
42. Tan, F.P.P.; Borghi, A.; Mohiaddin, R.H.; Wood, N.B.; Thom, S.; Xu, X.Y. Analysis of flow patterns in a patient-specific thoracic aneurysm model. *Comput. Struct.* **2009**, *87*, 680–690. [CrossRef]
43. Tzirakis, K.; Kamarianakis, Y.; Metaxa, E.; Kontopodis, N.; Ioannou, C.V.; Papaharilaou, Y. A robust approach for exploring hemodynamics and thrombus growth associations in abdominal aortic aneurysms. *Med. Biol. Eng. Comput.* **2017**, *55*, 1493–1506. [CrossRef]
44. Doyle, B.J.; McGloughlin, T.M.; Kavanagh, E.G.; Hoskins, P.R. From detection to rupture: A serial computational fluid dynamics case study of a rapidly expanding, patient-specific, ruptured abdominal aortic aneurysm. In *Computational Biomechanics for Medicine: Fundamental Science and Patient-Specific Applications*; Doyle, B., Miller, K., Wittek, A., Eds.; Springer: Berlin/Heidelberg, Germany, 2014; pp. 53–68. [CrossRef]
45. Koole, D.; Zandvoort, H.J.A.; Schoneveld, A.; Vink, A.; Vos, J.A.; Hoogen, L.L.V.D.; de Vries, J.-P.P.; Pasterkamp, G.; Moll, F.L.; van Herwaarden, J.A. Intraluminal abdominal aortic aneurysm thrombus is associated with disruption of wall integrity. *J. Vasc. Surg.* **2013**, *57*, 77–83. [CrossRef]
46. Boniforti, M.A.; Di Bella, L.; Magini, R. On the role of hemodynamics in predicting rupture of the abdominal aortic aneurysm. *J. Zhejiang Univ. Sci. A* **2021**, *22*, 957–978. [CrossRef]
47. Soudah, E.; Ng, E.Y.K.; Loong, T.H.; Bordone, M.; Pua, U.; Narayanan, S. CFD modelling of abdominal aortic aneurysm on hemodynamic loads using a realistic geometry with CT. *Comput. Math. Methods Med.* **2013**, *2013*, 472564. [CrossRef]
48. Patel, S.; Usmani, A.Y.; Muralidhar, K. Effect of aorto-iliac bifurcation and iliac stenosis on flow dynamics in an abdominal aortic aneurysm. *Fluid Dyn. Res.* **2017**, *49*, 035513. [CrossRef]
49. Canchi, T.; Kumar, S.D.; Ng, E.Y.K.; Narayanan, S. A review of computational methods to predict the risk of rupture of abdominal aortic aneurysms. *BioMed Res. Int.* **2015**, *2015*, 861627. [CrossRef]

fluids

MDPI

Article

Blood Flow Simulation of Aneurysmatic and Sane Thoracic Aorta Using OpenFOAM CFD Software

Francesco Duronio * and Andrea Di Mascio

Department of Industrial Engineering, Information and Economics, Università degli Studi dell'Aquila, Piazzale Ernesto Pontieri, Monteluco di Roio, 67100 L'Aquila, Italy; andrea.dimascio@univaq.it
* Correspondence: francesco.duronio@univaq.it; Tel.: +39-0862-434317

Abstract: Cardiovascular diseases still represent one of the most deadly pathologies worldwide. Knowledge of the blood flow dynamics within the cardio-vascular system is crucial in preventing these diseases and analysing their physiology and physio-pathology. CFD simulations are highly effective in guiding clinical predictions and, more importantly, allow the evaluation of physical and clinical parameters that are difficult to measure with common diagnostic techniques. Therefore, in particular, this study is focused on investigating the hemodynamics of the thoracic aorta. Real aortic geometries regarding a sane and diseased patient presenting an aneurysm were considered. CFD simulations were performed with the OpenFOAM C++ library using patient-specific pulsatile blood flow waveforms and implementing the Windkessel pressure boundary condition for the artery outflow. The adopted methodology was preliminarily verified for assessing the numerical uncertainty and convergence. Then, the CFD results were evaluated against experimental data concerning pressure and velocity of the thoracic aorta measured with standard diagnostic techniques. The normal aorta's blood flow was also compared against the pattern regarding the patient-specific aortic aneurysm. Parameters such as wall pressure, wall shear stress (WSS) and velocity distribution were investigated and discussed. The research highlighted that the blood flow in the aorta is strongly affected by the aneurysm onset, with the growth of recirculation zones being potentially hazardous. The outcomes of the investigation finally demonstrate how CFD simulation tools, capturing the detailed physics of the aortic flow, are powerful tools for supporting clinical activities of the cardio-vascular system.

Keywords: CFD; aorta aneurysm; cardiovascular flow; patient-specific simulation

Citation: Duronio, F.; Di Mascio, A. Blood Flow Simulation of Aneurysmatic and Sane Thoracic Aorta Using OpenFOAM CFD Software. *Fluids* **2023**, *8*, 272. https://doi.org/10.3390/fluids8100272

Academic Editors: Ricardo Ruiz Baier and D. Andrew S. Rees

Received: 6 August 2023
Revised: 30 August 2023
Accepted: 28 September 2023
Published: 2 October 2023

1. Introduction

The aorta, the largest artery in the human body, plays a vital role in systemic circulation, carrying oxygenated blood from the left ventricle to various organs. The thoracic aorta includes the ascending aorta, aortic arch and descending aorta, with three main branches originating from the arch to supply blood to the upper body. Despite significant advancements in clinical care and public awareness, cardiovascular diseases (CVD) remain the most important cause of mortality worldwide [1]. The aorta faces various diseases, such as atherosclerosis, aortic aneurysm and dissection, posing life-threatening risks due to potential rupture at weakened sections. Understanding the mechanisms behind the development and progression of these diseases is critical, leading to active research in this area.

The thoracic aorta exhibits a complex anatomy characterised by severe bending, non-planarity, branching, tapering lumen and elastic arterial walls. Blood flow in the aorta exhibits intricate patterns. During systole, as blood is accelerated, it leaves the left ventricle through the aortic valve. Hot-film anemometry measurements have demonstrated that velocity profiles are nearly flat in the ascending aorta, and reversed flow occurs during systolic deceleration and the diastolic phases [2]. These complex flow patterns expose large areas of the vessel walls to high and low shear stress, promoting the development of atherosclerotic lesions [3].

Extensive experimental and numerical studies were conducted to investigate the dynamics of aortic flow. While idealised geometries provided a fundamental understanding of flow phenomena [4,5], more detailed and realistic in vivo anatomical models become essential. Advancements in non-invasive methods, such as magnetic resonance imaging (MRI) technology, allow assessing the morphology and function of the cardiovascular system. Even more, MRI allows the measurement of complex arterial geometries, including their movement and blood velocities throughout the cardiac cycle [6–9]. Consequently, these techniques provide the anatomical and hemodynamic input necessary for computational fluid dynamics (CFD) simulations to obtain comprehensive fluid-dynamic information. By combining these two methods, CFD simulations can predict crucial hemodynamic quantities, such as wall shear stress, which cannot be directly measured in the human vascular system.

However, it must be underlined that boundary conditions play a critical role in obtaining accurate and reliable results from CFD simulations. At the aorta, inlet velocity boundary conditions are usually prescribed. Pulsatile waveforms, based on experimental data, are frequently used to represent accelerating, decelerating, reversed and zero flow regions [10,11].

The downstream side of the artery of interest presents challenges in determining appropriate boundary conditions for computational fluid dynamics (CFD) simulations. Arteries are branched and connected to smaller vessels, making it impractical to trace all the branching in the simulation. Therefore, the model needs to be terminated at some point, and the branches must be lumped into a suitable terminal description to accurately represent wave propagation and artery impedance characteristics.

Previous studies used various outlet boundary conditions, including constant or time-varying pressures and velocity profiles. However, these conditions have limitations, as they do not accurately replicate the fluid impedance of the downstream vasculatures. Prescribing zero or equal pressures for different outlets neglects the dominant effect of the resistance of the downstream vasculatures, leading to inaccurate flow [11–14]. Additionally, prescribing constant pressure or velocity may not result in physiologically realistic blood pressure values, limiting the accuracy of the simulations.

Outflow conditions can be prescribed also as constant fractions of the in-flowing blood. However, these methods may not fully account for patient-specific characteristics and flow dynamics during the cardiac cycle. Alternative approaches also involve using personalised PC-MRI-measured blood flow rates as outlet boundary conditions.

More sophisticated zero-dimensional (lumped parameter) models can be used to overcome these shortcomings. These models provide boundary conditions for 3D computational simulations and consider branching patterns and vasculature properties. They allow for a more dynamic description of the blood flow downstream of the computational domain and can be coupled explicitly or embedded into CFD codes [14–16].

In this study, the OpenFOAM C++ library was employed to simulate blood flow through two different aorta geometries: a normal subject's aorta and that of a patient with an aortic aneurysm. The convergence and the grid sensibility were initially evaluated with a commonly adopted procedure. Next, the validity of the results was assessed by relying on diagnostic data.

The blood flow patterns of the normal aorta model and the aneurysmatic model were compared. Important hemodynamic parameters such as blood pressure, wall shear stress (WSS) and velocity distribution were estimated during the analysis. This comparison aims to gain insights into the differences in flow behaviour between the two models and their potential implications for cardiovascular health.

2. Mathematical and Numerical Method

2.1. Geometries of Aorta and Generation of the Computational Domain

The aorta of two young male subjects of comparable age (18 and 23 years old) were considered in this study. The two geometries regard a sane and a diseased aorta. The latter presents an aneurysm on the descending section due to Marfan syndrome. Marfan syndrome is a genetic disorder that affects the body's connective tissue. Connective tissue provides support and strength to various structures in the body, including bones, joints, ligaments, blood vessels and the heart. This condition is named after Antoine Marfan, the French paediatrician who first described it in the late 19th century. Among cardiovascular issues, Marfan syndrome affects the heart and the blood vessels, originating aortic aneurysms (bulging of the aorta) and mitral valve prolapse.

Computed tomography (CT) scans of the aorta were used to create the 3D patient-specific anatomical models used in computational fluid dynamics (CFD) simulations. All the required data were obtained from the public repository of the National Institutes of Health [17].

The aorta geometries were discretised using OpenFOAM's utility snappyHexMesh. The computational grid so obtained was further refined and improved, creating boundary layer cells. This process refines the mesh near the boundaries, adding additional layers of hexahedral cells aligned to the boundary surface and removing irregular cells.

Three different meshes were computed to perform grid sensitivity analysis: a fine grid featuring elements of 0.5 mm average dimension, an intermediate grid of 1 mm and a coarse grid of 2 mm.

Figure 1 shows the computational grid.

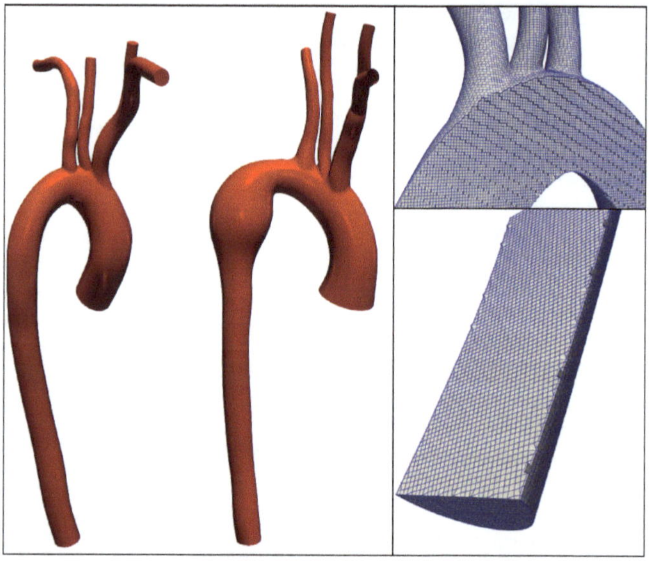

Figure 1. Sane and diseased aorta geometries are shown on the left while, on the right, some details of the computational grid used are shown.

For the finer grid, the boundary layer cells are 0.15 mm high. A posteriori check, using the average wall shear stress, showed that the y+ was everywhere lower than 1.

2.2. Model Setup

Governing equations of the blood flow within the aorta are essentially the continuity and the Navier–Stokes equations that express mass and momentum conservation of an incompressible flow. This assumption was deemed true considering the blood density and low velocities typical of this kind of flow [18,19].

$$\nabla \cdot \mathbf{u} = 0 \tag{1}$$

$$\rho \frac{D\mathbf{u}}{Dt} = -\nabla p + \rho \mathbf{f} + \mu \nabla^2 \mathbf{u} \tag{2}$$

This is a system of four partial differential equations. The analytical solution of them is possible only in very few cases. A numerical solution is required due to the aorta's complex geometry and unsteadiness of the blood flow. The equations were discretised by adopting the upwind scheme, while temporal integration was performed with the Euler method. The PIMPLE algorithm, featuring an implicit solution of the pressure–velocity coupling, was adopted to compute the solution [20–23].

A variable time step was chosen to keep the Courant number equal to 0.8.

The aorta walls are considered solid, rigid and impermeable boundaries. The blood flow near the wall was assumed to have zero velocity (no-slip boundary condition). Typically, blood exhibits non-Newtonian behaviour due to its complex composition and interactions between its components. However, it can be considered Newtonian when the shear rate exceeds 100 s^{-1} [24]. This happens particularly in large vessels such as the aorta [25]. Numerous literature studies adopted Newtonian behaviour for blood in large arteries [26–28]. In this particular study, the focus is on the aorta and its three major branches, where the diameters are larger than 0.1 mm. As a result, the shear rate in these large arteries is well above 100 s^{-1}, allowing for the acceptance of blood as a Newtonian fluid, so a constant kinematic viscosity of 4×10^{-6} m^2/s and a density of 1050 kg/m^3 were considered [29–31].

Based on the nominal diameter of the aorta, the peak pulsatile flow resulted in a Reynolds number ranging from 5000 to 6000, indicating turbulent flow conditions [18,32]. The Spalart–Allmaras detached eddy simulation (DES) [33,34] was chosen to model this turbulent flow. This hybrid model switches between a pure LES approach in the core turbulent region, where large unsteady turbulence scales can be resolved by grid size, and a RANS Spalart–Allmaras model near solid walls, where the typical length scale of the turbulent eddies is significantly smaller than the grid dimensions.

In the study of blood flow in the arteries, boundary conditions play a critical role in accurately representing the complex nature of the flow. At the upstream side of the artery of interest, a velocity boundary condition was prescribed.

Figures 2 and 3 report the flow-rate waveforms used for the present study regarding the sane and diseased patients obtained from specific measurements [17].

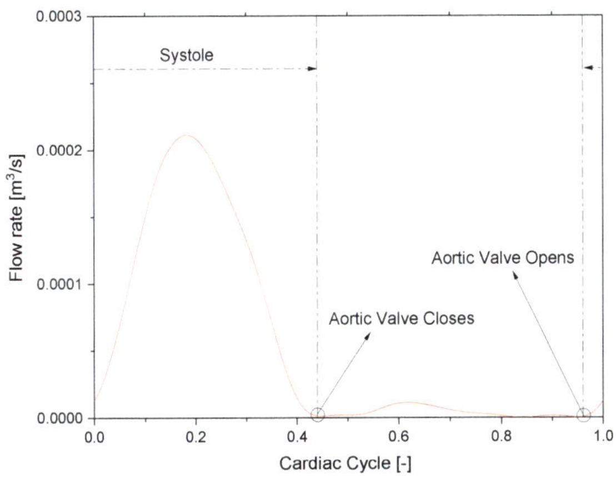

Figure 2. Pulsatile waveform used as Dirichlet inlet boundary condition for the sane patient.

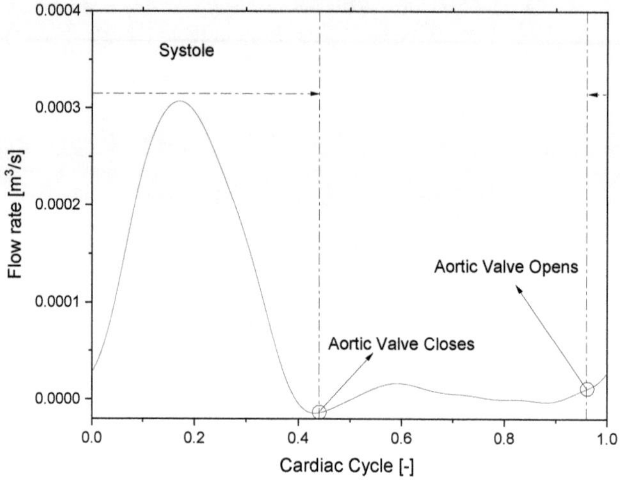

Figure 3. Pulsatile waveform used as Dirichlet inlet boundary condition for the diseased patient (aorta with aneurysm).

A three-element Windkessel model was adopted as boundary condition for the outflow patches. The Windkessel model is a simplified model that considers the downstream arterial system as a combination of a capacitance vessel (arterial compliance) and a resistance vessel (arterial resistance). The compliance of the arteries allows them to stretch during systole, when blood is ejected from the heart, and to store part of the energy. During diastole, when the heart is relaxed, the stored energy is released, helping to maintain blood flow and pressure in the arteries. The three-element Windkessel model features a resistor to account for the resistance to blood flow caused by the aortic valve and a parallel combination of another resistor and a capacitor representing the total arterial compliance and peripheral resistance. Exploiting an electrical analogy, the three-element Windkessel model is represented in Figure 4.

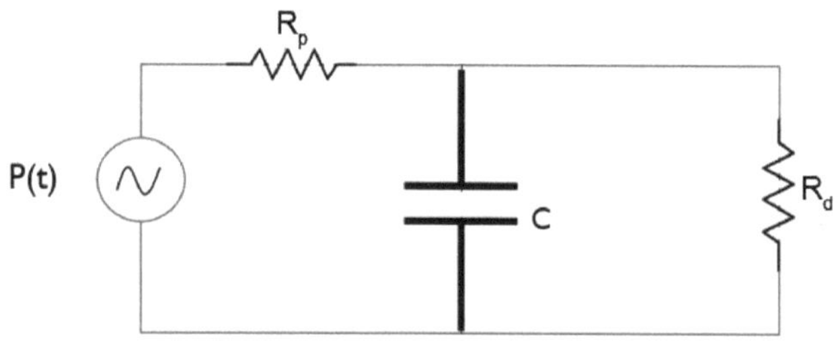

Figure 4. Electrical analog of the 3-element Windkessel model.

The dynamics of this circuit can be mathematically expressed by the following Equation (3):

$$\left(1 + \frac{R_d}{R_p}\right)Q(t) + CR_d\frac{dQ(t)}{dt} = \frac{P(t)}{R_p} + C\frac{dP(t)}{dt} \tag{3}$$

A new OpenFOAM solver called pimpleWKFoam was developed to include the three-element Windkessel model as a pressure boundary condition. Further details regarding the solution procedure are not reported here for the sake of brevity and can be found in the literature [15].

The numerical values of R_p, R_d and C can be defined relying on specific patient measurements regarding mainly cardiac output and mean aortic pressure [16,35–37]. Following the various literature references, the proximal resistance results to be 10% of the overall vascular resistance while the distal resistance is 90% [38].

All the values adopted are reported in Tables 1 and 2.

Table 1. Resistance and compliance values for healthy aorta. LCCA = left common carotid artery; LSA = left subclavian artery; RCCA = right common carotid artery; RSA = right subclavian artery.

	R_p [Pa · s/m^5]	R_d [Pa · s/m^5]	C [m^5/Pa]
OUTFLOW	4.09×10^7	1.29×10^8	1.29×10^{-8}
LCCA	3.23×10^8	1.02×10^9	1.64×10^{-9}
LSA	2.72×10^8	8.58×10^8	1.94×10^{-9}
RCCA	2.81×10^8	8.86×10^8	1.88×10^{-9}
RSA	1.52×10^8	4.80×10^8	3.46×10^{-9}

Table 2. Resistance and compliance values for aneurysmatic aorta. LCCA = left common carotid artery; LSA = left subclavian artery; RCCA = right common carotid artery; RSA = right subclavian artery.

	R_p [Pa · s/m^5]	R_d [Pa · s/m^5]	C [m^5/Pa]
OUTFLOW	3.03×10^7	9.13×10^7	1.74×10^{-8}
LCCA	2.18×10^8	6.58×10^8	2.42×10^{-9}
LSA	1.74×10^8	5.24×10^8	3.04×10^{-9}
RCCA	1.71×10^8	5.14×10^8	3.09×10^{-9}
RSA	1.32×10^8	3.98×10^8	4.00×10^{-9}

3. Results

3.1. Convergence and Grid Sensibility Study: Preliminary Discussion

The assessment of the simulations' accuracy and stability is crucial to ensure reliable results.

Figure 5 reports the pressure and the velocity sampled along the aorta path line, together with the computed uncertainty during the systolic peak.

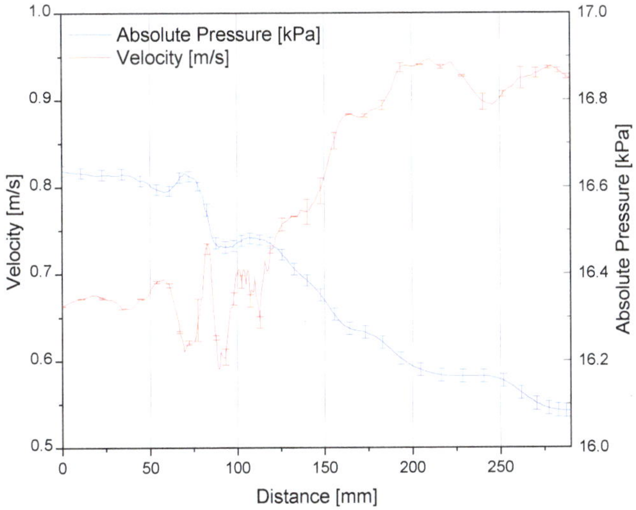

Figure 5. Pressure and velocity sampled on the aorta path-line during the systolic peak.

The uncertainty of the numerical results was evaluated in accordance with the criterion illustrated by Roache in [20,39], where the assessment procedures adopted by the AIAA, ITTC and IEEE are described and discussed in detail. Table 3 reports the grid convergence index and the order of convergence for the average value of pressure and velocity fields sampled along the aorta path-line.

Table 3. Grid convergence index and the order of convergence for the pressure and velocity fields.

	Grid Convergence Index	**Order of Convergence**
U	0.72%	2.03
p	0.23%	2.83

To verify the adequacy of the adopted grid far from the aorta surfaces, where the turbulence model reduces to a large-eddy simulation, the modelled kinetic energy was evaluated and compared with the total energy. To do that, we followed Di Mascio et al. [40]. Figure 6 shows the ratio between the modelled kinetic energy and the total kinetic energy on various cross-sections of the aorta geometry for a specific temporal instant (during the systolic phase).

Figure 6. $K_{mod}/(K_{mod}+K_{res})$ horizontal sections.

The modelled kinetic energy is reasonably small in the bulk flow compared to the resolved one; the ratio between the modelled and total kinetic energy is almost everywhere smaller than the value of 0.3, and therefore, the grid can be considered adequate for LES resolution, according to the Pope criterion [41] (there are only a few spots where the latter exceeds 0.3). Of course, the ratio becomes larger than 0.3 in the boundary layer on the aorta walls, where we have a Reynolds-averaged Navier–Stokes equation simulation.

Figure 7 shows a 3D volumetric representation of the velocity field captured during early, peak and late systole, 0.08, 0.18 and 0.3 s, respectively.

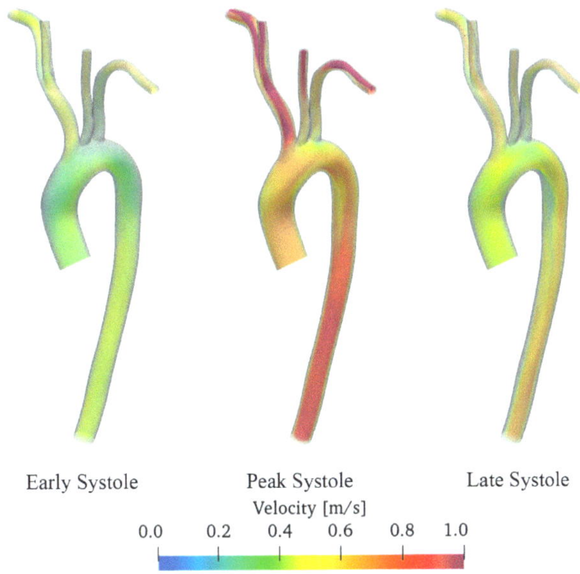

Figure 7. Computed velocity flow field of the sane aorta.

Unfortunately, flow measurements regarding the specific aorta here numerically investigated are not available, but the magnitude of the velocities predicted by CFD simulations are physiologically reasonable, being totally comparable with the ones measured with 4D magnetic resonance imaging (4D-MRI) in various literature works [7,42].

The same can be stated for pressure. The pressure field is reported in Figure 8, where the colour map shows the relative pressure computed with respect to a reference point chosen at the sino-tubular junction (inlet patch).

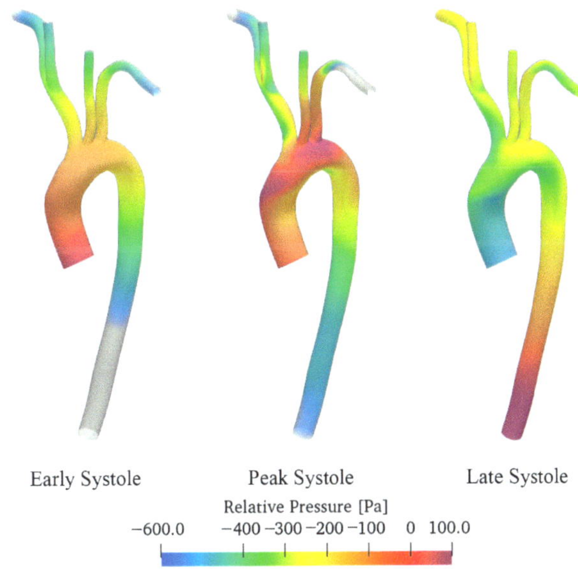

Figure 8. Computed pressure field of the sane aorta.

The range of variation is of the order of some hundreds of Pa, or, in mmHg, from approximately −10 mmHg to +1/+2 mmHg as usually experimentally measured with diagnostic techniques [9,43]. The pressure peak zone travels from the ascending to the descending as moving from the early to late systole phases.

3.2. Discussion of Aneurysmatic and Sane Aorta Results

Once having assessed the reliability of the computational methodology developed, further simulations regarding the aorta with the aneurysm were performed, and the results here were reported.

Figure 9 shows the velocity streamline of the sane and aneurysmatic aorta.

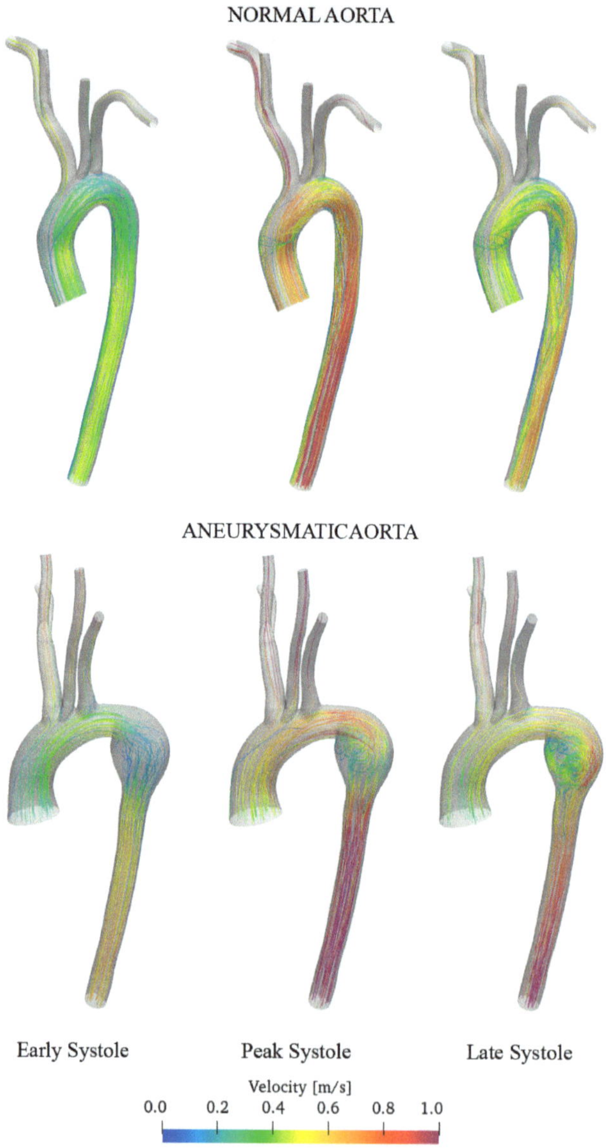

Figure 9. Comparison of sane and diseased aorta. Streamline coloured accordingly with the velocity magnitude.

The flow in the normal aorta model appears to be highly stable with smooth and continuous streamlines. In contrast, in the aneurysmatic aorta model, the flow is unstable and recirculation regions are observed within the aneurysm region.

Figure 10 highlights this fact, reporting the vorticity vector. It is the curl of the velocity field. It is fundamental in understanding the behaviour of turbulent flows and the vortex formation describing the local rotation or swirling motion of fluid particles.

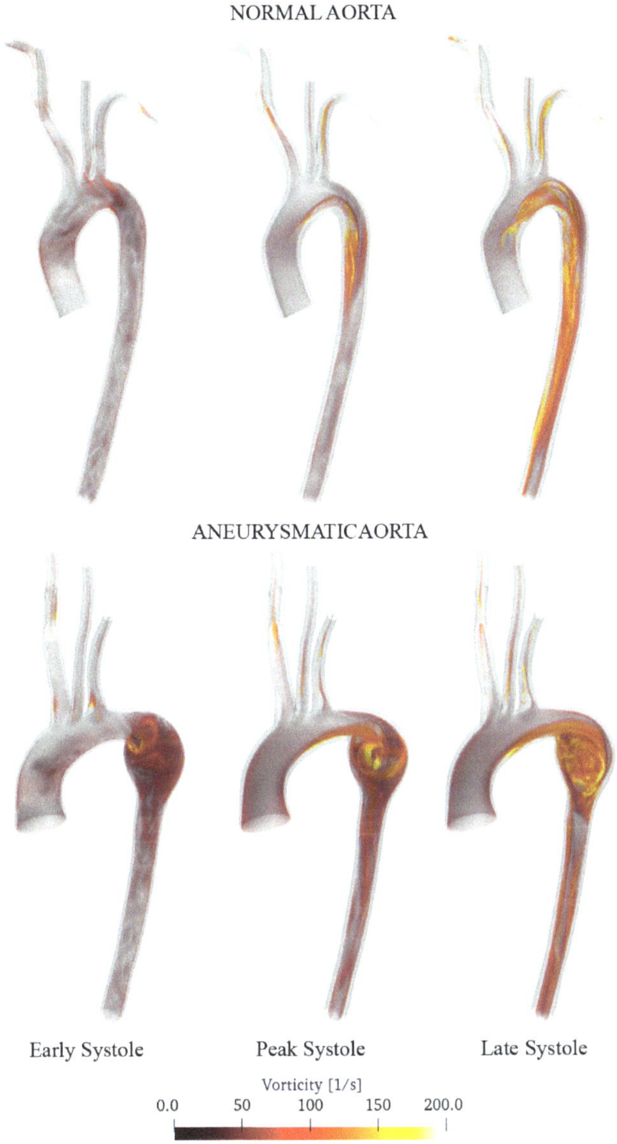

Figure 10. Comparison of sane and diseased aorta. 3D visualisation of the vorticity vector.

Observing the descending aorta, it can be noted how the flow is well organised and almost laminar, while the aneurysm enlargement creates chaotic vortical structures. The flow becomes unorganised, and the recirculation causes the blood particles to remain in prolonged contact with the aneurysm lumen surface, which can lead to the deposition of platelets on the surface [44]. The deposition of platelets on the lumen surface contributes to thrombus formation.

Figure 11 depicts the relative pressure distribution at the systole peak.

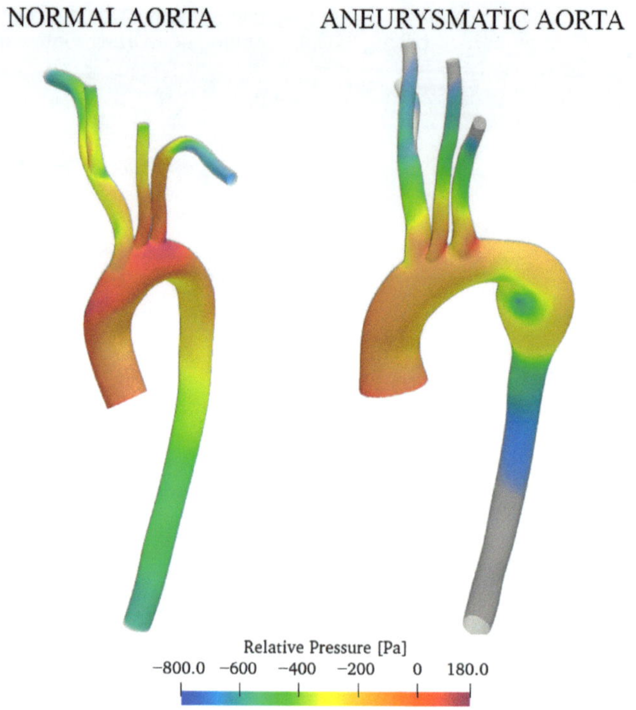

Figure 11. Relative pressure with reference to the inlet section. Comparison of sane and aneurysmatic aorta.

Similar pressure values were obtained for the two aorta models.

The maximum pressure is located at the aortic arch. Then, in the descending region of the aorta, the values decrease. The overall pressure variation is greater for the aneurysmatic aorta as a consequence of the structure of the flow field [45]. The aneurysm cavity presents a low-pressure region. A pressure minimum can be observed at the centre of the vortex as a consequence of the turbulent energy here dissipated.

The distribution of wall shear stress (WSS) on the artery walls is an important parameter that can be used to investigate risks related to the aneurysms of the aortic arch [45,46]. Abnormal low and high WSS patterns can have significant implications for the development of vascular diseases on the artery walls. The advantage of CFD simulations in studying aortic flow is that they can help identify regions with abnormal WSS levels, which may be more prone to dilatation or aneurysm formation.

Figure 12 reports the WSS at the peak systole for both the aorta.

The aneurysm zone and the region where the aorta branches off in arteries are subject to more intensity compared to the normal aorta's relative areas. The descending section of the diseased aorta, downstream of the enlargement, also presents spots of high WSS. The differences in WSS values between subjects with aneurysms and normal subjects can be attributed to many factors. However, among them, the variation of the velocity field should be cited as a consequence of the distorted aorta geometry.

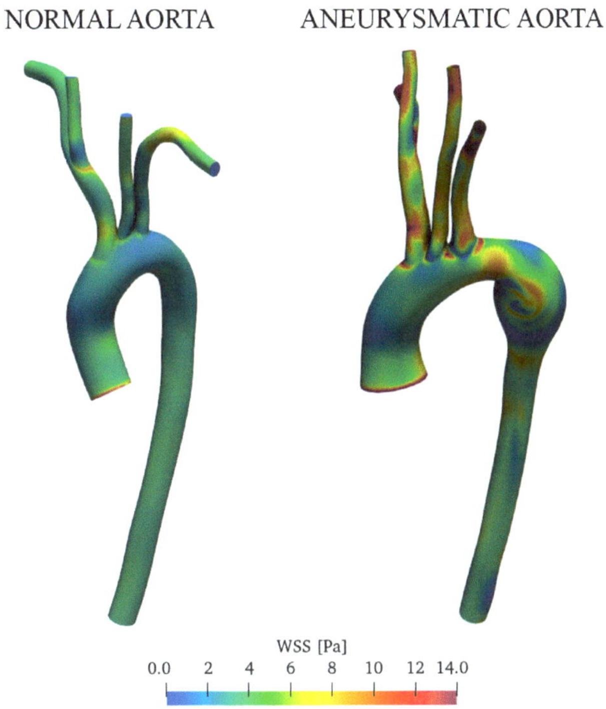

Figure 12. WSS acting on the boundary surfaces of the computational fluid domain. Comparison of sane and aneurysmatic aorta.

4. Conclusions

This study aims to contribute to the understanding of thoracic aortic flow characteristics. CFD simulations were performed using real-patient flow characteristics for both a normal subject and a subject with an aneurysm of the thoracic aorta. Real 3D geometries of the aorta were exploited with patient-specific pulsatile waveform used as the inlet boundary condition to reflect the real cardiac cycle. A three-element Windkessel lumped parameter model was implemented in a new OpenFOAM solver called pimpleWKFoam to set the pressure outlet boundary conditions properly. The main findings are:

- The developed methodology with the implementation of the WK model is capable of reproducing the fluid-dynamic characteristics of the aortic flow, providing realistic pressure and velocity field values.
- In the thoracic aorta, blood velocity is on the order of 1 m/s, while the pressure varies by about 500/700 Pa crossing the aorta itself.
- Aneurysm onset causes the flow field to become unstable, and recirculation zones grow in the enlargement section with the consequent deposition of platelets and thrombus formation.
- CFD simulations allow identifying regions with WSS levels that may be more prone to dilatation or aneurysm formation. The magnitude of the WSS reaches the maximum values in the enlarged zone.

Future developments of this work will regard the implementation of elastic walls with a fluid-structure-interaction (FSI) model.

Author Contributions: Conceptualisation, F.D.; methodology, F.D.; software, F.D.; validation, F.D. and A.D.M.; investigation, F.D.; resources, A.D.M.; writing—original draft preparation, F.D.; writing—review and editing, A.D.M.; supervision, A.D.M. All authors have read and agreed to the published version of the manuscript.

Funding: This research received no funding.

Institutional Review Board Statement: Not applicable.

Informed Consent Statement: Not applicable.

Data Availability Statement: Data is contained within the article or in the supplementary material.

Acknowledgments: All the simulations were performed with the developed solver on the Galileo100 cluster of HPC CINECA facilities within the agreement between DIIIE—Università degli Studi dell'Aquila and CINECA.

Conflicts of Interest: The authors declare no conflicts of interest.

Abbreviations

The following abbreviations are used in this manuscript:

CFD	Computational Fluid Dynamic
CVD	Cardio-Vascular Diseases
CT	Computed Tomography
DES	Detached Eddy Simulation
MRI	Magnetic Resonance Imaging
PISO	Pressure Implicit with Splitting of Operator
LCCA	Left Common Carotid Artery;
LES	Large Eddy Simulation
LSA	Left Subclavian Artery
RCCA	Right Common Carotid Artery
RANS	Reynolds-Averaged Navier–Stokes
RSA	Right Subclavian Artery
WK	Windkessel
WSS	Wall Shear Stress

References

1. Gaidai, O.; Cao, Y.; Loginov, S. Global Cardiovascular Diseases Death Rate Prediction. *Curr. Probl. Cardiol.* **2023**, *48*, 101622. [CrossRef] [PubMed]
2. Nerem, R. Hot film measurement of arterial blood flow and observation of flow disturbances. In *Cardiovascular Flow Dynamics and Measurement*; University Park Press: University Park, PA, USA, 1977; pp. 191–215.
3. Smedby, Ö. Geometric risk factors for atherosclerosis in the aortic bifurcation: A digitized angiography study. *Ann. Biomed. Eng.* **1996**, *24*, 481–488. [CrossRef] [PubMed]
4. Menichini, C.; Xu, X.Y. Mathematical modeling of thrombus formation in idealized models of aortic dissection: Initial findings and potential applications. *J. Math. Biol.* **2016**, *73*, 1205–1226. [CrossRef] [PubMed]
5. Vasava, P.; Jalali, P.; Dabagh, M.; Kolari, P.J. Finite Element Modelling of Pulsatile Blood Flow in Idealized Model of Human Aortic Arch: Study of Hypotension and Hypertension. *Comput. Math. Methods Med.* **2012**, *2012*, 861837. [CrossRef] [PubMed]
6. Voges, I.; Jerosch-Herold, M.; Hedderich, J.; Pardun, E.; Hart, C.; Gabbert, D.D.; Hansen, J.H.; Petko, C.; Kramer, H.H.; Rickers, C. Normal values of aortic dimensions, distensibility, and pulse wave velocity in children and young adults: A cross-sectional study. *J. Cardiovasc. Magn. Reson.* **2012**, *14*, 77. [CrossRef] [PubMed]
7. Bouaou, K.; Bargiotas, I.; Dietenbeck, T.; Bollache, E.; Soulat, G.; Craiem, D.; Houriez-Gombaud-Saintonge, S.; De Cesare, A.; Gencer, U.; Giron, A.; et al. Analysis of aortic pressure fields from 4D flow MRI in healthy volunteers: Associations with age and left ventricular remodeling. *J. Magn. Reson. Imaging* **2019**, *50*, 982–993. [CrossRef]
8. Lamata, P.; Pitcher, A.; Krittian, S.; Nordsletten, D.; Bissell, M.M.; Cassar, T.; Barker, A.J.; Markl, M.; Neubauer, S.; Smith, N.P. Aortic relative pressure components derived from four-dimensional flow cardiovascular magnetic resonance. *Magn. Reson. Med.* **2014**, *72*, 1162–1169. [CrossRef]
9. Rengier, F.; Delles, M.; Eichhorn, J.; Azad, Y.J.; von Tengg-Kobligk, H.; Ley-Zaporozhan, J.; Dillmann, R.; Kauczor, H.U.; Unterhinninghofen, R.; Ley, S. Noninvasive 4D pressure difference mapping derived from 4D flow MRI in patients with repaired aortic coarctation: Comparison with young healthy volunteers. *Int. J. Cardiovasc. Imaging* **2015**, *31*, 823–830. [CrossRef]

10. Yull Park, J.; Young Park, C.; Mo Hwang, C.; Sun, K.; Goo Min, B. Pseudo-organ boundary conditions applied to a computational fluid dynamics model of the human aorta. *Comput. Biol. Med.* **2007**, *37*, 1063–1072. [CrossRef]
11. Madhavan, S.; Kemmerling, E.M.C. The effect of inlet and outlet boundary conditions in image-based CFD modeling of aortic flow. *Biomed. Eng. Online* **2018**, *17*, 66. [CrossRef]
12. Pirola, S.; Cheng, Z.; Jarral, O.; O'Regan, D.; Pepper, J.; Athanasiou, T.; Xu, X. On the choice of outlet boundary conditions for patient-specific analysis of aortic flow using computational fluid dynamics. *J. Biomech.* **2017**, *60*, 15–21. [CrossRef] [PubMed]
13. Hardman, D.; Semple, S.I.; Richards, J.M.; Hoskins, P.R. Comparison of patient-specific inlet boundary conditions in the numerical modelling of blood flow in abdominal aortic aneurysm disease. *Int. J. Numer. Methods Biomed. Eng.* **2013**, *29*, 165–178. [CrossRef] [PubMed]
14. Antonuccio, M.N.; Mariotti, A.; Fanni, B.M.; Capellini, K.; Capelli, C.; Sauvage, E.; Celi, S. Effects of Uncertainty of Outlet Boundary Conditions in a Patient-Specific Case of Aortic Coarctation. *Ann. Biomed. Eng.* **2021**, *49*, 3494–3507. [CrossRef]
15. Catanho, M.; Sinha, M.; Vijayan, V. *Model of Aortic Blood Flow Using the Windkessel Effect*; University of California of San Diego: San Diego, CA, USA, 2012.
16. Westerhof, N.; Lankhaar, J.W.; Westerhof, B.E. The arterial Windkessel. *Med Biol. Eng. Comput.* **2009**, *47*, 131–141. [CrossRef] [PubMed]
17. Wilson, N.M.; Ortiz, A.K.; Johnson, A.B. The Vascular Model Repository: A Public Resource of Medical Imaging Data and Blood Flow Simulation Results. *J. Med. Device* **2013**, *7*, 0409231. [CrossRef] [PubMed]
18. Caballero, A.D.; Laín, S. A Review on Computational Fluid Dynamics Modelling in Human Thoracic Aorta. *Cardiovasc. Eng. Technol.* **2013**, *4*, 103–130. [CrossRef]
19. Etli, M.; Canbolat, G.; Karahan, O.; Koru, M. Numerical investigation of patient-specific thoracic aortic aneurysms and comparison with normal subject via computational fluid dynamics (CFD). *Med. Biol. Eng. Comput.* **2021**, *59*, 71–84. [CrossRef]
20. Duronio, F.; Di Mascio, A.; De Vita, A.; Innocenzi, V.; Prisciandaro, M. Eulerian–Lagrangian modeling of phase transition for application to cavitation-driven chemical processes. *Phys. Fluids* **2023**, *35*, 053305. [CrossRef]
21. Dutta, H. *Mathematical Methods in Engineering and Applied Sciences*; Mathematics and Its Applications; CRC Press: Boca Raton, FL, USA, 2020.
22. De Vita, M.; Duronio, F.; De Vita, A.; De Berardinis, P. Adaptive Retrofit for Adaptive Reuse: Converting an Industrial Chimney into a Ventilation Duct to Improve Internal Comfort in a Historic Environment. *Sustainability* **2022**, *14*, 3360. [CrossRef]
23. Duronio, F.; Mascio, A.D.; Villante, C.; Anatone, M.; Vita, A.D. ECN Spray G: Coupled Eulerian internal nozzle flow and Lagrangian spray simulation in flash boiling conditions. *Int. J. Engine Res.* **2023**, *24*, 1530–1544. [CrossRef]
24. Berger, S.A.; Jou, L.D. Flows in Stenotic Vessels. *Annu. Rev. Fluid Mech.* **2000**, *32*, 347–382. [CrossRef]
25. Pedley, T.J. *The Fluid Mechanics of Large Blood Vessels*; Cambridge University Press: Cambridge, UK, 1980. [CrossRef]
26. Morris, L.; Delassus, P.; Callanan, A.; Walsh, M.; Wallis, F.; Grace, P.; McGloughlin, T. 3-D Numerical Simulation of Blood Flow through Models of the Human Aorta. *J. Biomech. Eng.* **2005**, *127*, 767–775. [CrossRef] [PubMed]
27. Youssefi, P.; Gomez, A.; Arthurs, C.; Sharma, R.; Jahangiri, M.; Alberto Figueroa, C. Impact of Patient-Specific Inflow Velocity Profile on Hemodynamics of the Thoracic Aorta. *J. Biomech. Eng.* **2017**, *140*, 011002. [CrossRef]
28. Qian, Y.; Liu, J.L.; Itatani, K.; Miyaji, K.; Umezu, M. Computational Hemodynamic Analysis in Congenital Heart Disease: Simulation of the Norwood Procedure. *Ann. Biomed. Eng.* **2010**, *38*, 2302–2313. [CrossRef] [PubMed]
29. Zakaria, M.S.; Ismail, F.; Tamagawa, M.; Abdul Azi, A.F.; Wiriadidjaya, S.; Basri, A.A.; Ahmad, K.A. Computational Fluid Dynamics Study of Blood Flow in Aorta using OpenFOAM. *J. Adv. Res. Fluid Mech. Therm. Sci.* **2020**, *43*, 81–89.
30. Boyd, J.; Buick, J.M. Comparison of Newtonian and non-Newtonian flows in a two-dimensional carotid artery model using the lattice Boltzmann method. *Phys. Med. Biol.* **2007**, *52*, 6215. [CrossRef]
31. Otani, T.; Nakamura, M.; Fujinaka, T.; Hirata, M.; Kuroda, J.; Shibano, K.; Wada, S. Computational fluid dynamics of blood flow in coil-embolized aneurysms: Effect of packing density on flow stagnation in an idealized geometry. *Med. Biol. Eng. Comput.* **2013**, *51*, 901–910. [CrossRef]
32. Antón, R.; Chen, C.Y.; Hung, M.Y.; Finol, E.; Pekkan, K. Experimental and computational investigation of the patient-specific abdominal aortic aneurysm pressure field. *Comput. Methods Biomech. Biomed. Eng.* **2015**, *18*, 981–992. [CrossRef]
33. Spalart, P.R.; Deck, S.; Shur, M.L.; Squires, K.D.; Strelets, M.K.; Travin, A. A new version of detached-eddy simulation, resistant to ambiguous grid densities. *Theor. Comput. Fluid Dyn.* **2006**, *20*, 181–195. [CrossRef]
34. Di Angelo, L.; Duronio, F.; De Vita, A.; Di Mascio, A. Cartesian Mesh Generation with Local Refinement for Immersed Boundary Approaches. *J. Mar. Sci. Eng.* **2021**, *9*, 572. [CrossRef]
35. Safar, M.E.; Lévy, B.I. Chapter 13—Resistance Vessels in Hypertension. In *Comprehensive Hypertension*; Lip, G.Y., Hall, J.E., Eds.; Mosby: Philadelphia, PA, USA, 2007; pp. 145–150. [CrossRef]
36. Pochet, T.; Gerard, P.; Marnette, J.M.; D'orio, V.; Marcelle, R.; Fatemi, M.; Fossion, A.; Juchmes, J. Identification of three-element windkessel model: Comparison of time and frequency domain techniques. *Arch. Int. Physiol. Biochim. Biophys.* **1992**, *100*, 295–301. [CrossRef] [PubMed]
37. Tricarico, R.; Berceli, S.A.; Tran-Son-Tay, R.; He, Y. Non-invasive estimation of the parameters of a three-element windkessel model of aortic arch arteries in patients undergoing thoracic endovascular aortic repair. *Front. Bioeng. Biotechnol.* **2023**, *11*, 1127855. [CrossRef] [PubMed]

38. Lungu, A.; Wild, J.; Capener, D.; Kiely, D.; Swift, A.; Hose, D. MRI model-based non-invasive differential diagnosis in pulmonary hypertension. *J. Biomech.* **2014**, *47*, 2941–2947. [CrossRef]
39. Roache, P.J. Quantification of uncertainty in computational fluid dynamics. *Annu. Rev. Fluid Mech.* **1997**, *29* , 123–160. [CrossRef]
40. Di Mascio, A.; Dubbioso, G.; Muscari, R. Vortex structures in the wake of a marine propeller operating close to a free surface. *J. Fluid Mech.* **2022**, *949*, A33. [CrossRef]
41. Pope, S.B.; Pope, S.B. *Turbulent Flows*; Cambridge University Press: Cambridge, UK, 2000.
42. Garcia, J.; Barker, A.J.; Markl, M. The Role of Imaging of Flow Patterns by 4D Flow MRI in Aortic Stenosis. *JACC Cardiovasc. Imaging* **2019**, *12*, 252–266. [CrossRef]
43. Soulat, G.; McCarthy, P.; Markl, M. 4D Flow with MRI. *Annu. Rev. Biomed. Eng.* **2020**, *22*, 103–126. [CrossRef]
44. Bluestein, D.; Niu, L.; Schoephoerster, R.T.; Dewanjee, M.K. Steady Flow in an Aneurysm Model: Correlation Between Fluid Dynamics and Blood Platelet Deposition. *J. Biomech. Eng.* **1996**, *118*, 280–286. [CrossRef]
45. Vinoth, R.; Kumar, D.; Adhikari, R.; Vijayapradeep, S.; Geetha, K.; Ilavarasi, R.; Mahalingam, S. Steady and Transient Flow CFD Simulations in an Aorta Model of Normal and Aortic Aneurysm Subjects. In *Proceedings of the International Conference on Sensing and Imaging*; Jiang, M., Ida, N., Louis, A.K., Quinto, E.T., Eds.; Springer International Publishing: Cham, Switzerland, 2019; pp. 29–43.
46. Febina, J.; Sikkandar, M.Y.; Sudharsan, N.M. Wall Shear Stress Estimation of Thoracic Aortic Aneurysm Using Computational Fluid Dynamics. *Comput. Math. Methods Med.* **2018**, *2018*, 7126532. [CrossRef]

fluids

MDPI

Article

Evaluating the Impact of Domain Boundaries on Hemodynamics in Intracranial Aneurysms within the Circle of Willis

Pablo Jeken-Rico [1], Aurèle Goetz [1], Philippe Meliga [1], Aurélien Larcher [1], Yigit Özpeynirci [2] and Elie Hachem [1,*]

[1] Mines Paris, Université PSL, Centre de Mise en Forme des Matériaux (CEMEF), UMR7635 CNRS, 06904 Sophia Antipolis, France; pablo.jeken_rico@minesparis.psl.eu (P.J.-R.)

[2] Institute of Neuroradiology, University Hospital LMU Munich, 81377 Munich, Germany; yigit.oezpeynirci@med.uni-muenchen.de

* Correspondence: elie.hachem@minesparis.psl.eu

Abstract: Hemodynamic simulations are increasingly used to study vascular diseases such as Intracranial Aneurysms (IA) and to further develop treatment options. However, due to limited data, certain aspects must rely on heuristics, especially at the simulation's distal ends. In the literature, Murray's Law is often used to model the outflow split based on vessel cross-section area; however, this poses challenges for the communicating arteries in the Circle of Willis (CoW). In this study, we contribute by assessing the impact of Murray's Law in patient-specific geometries featuring IA at the posterior communication. We simulate different domain extensions representing common modelling choices and establish Full CoW simulations as a baseline to evaluate the effect of these modelling assumptions on hemodynamic indicators, focusing on IA growth and rupture-related factors such as the Wall Shear Stress (WSS) and Oscillatory Shear Index (OSI). Our findings reveal qualitative alterations in hemodynamics when not modeling posterior communication. Comparisons between computing the anterior circulation and computing the whole Circle of Willis reveal that quantitative changes in WSS may reach up to 80%, highlighting the significance of modelling choices in assessing IA risks and treatment strategies.

Keywords: computational hemodynamics; intracranial aneurysms; Circle of Willis; wall shear stress; boundary conditions

Citation: Jeken-Rico, P.; Goetz, A.; Meliga, P.; Larcher, A.; Özpeynirci, Y.; Hachem, E. Evaluating the Impact of Domain Boundaries on Hemodynamics in Intracranial Aneurysms within the Circle of Willis. *Fluids* **2024**, *9*, 1. https://doi.org/10.3390/fluids9010001

Academic Editors: Fang-Bao Tian and D. Andrew S. Rees

Received: 21 October 2023
Revised: 23 November 2023
Accepted: 18 December 2023
Published: 21 December 2023

Correction Statement: This article has been republished with a minor change. The change does not affect the scientific content of the article and further details are available within the backmatter of the website version of this article.

1. Introduction

Intracranial Aneurysms (IA) are estimated to affect about 3.2% of the adult population. While the annual risk of rupture is moderate at 2.2%, they impose a significant burden on patients, physicians, and the healthcare system [1,2]. To prevent ruptures, considerable effort is invested in early diagnosis, growth prediction, and treatment of IA. Advances in non-invasive imaging techniques such as Magnetic Resonance Angiography (MRA) have made regular preventive screenings increasingly feasible. Detecting IA at an early stage provides neuroradiologists with the opportunity to weigh treatment options or continue monitoring the aneurysm's evolution. In this decision process, physicians rely on geometrical and topological characterization of the IA along with other patient records. While scoring methods based on statistics such as the PHASES score may contribute to the decision-making process, they have been shown to be weak overall predictors [3].

Instead of relying solely on statistics, Computational Fluid Mechanics (CFD) has been proposed as a promising complementary tool. The goal is to simulate the hemodynamics of IA and extract risk indicators in order to assess the severity of the case and potentially predict a rupture site. Despite substantial progress in this field [4–6], a latent dissent exists in the research community, highlighted during numerous CFD challenges. It has been confirmed that due to their various modeling strategies the participating teams obtained dissimilar results that, in certain cases, could point towards different interpretations [7–10]. For a consolidated use of CFD in IA research, it is imperative to address open modelling questions in advance [11].

Outflow Boundary Conditions (BCs) represent one recurring source of uncertainty in vascular fluid dynamics. Reliable measurements are challenging to obtain, and as such are rarely employed [11]. Instead, the Principle of Minimum Work can be used to prescribe the flow split among the distal ends of the simulated network based on the relationship of their cross-sectional areas [12]. This heuristic, arising from the pursuit of minimizing the energy spent on the transport and storage of blood, has been confirmed through ex vivo analysis by analyzing the regularity of arterial branching patterns [13]. The Law of Minimum Work, known as Murray's law, is not only more realistic than plain stress-free outflows [14,15], it offers the advantage that it solely depends on the geometrical features of the network. This is advantageous because it makes boundary conditions reproducible across varying modeling assumptions such as rheology laws, solver schemes, and boundary extrusions.

Nevertheless, Murray's law has an unstudied implication when applied to the communicating arteries in the Circle of Willis (CoW), a loop of arteries found at the base of the brain (see Figure 1). From an anatomical point of view, these vessels serve the purpose of linking the anterior and posterior circulations, thereby providing alternative pathways for the blood in case of ischemia [16]. Frequently, the PCom carries a lesser net flow from the ICA to the PCA, although different CoW configurations may perturb this rule [17–19]. The majority of CFD studies that examine IA at the carotid arteries only consider one part of the anterior circulation, either defining the PCom as a regular system outlet [14,20] or neglecting it [21,22]. These two strategies pose two extreme cases, and inevitably raise the question of whether or not they lead to physiological conditions.

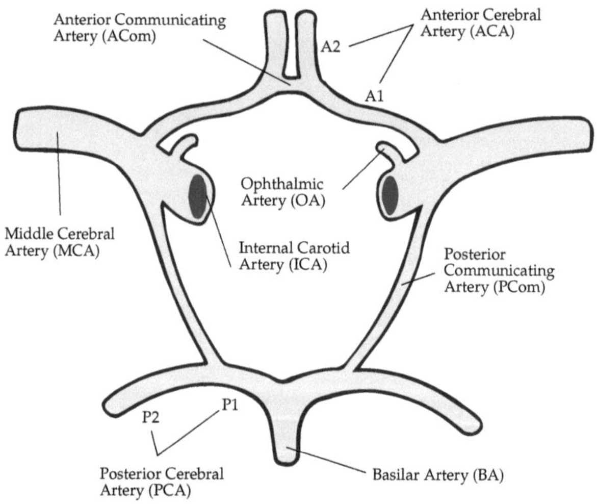

Figure 1. Schematic view of the complete CoW and the ophthalmic arteries.

The uncertainty surrounding the simulation domain of IA and the uncertainty in treating the PCom artery in CFD, as per the existing literature, underscores the motivation for this study. Our aim is to evaluate changes in intra-aneurysmal hemodynamics in ICA-PCom bifurcation aneurysms by varying the extent of the simulated vasculature. These extensions are selected based on common observations from the IA literature, and are compared with full CoW simulations. The primary focus of the comparison lies in intra-aneurysmal flow patterns and the exposure of the lumen to shear stresses.

2. Materials and Methods

2.1. Medical Imaging and Segmentation

Angiographic images were obtained using time-of-flight angiography on 1.5 T and 3 T MRI scanners with isotropic imaging and 0.6 mm slice thickness. The lumen was segmented using 3DSlicer (https://www.slicer.org/, accessed on 5 May 2023) under the supervision of neuroradiologists of the cooperating medical institution. Patient A has a complete CoW with one aneurysm of diameter 7.5 mm at the left ICA-PCom bifurcation. Patient B has a bilateral P1 hypoplasia, and as such is missing a connection of the basilar artery with both anterior circulations. The PComs are defined in this case as fetal, as their size is larger than usual and they are in fact the main supplier of blood to the PCA [23]. The aneurysm of the latter is located on the right side ($d = 5$ mm) and features a pronounced lateral daughter sack. The complex geometry of this formation was later confirmed through image recordings during the operation procedure. Lastly, patient C has an incomplete CoW due to an absent right PCom. In this case, two IAs can be found, one at the left ICA-PCom bifurcation and one at the right MCA bifurcation. The former, which was the only one considered during analysis, is classified as bilobular due to the presence of two rounded sacks.

2.2. Spatial Discretization

For each of the segmented geometries, three extensions of ascending complexity (see Figure 2) were generated to assess the implications of Murray's law on simulated hemodynamics. Geometries with the keyword Simplified are constrained to the main supplying artery that leads to the IA. Communicating arteries and A2 segments (see Figure 1) are neglected in this case. Middle extends the prior Simplified complexity by adding the PCom. Finally, the keyword Full denotes the full CoW. To minimize perturbations caused by inlet and outlet models, extrusions were made along these boundary patches [11,21]. Circular profiles were adapted by a least-square fit to the irregular vessel cross-sections and extended along the mean vessel direction using transfinite interpolations. This procedure additionally simplifies the imposition of circular inflow profiles and facilitates geometrical information required for the outflow boundary conditions.

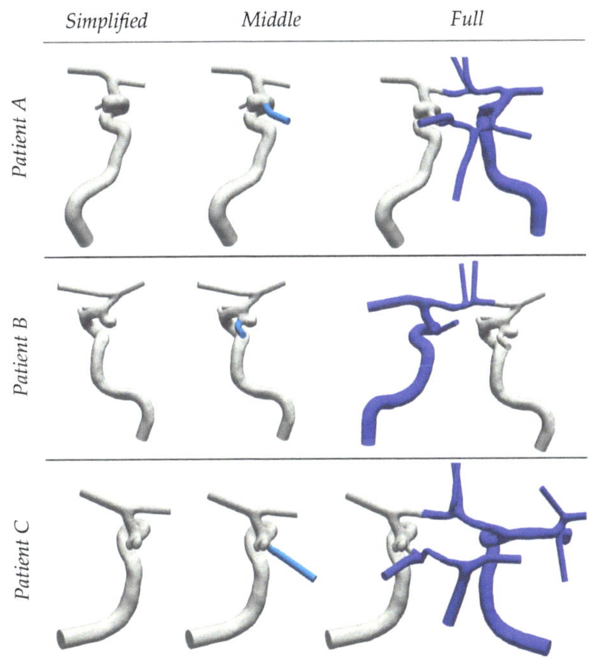

Figure 2. Overview of cases A–C (**top** to **bottom**) in the different extensions Simplified–Full (**left** to **right**). Simplified—Neglection of the adjacent PCom. Middle—Simulation of ICA and proximal vessels. Full—Full computation of the CoW.

The open 2D surface meshes were parametrized using conformal maps, remeshed, and extruded inwards to generate a set of tetrahaedral boundary layers (growth factor 1.2) [24]. These boundary layers serve two crucial purposes in our fluid simulations of arteries. First, the incorporation of boundary layers is essential for a more accurate resolution at the wall regions. Second, maintaining consistency in the structure at the boundaries permits comparison between different cases without disturbances from mesh-induced boundary effects. The layer thickness was determined based on the dimensionless wall distance of $y^+ = 1$, an overestimated Reynolds number of $Re = 1000$ and the averaged diameter of the ICAs. As proposed in [4], we fixed the mesh size $h = 0.2$ mm based on a preliminary mesh convergence, which was progressively refined to $h = 0.1$ mm at the aneurysm periphery (see Figure 3) utilizing gmsh (https://gmsh.info/, accessed on 30 May 2023).

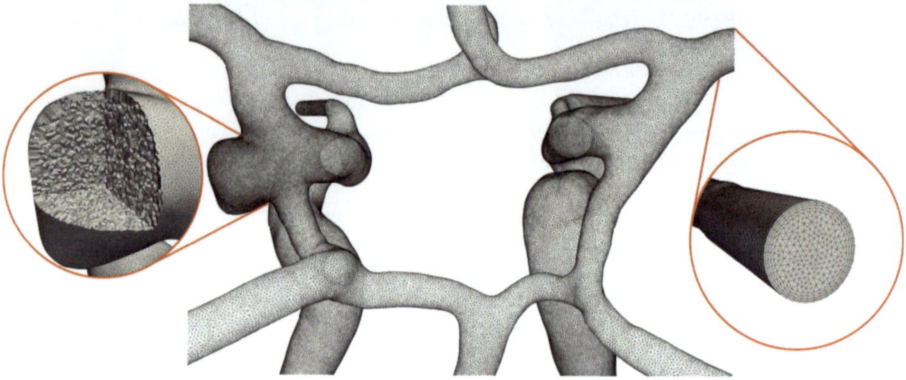

Figure 3. Visualization of the CoW of patient A with highlighted mesh features (orange circles) showing the linear isotropic refinement (**left**) and structured tetrahaedral boundary layers (**right**).

2.3. *Hemodynamics Simulations*

2.3.1. Navier–Stokes

Simulations were carried out by numerically solving the transient incompressible Navier–Stokes equations using an in-house Finite Element (FE) solver. The discrete system is set up by linear elements for both pressure and velocity, and is consequently stabilized by a residual-based Variational Multiscale-type method [25,26]. The weak formulation (see Equations (1) and (2)) is enriched with the residuals of continuity $\mathcal{R}_C$ and residuals of momentum $\mathcal{R}_M$. For details on the stabilization parameters (τ_C and τ_M), see [25] and references therein. Time integration was carried out with a second-order semi-implicit backwards scheme, which has been shown to provide a good balance between memory and consistency [27,28].

$$\int_\Omega \underbrace{\rho(\partial_t \vec{u}_h + (\vec{u}_h \cdot \nabla)\vec{u}_h) \cdot \vec{w} + \sigma : \nabla \vec{w}}_{\text{Galerkin terms}} \, d\Omega + \sum_{\Omega_e \in \Omega} \int_{\Omega_e} \underbrace{\tau_M \rho (\vec{u}_h \cdot \nabla)\vec{w} \cdot \mathcal{R}_M}_{\text{Upwind stabilization}} \, d\Omega$$
$$+ \sum_{\Omega_e \in \Omega} \int_{\Omega_e} \underbrace{\tau_C \mathcal{R}_C \rho \nabla \cdot \vec{w}}_{\text{Grad-div stabilization}} \, d\Omega = \int_{\Gamma_h} \underbrace{\vec{w} \cdot \sigma \cdot \vec{n}}_{\text{Boundary stress}} \, d\Gamma, \quad \forall \vec{w} \in H^1 \tag{1}$$

$$\sum_{\Omega_e \in \Omega} \int_{\Omega_e} \underbrace{\tau_M \nabla q \cdot \mathcal{R}_M}_{\text{Pressure stabilization}} \, d\Omega + \int_\Omega \underbrace{q \nabla \cdot \vec{u}_h}_{\text{Galerkin term}} \, d\Omega = 0, \quad \forall q \in H^1 \tag{2}$$

As mentioned previously, several benchmark validations with both space and time convergences analyses were made in [25,27], and more recently in the context of IA in [29]. The obtained parameters yielded stable simulations and matched the proposed resolutions in the IA and hemodynamics literature [4,22,30]. For completeness, we added the effects of different timesteps and mesh resolutions (see Figures 4 and 5) on a representative test case, patient B, allowing us to finally fix the time step to 1 ms and the mesh size graduation between $h = 0.2$ mm and $h = 0.1$ mm at the aneurysm periphery for for all our simulations.

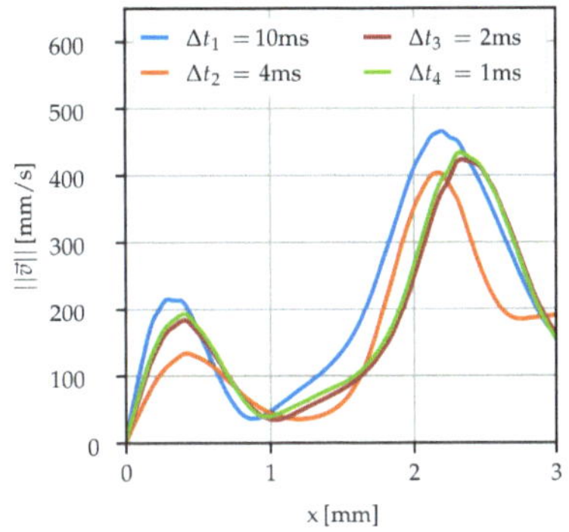

Figure 4. Velocity profile over aneurysm B for different time resolutions using the described mesh.

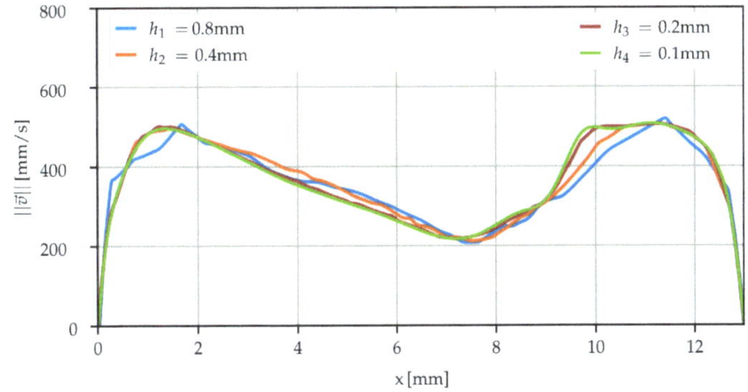

Figure 5. Velocity profile plotted over the C6-C7 ICA sections (Patient B) for different mesh sizes.

2.3.2. Rheology

The shear-thinning rheology was modelled through the Carreau model describing its behaviour [11,31,32], as provided in Equation (3). The constants $\rho = 1056\,\mathrm{kg/m^3}$, $\mu_0 = 0.0456$ Pa·s, $\mu_\infty = 0.0032$ Pa·s, $\lambda = 10.03$ s, and $n = 0.344$ and shear rate $\dot{\gamma}$ were employed [33].

$$\mu(\dot{\gamma}) = \mu_\infty + (\mu_0 - \mu_\infty)\left(1 + (\lambda\dot{\gamma})^2\right)^{(n-1)/2} \tag{3}$$

2.3.3. Boundary Conditions

Inflow conditions were described with the help of a generalized volumetric flow curve which was scaled and split among the supplying arteries (see Figure 6). More precisely, in case A each ICA contributes 44% of the total volumetric flow, while the BA carries only 12% due to its small size [34]. Cases B and C follow a split of 40%–20% between each ICA and the BA. The flow rate Q was imposed via parabolic velocity profiles at the base of the arterial system. The walls were considered fully rigid, and as such were set through no-slip conditions.

The outflow of the system followed Murray's Law, guiding the flow distribution based on the cubed outlet radius, as expressed in Equation (4) [14,15]. The exponent three ($n = 3$) aligns with Poiseuille flow assumptions on the long-term cost of blood transport in brain arteries [12]. This exponent was chosen for consistency with the linear relationship $P = QR$ used to impose the desired flow rates. The rule is derived from fully established Poiseuille flow principles [35], and is commonly used in brain arteries [28,36]. The resistances R were adjusted iteratively for each outlet, ensuring a systematic application of Murray's Law across the system.

$$Q_{out,i}(t) = Q_{in}(t)\frac{r_i^3}{\sum_j r_j^3} \tag{4}$$

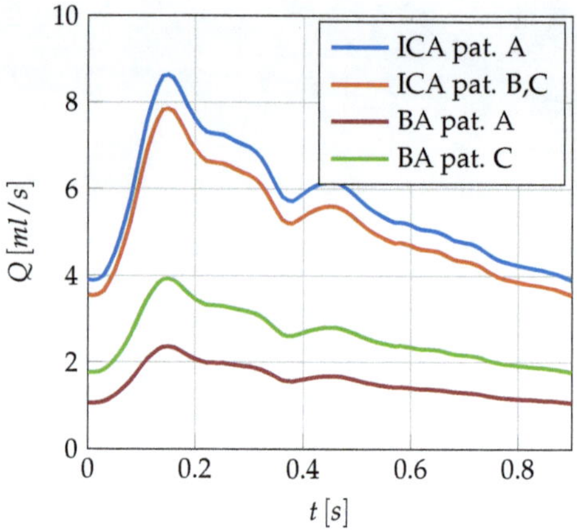

Figure 6. Volumetric inflow of patients A–C.

2.3.4. Computation Details

For the solution, the fully coupled system of pressure and velocity was solved in parallel using the ILU(4) block preconditioned Stabilized Bi-Conjugate Gradient method. The simulations were carried out on dual processors (32-Core AMD EPYC 64-bit Processor 7502, Advanced Micro Devices, Santa Clara, CA, USA) with a 2.5 GHz base clock rate and a HDR 100 interconnection. The computing times of each cardiac cycle can be extracted from Table 1. The number of elements, likewise provided in Table 1, was notably higher for case A due to its larger volume and aneurysm sack.

Table 1. Element count in millions of elements and computation time per cardiac cycle of different meshes.

Patient		Simplified	Middle	Full
A	n_{elem}	3.4	3.8	7.1
	T_{comp}	9:20 h	10:12 h	22:10 h
B	n_{elem}	2.0	2.1	3.6
	T_{comp}	4:41 h	5:19 h	8:23 h
C	n_{elem}	1.6	1.7	4.7
	T_{comp}	5:07 h	5:46 h	8:30 h

2.4. Hemodynamic Descriptors

The influence of different domain extensions was assessed through key hemodynamic indicators, including Wall Shear Stress(WSS), Oscillatory Shear Index (OSI), and velocity profiles. As a primal variable, velocity is inherently linked to any alterations in system dynamics, and as such was considered for initial assessments. WSS and OSI belong to a set of indicators related to vascular remodeling, and are recognized as risk factors in aneurysm formation [37–39]. According to Meng et al. [40], destructive remodeling phenomena of arterial walls can be categorized into mural cell-mediated events caused by abnormally high WSS and inflammatory remodeling induced by low WSS and high OSI. However, despite the identification of general trends in the literature [41], a consensus on threshold values for WSS and OSI in brain arteries remains elusive [42]. Recent advancements in hemodynamic research have introduced WSS-derived quantities such as the WSS gradient topology [43] and the relative residence time (see [44] and therein). These novel indicators provide insights into the pulsatility and topological structure of WSS distributions. Recognizing the critical role of WSS in these advancements, our work primarily focuses on its accurate computation. These quantities were calculated from the velocity fields through the stress tensor σ, as illustrated in Equations (5) and (6), respectively.

$$\vec{\tau}_{wss} = \vec{n} \times (\sigma \cdot \vec{n}) \times \vec{n} \tag{5}$$

$$\mathrm{OSI} = \frac{1}{2}\left(1 - \frac{\left\| \int_0^T \vec{\tau}_{wss}\, dt \right\|}{\int_0^T \left\| \vec{\tau}_{wss} \right\| dt}\right) \tag{6}$$

3. Results

3.1. Simulations

The overall flow distribution can be visually followed from the velocity magnitudes shown in Figure 7. Patient A stands out due to slower flow velocities despite similar tributary flows, which is a direct consequence of having larger vessels overall compared to B and C. The plot highlights regions of high flow speeds, for instance in the terminal ICA bifurcation. The highly inertia-driven flow turns mostly into the MCAs, which due to both their size and the Murray exponent (n = 3) carry the largest outflow. When present, the BA supplies the posterior circulation of the brain without directly irrigating the anterior circulation. Therefore, the flux traversing the PCom is strictly unidirectional, carrying blood away from the ICA towards the PCA in accordance with the commonly observed flow direction [18,45]. According to our models, Patients B and C feed the distal PCA mostly (Patient C) or exclusively (Patient B) through the PCom due to their small or fetal P1 segments.

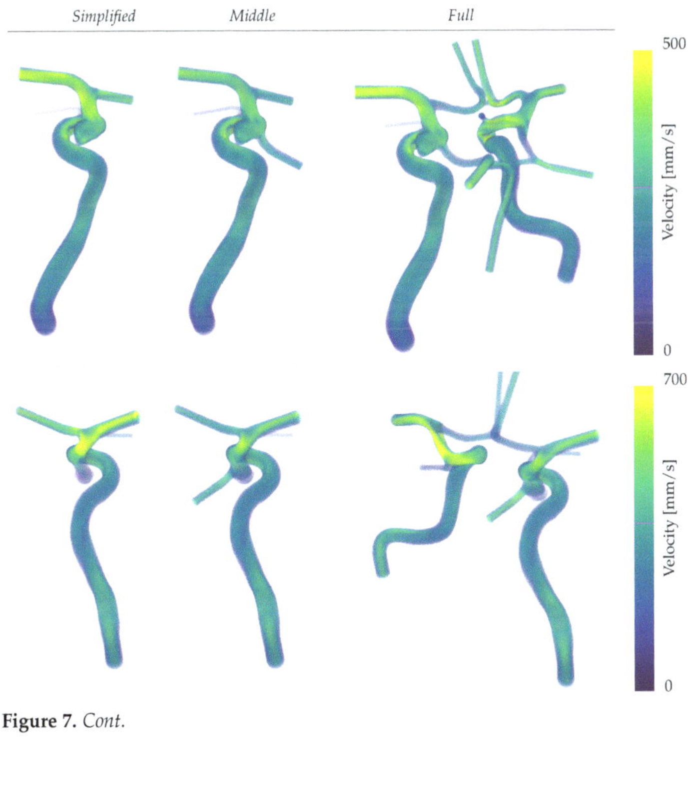

Figure 7. *Cont.*

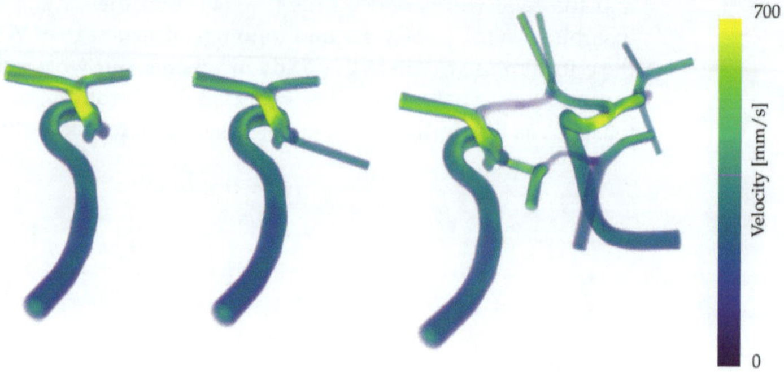

Figure 7. Maximum velocity projection of patients A–C (**top** to **bottom**) for all complexity levels Simplified–Full (**left** to **right**) at diastole.

The detailed presentation of the average blood transported through each vessel is provided in Table 2. Notably, in the Full configuration, the flow through the A2 segment of the ACA is consistently smaller, exhibiting a deviation of up to 5 percentage points compared to the Middle and Simplified configurations. The Middle configuration is characterized by guiding flow away from the MCA towards the A1 segment of the ACA, representing a distinct pattern in the distribution of the blood flow. These observations are relevant for the later discussion of the biases of Murray's law. Additionally, the volumetric flow was recorded at the vessels close to the aneurysms. Figure 8 shows this quantity normalized with respect to the maximum mean flow of the three extensions. It can be seen that in the Simplified simulations the flow that would leave through the PCom is distributed among the other outlets, causing their flow rates to be consistently higher. While Middle is distinguished from Full through lower outflows through the MCA and larger ones through the ACA, it exhibits no general trend on the PCom.

For verification purposes, the goodness of the boundary condition model is shown, for which the resistances R_i have been fitted in an iterative process to minimize the L_1 error between the targeted flow and the achieved one. In Figure 9, the volumetric flow of the left hemisphere's outlets (aneurysm A) are shown together with the targeted curves as a representative example of the fitting performance. The linear law enables the mass flow to be controlled through pressure BCs within a relative error of 5% per cardiac cycle despite non-Newtonian and transient flows. The largest relative deviations are observed at small vessels during peak systolic and peak diastolic times, for instance, the OA (green curves) in Figure 9.

Table 2. Outflow per cardiac cycle with respect to the in total inflow of each patient. In the Full configuration of case C, the flow in the right MCA is a consequence of imposing Murray's law on the vessels of the M1–M2 bifurcation. The massflow for the Middle and Simplified configurations is taken at the A1 ACA segment due to the inherent cut of the domain.

Case	Complexity	MCA		PCA		ACA		OA	
		Left	Right	Left	Right	Left	Right	Left	Right
A	Full	28.8%	25.7%	14.0%	7.7%	7.2%	9.3%	3.1%	4.2%
	Middle	21.8%	-	10.6%	-	9.2%	-	2.4%	-
	Simplified	28.8%	-	-	-	12.1%	-	3.1%	-
B	Full	29.3%	21.6%	4.9%	9.6%	4.3%	7.0%	-	3.1%
	Middle	-	20.9%	-	9.3%	-	6.8%	-	3.1%
	Simplified	-	27.2%	-	-	-	8.8%	-	4.0%
C	Full	27.1%	35.9%	12.0%	12.4%	4.0%	8.5%	-	-
	Middle	22.1%	-	7.5%	-	10.3%	-	-	-
	Simplified	27.3%	-	-	-	12.7%	-	-	-

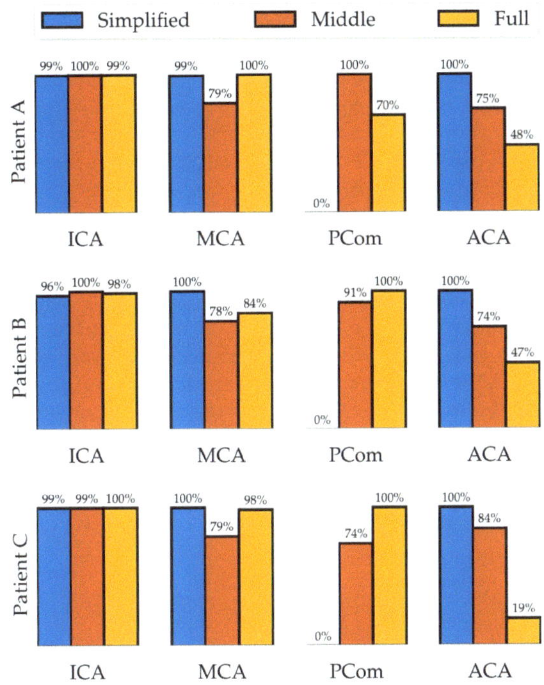

Figure 8. Mean volumetric flow per cardiac cycle in the IA proximity. The percentage values describe how much flow passes through the indicated vessel with respect to the other configurations. The cuts of the MCA and ACA are set at the M1 and A1 segments, respectively.

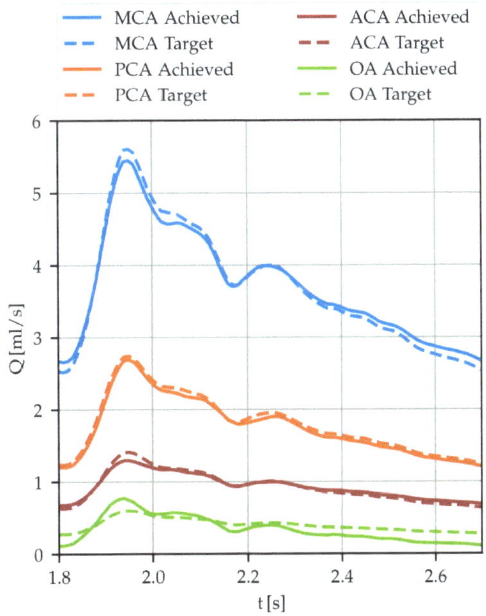

Figure 9. Volumetric flow through the left ICA outlets of patient A, comparing the values of the targeted flow (dashed lines) as dictated by Murray's Law (see Equation (4)) with those obtained using the linear pressure relationship $P = QR$ after fitting the resistances.

3.2. Intra-Aneurysmal Dynamics

The flow structures in the domes are visualized in Figure 10 for each patient (rows) and configuration (columns). The aneurysms are cut with a plane at the neck, and velocity streamlines of the intra-aneurysmal flow are inscribed lightly to indicate the structure of the present vortices at peak diastolic time. For the inertial flow of aneurysm A (first row), all three configurations show a large centred vortex traversing the dome that can be regarded as consistent across the three sizes. On the contrary, Patients B and C in the second and third rows of Figure 10 display clear discrepancies. Starting with aneurysm B (second row), it is apparent that in the simple configuration the main inflow enters the dome focused as a small jet, then later decays into a stable weaker recirculation. For the Middle and Full extensions of the same case, the inflow hits the neck of the daughter sack, splitting its momentum among two vortices that traverse the main lobe and the daughter sack, respectively. In the Full simulation of B (second row, third column), the daughter sack's recirculation is located farther up inside the dome, leading to larger in-plane velocities. Lastly, Patient C (the third row of Figure 10) presents the largest disparities on both the quantitative and qualitative levels. One of the most noticeable differences in Figure 10 is the inverted circulation of the right aneurysm dome with respect to the other two extensions in the *Full* simulation. At the same time, the average velocity of the impinging jet exceeds those in the other cases by 31% and 19%, respectively.

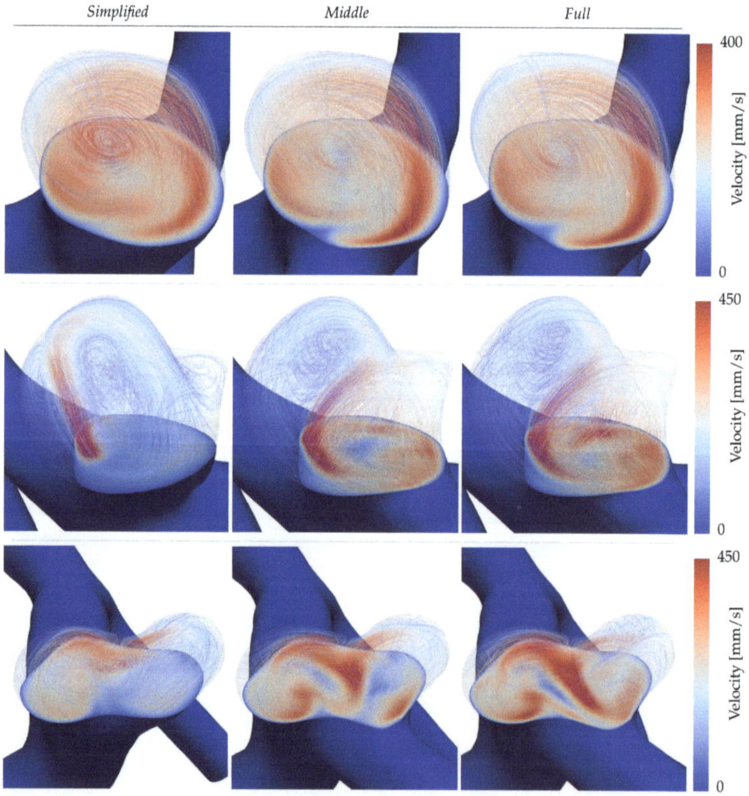

Figure 10. Cross-section cuts at the IA neck of Patients A–C (**top** to **bottom**) with the inscribed velocity magnitudes at diastole. The velocity streamlines are superposed with the aneurysm necks to highlight the structure of the flow.

3.3. Wall Shear Stress

The WSS magnitudes for peak diastole and systole are shown in Figure 11a,b, respectively. Patient A (first row) is characterized by a stable and qualitatively similar WSS pattern across all extensions. Here, only the Simplified case demonstrates smaller patches of higher shearing that are not present in the Middle and Full systems. Patients B and C instead exhibit persistently altered WSS patterns for all of the domain extensions. Setting the Simplified system aside, it is possible to observe shifts in patterns between the Middle and Full extensions in the second and third columns, respectively. Most notable here is the daughter sack in patient B's aneurysm (second row of Figure 11). During diastole, the WSS is relatable between the Middle and Full systems. On the contrary, at systole the shearing increases for the Middle system while remaining low for the Full system, leading to two distinct characterizations. Following the Full system, the daughter sack could be identified as a risk factor due to low WSS exposure, whereas the Middle simulation would not. Daughter sacks play an important role here, as their appearance is linked to an augmented risk of rupture. The last patient case in the bottom row of Figure 11a,b, again shows an inconsistent WSS among the different extensions, which is especially aggravated at the dent separating the two sacks.

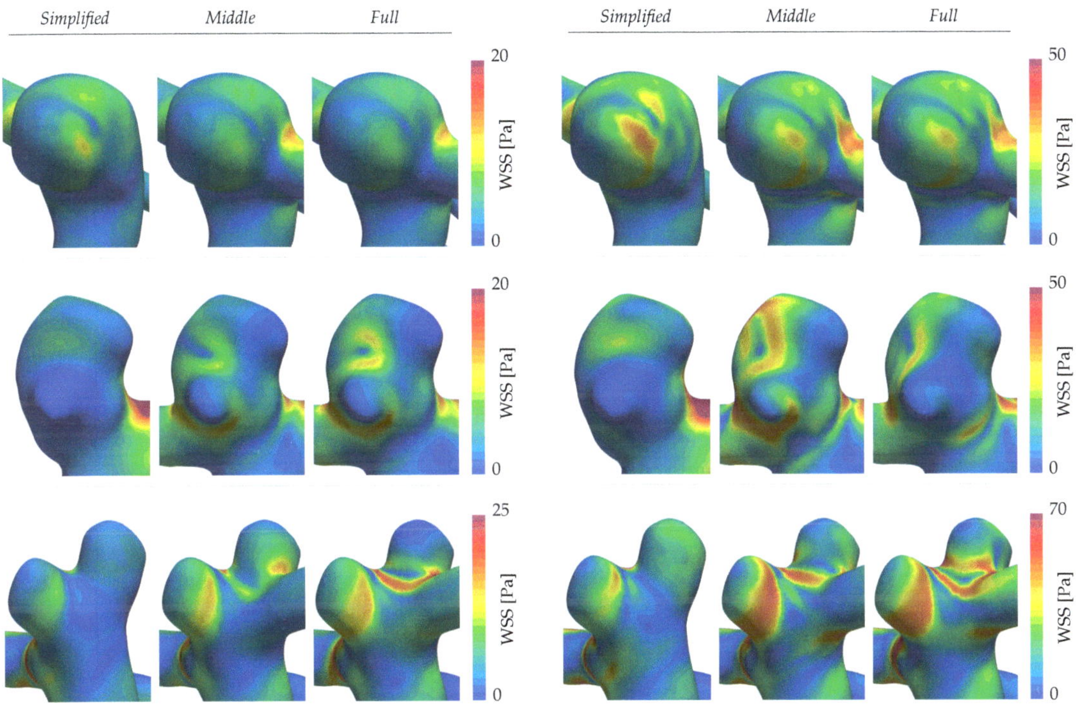

(**a**) Diastolic WSS (**b**) Systolic WSS

Figure 11. WSS distributions on aneurysms A–C (**top** to **bottom**) for the different complexity levels at diastole (**a**) and peak systole (**b**).

3.4. Oscillatory Shearing

Following the same structure as in Figure 11a,b, the OSI is displayed in Figure 12. The observed stable WSS profile of aneurysm A inevitably leads to nearly absent oscillatory shearing with only small localized peaks (see the first row in Figure 12). Large dissent of the OSI distributions is again found in patients B and C (the second and third rows of Figure 12), which is especially apparent in the distal lobe of Patient C. There, in contrast to the other cases, high overall values are found on the visible face in the Full CoW simulation.

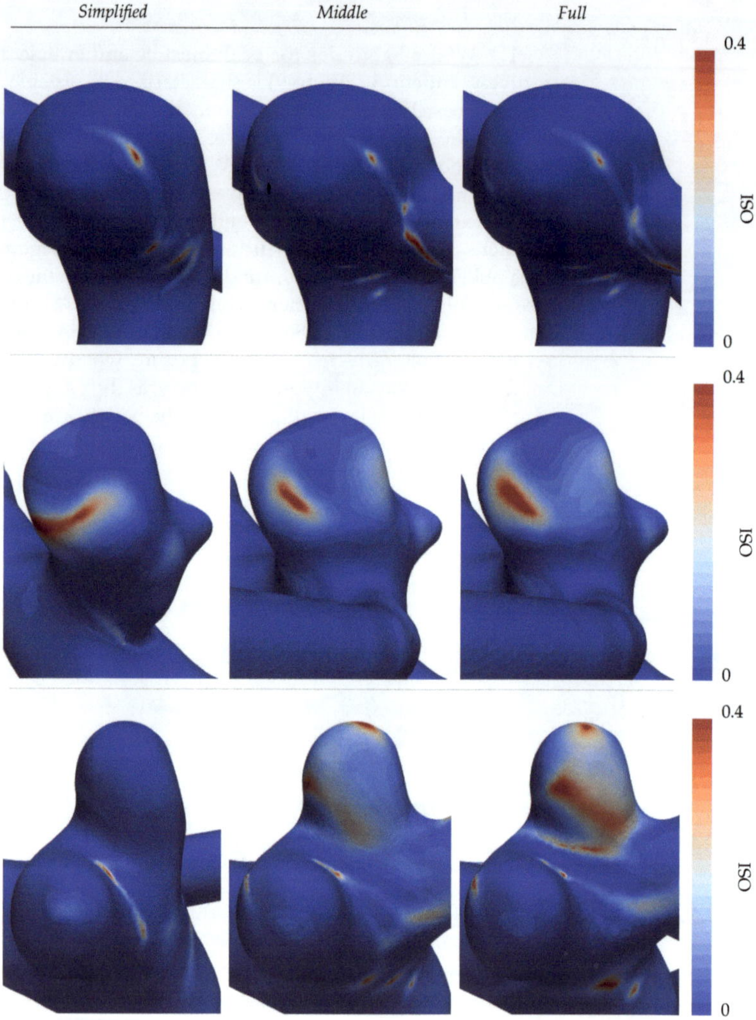

Figure 12. OSI distributions for cases A–C (**top** to **bottom**) under different geometry extensions.

4. Discussion

The results show a satisfactory application of resistive boundary conditions to the simulation of hemodynamics, providing accuracies above 95% with respect to the targeted flows of Murray's law. The resulting flow in the CoW resembles the values from the literature [46], though with a tendency to overestimate the flow through the MCAs. Using an exponent of ($n = 2$) in Murray's law (see Equation (4)) reduces the discrepancy for the flow in the MCA, though at the cost of overestimating the flows in smaller vessels such as the OA. Detailed measurements of brain arterial trees [13] suggests that Murray's law is best fit by exponents that grow from 2 in the proximal sections towards 3 in the more distal sections. While this factor is important to consider in general, we opted for a consistent modelling of the distal systems based on the Poiseulle flow. For a more detailed analysis of this matter, refer to [15,20], which studied the different exponents, area estimations, and bifurcation rules in relationship with Murray's law.

The changes in the flow through the ACA between the different computational domain extensions are a direct consequence of Murray's law being applied to vessels of varying diameter. In all cases, the proximal ACA (A1) has a larger radius than its post-communicating segment (A2), leading to different flow rates according to Equation (4). However, this apparent inconsistency is caused by perforator vessels around the ACom, such as the Recurrent Artery of Heubner, which due to their small size cannot be properly simulated. Experimental studies [46] support the observed decrease in flow rates. In our cases, applying Murray's law shows better agreement with the aforementioned study in both ACA and MCA flow rates when setting the domain boundary in the A1 segment.

Further analysis of the hemodynamics in the IA shows differences that exceed expectations, comparable to those seen in simulation challenges [7,8,10]. The Simplified strategy lacks physical reasoning, as all of the patient geometries in this study exhibit PCom arteries. Simulations confirm that neglecting these arteries significantly alters the intrasaccular hemodynamics, confirming the initial intuition. Without further exploration of this strategy, we conclude that it is unsuitable for the presented vasculatures.

The Middle extension, commonly found in the literature, presents more similarities with the Full CoW simulations. However, moderate changes in hemodynamics sometimes result in magnified alterations in WSS. These differences may lead to varied correlations between the flow dynamics and aneurysm evolution [28]. Causes for these observations include changed boundary flow rates on one side and a lack of interactions with the opposing hemisphere and posterior circulation on the other.

The additional blood suppliers in the case of the Full simulations, such as the neighbouring ICA and the BA, are participants in causing blood to traverse communicating vessels into other circulation segments. Although most of the interhemispheric flow was observed at the ACom, making it distal to the aneurysm fundus, the importance of the inflow streams' influence cannot be generally disregarded [21]. Due to a lack of measurement data, we imposed equal flow rates for the ICAs, aligning with a popular choice demonstrated in [7]. As stated before, this is irrefutably one of the biases taken into consideration that could be lifted by measurements in follow-up work. The second factor causing altered results between the Full CoW and Middle simulations, as stated earlier, is the interaction between the anterior and posterior circulations. By not imposing a midway boundary in a posterior communicating artery, more slack exists for the flow to unravel its natural characteristics, thereby providing a more physiological hemodynamic profile. Examples of this, among others, are pressure drops at posterior arterial junctions and bifurcations that effectively alter upstream pressure distributions, and are of notable relevance [14,47].

The flow in these communicating vessels holds significant medical importance, considering that 60–70% of all IAs are located on the Circle of Willis, which itself constitutes 85% of all IAs [48]. These vessels are both abundant and intricate, often presenting challenges in treatment due to their small size and contributing substantially to the incidence of ruptures [49]. While the role of CFD in diagnosing and analyzing IA growth and remodeling processes remains uncertain, it is crucial to acknowledge that modeling choices significantly impact the hemodynamics of communicating vessels [50]. Moreover, statistical evidence indicates that the absence or hypoplasia of certain communicating segments is associated with a higher likelihood of developing IAs [51,52]. As such, it becomes imperative to continually evaluate and refine CFD models in order to capture the hemodynamics of the CoW more effectively. This ongoing consideration is essential for advancing research and addressing the multitude of unanswered questions surrounding intracranial aneurysms.

5. Conclusions

In this study, we have conducted a detailed analysis of how different domain extensions and complexity levels affect hemodynamics in the brain's Circle of Willis. Using patient-specific geometries and computational fluid dynamics, we simulated three models of increasing complexity inspired by common literature findings for each patient. All of these simulations considered non-Newtonian pulsatile fluid flow, and outflow splits were determined using Murray's law.

The obtained results showed notable variations in flow patterns throughout the arterial network, particularly in two of the three cases. This highlights the importance of carefully analyzing topology in order to avoid modelling errors and conflicting results.

We conclude that studying the patient-specific Circle of Willis anatomy or utilizing measurement data is important for analysis of hemodynamic risk indicators in order to ensure physiological hemodynamics. To balance computational demands and accuracy, we suggest a two-step approach: a coarse large-scale simulation of the CoW to provide boundary conditions, and a highly resolved local simulation focused on the region of interest, ensuring both reliability and high-fidelity hemodynamics. This approach can be applied to other vascular pathologies as well.

This study deals with the modelling challenge in IA research, which, if addressed comprehensively, could reduce discrepancies in computational fluid dynamics results, enhancing our understanding of IAs and their treatment principles.

Author Contributions: Conceptualization, P.J.-R. and A.G.; methodology, P.J.-R.; software, A.L.; validation, P.J.-R., P.M. and A.G.; formal analysis, P.J.-R.; investigation, P.J.-R.; resources, Y.Ö.; data curation, Y.Ö.; writing—original draft preparation, P.J.-R.; writing—review and editing, A.G. and P.M.; visualization, A.G.; supervision, E.H.; project administration, E.H.; funding acquisition, E.H. All authors have read and agreed to the published version of the manuscript.

Funding: This research was funded by a Horizon ERC (European Union) grant (number 101045042) as part of Project CURE. The views and opinions expressed herein are solely those of the authors, and do not necessarily reflect those of the European Union or the European Research Council Executive Agency; neither the European Union nor the granting authority can be held responsible for them.

Informed Consent Statement: Specific patient consent was waived due to the retrospective study design.

Data Availability Statement: Data are contained within the article.

Conflicts of Interest: The authors declare no conflicts of interest.

Abbreviations

The following abbreviations are used in this manuscript:

IA	Intracranial Aneurysm
CFD	Computational Fluid Dynamics
CoW	Circle of Willis
WSS	Wall Shear Stress
OSI	Oscillatory Shear Index
ICA	Internal Carotid Artery
MCA	Middle Cerebral Artery
PCA	Posterior Cerebral Artery
ACA	Anterior Cerebral Artery
ACom	Anterior Communicating Artery
PCom	Posterior Communicating Artery

References

1. Wójtowicz, K.; Przepiorka, L.; Kujawski, S.; Marchel, A.; Kunert, P. Unruptured Anterior Communicating Artery Aneurysms: Management Strategy and Results of a Single-Center Experience. *J. Clin. Med.* **2023**, *12*, 4619. [CrossRef] [PubMed]
2. Juvela, S. Outcome of Patients with Multiple Intracranial Aneurysms after Subarachnoid Hemorrhage and Future Risk of Rupture of Unruptured Aneurysm. *J. Clin. Med.* **2021**, *10*, 1712. [CrossRef] [PubMed]
3. Pagiola, I.; Mihalea, C.; Caroff, J.; Ikka, L.; Chalumeau, V.; Iacobucci, M.; Ozanne, A.; Gallas, S.; Marques, M.; Nalli, D.; et al. The PHASES score: To treat or not to treat? Retrospective evaluation of the risk of rupture of intracranial aneurysms in patients with aneurysmal subarachnoid hemorrhage. *J. Neuroradiol.* **2020**, *47*, 349–352. [CrossRef]
4. Cebral, J.R.; Mut, F.; Raschi, M.; Scrivano, E.; Ceratto, R.; Lylyk, P.; Putman, C.M. Aneurysm rupture following treatment with flow-diverting stents: Computational hemodynamics analysis of treatment. *Am. J. Neuroradiol.* **2011**, *32*, 27–33. [CrossRef] [PubMed]
5. Janiga, G.; Berg, P.; Sugiyama, S.; Kono, K.; Steinman, D.A. The computational fluid dynamics rupture challenge 2013—Phase I: Prediction of rupture status in intracranial aneurysms. *Am. J. Neuroradiol.* **2015**, *36*, 530–536. [CrossRef] [PubMed]

6. Berg, P.; Roloff, C.; Beuing, O.; Voß, S.; Sugiyama, S.; Aristokleous, N.; Anayiotos, A.S.; Ashton, N.; Revell, A.; Bressloff, N.W.; et al. The Computational Fluid Dynamics Rupture Challenge 2013—Phase II: Variability of Hemodynamic Simulations in Two Intracranial Aneurysms. *J. Biomech. Eng.* **2015**, *137*, 121008. [CrossRef] [PubMed]
7. Valen-Sendstad, K.; Bergersen, A.W.; Shimogonya, Y.; Goubergrits, L.; Bruening, J.; Pallares, J.; Cito, S.; Piskin, S.; Pekkan, K.; Geers, A.J.; et al. Real-World Variability in the Prediction of Intracranial Aneurysm Wall Shear Stress: The 2015 International Aneurysm CFD Challenge. *Cardiovasc. Eng. Technol.* **2018**, *9*, 544–564. [CrossRef]
8. Berg, P.; Voß, S.; Saalfeld, S.; Janiga, G.; Bergersen, A.; Valen-Sendstad, K.; Bruening, J.; Goubergrits, L.; Spuler, A.; Cancelliere, N.; et al. Multiple Aneurysms AnaTomy CHallenge 2018 (MATCH): Phase I: Segmentation. *Cardiovasc. Eng. Technol.* **2018**, *9*, 565–581. [CrossRef]
9. Voß, S.; Beuing, O.; Janiga, G.; Berg, P. Multiple Aneurysms AnaTomy CHallenge 2018 (MATCH)-Phase Ib: Effect of morphology on hemodynamics. *PLoS ONE* **2019**, *14*, e0216813. [CrossRef]
10. Berg, P.; Voß, S.; Janiga, G.; Saalfeld, S.; Bergersen, A.W.; Valen-Sendstad, K.; Bruening, J.; Goubergrits, L.; Spuler, A.; Chiu, T.L.; et al. Multiple Aneurysms AnaTomy CHallenge 2018 (MATCH)—phase II: Rupture risk assessment. *Int. J. Comput. Assist. Radiol. Surg.* **2019**, *14*, 1795–1804. [CrossRef]
11. Berg, P.; Saalfeld, S.; Voß, S.; Beuing, O.; Janiga, G. A review on the reliability of hemodynamic modeling in intracranial aneurysms: Why computational fluid dynamics alone cannot solve the equation. *Neurosurg. Focus* **2019**, *47*, E15. [CrossRef] [PubMed]
12. Marsden, A.L.; Feinstein, J.A.; Taylor, C.A. A computational framework for derivative-free optimization of cardiovascular geometries. *Comput. Methods Appl. Mech. Eng.* **2008**, *197*, 1890–1905. [CrossRef]
13. Helthuis, J.H.; van Doormaal, T.P.; Hillen, B.; Bleys, R.L.; Harteveld, A.A.; Hendrikse, J.; van der Toorn, A.; Brozici, M.; Zwanenburg, J.J.; van der Zwan, A. Branching Pattern of the Cerebral Arterial Tree. *Anat. Rec.* **2019**, *302*, 1434–1446. [CrossRef] [PubMed]
14. Chnafa, C.; Valen-Sendstad, K.; Brina, O.; Pereira, V.M.; Steinman, D.A. Improved reduced-order modelling of cerebrovascular flow distribution by accounting for arterial bifurcation pressure drops. *J. Biomech.* **2017**, *51*, 83–88. [CrossRef] [PubMed]
15. Saalfeld, S.; Voß, S.; Beuing, O.; Preim, B.; Berg, P. Flow-splitting-based computation of outlet boundary conditions for improved cerebrovascular simulation in multiple intracranial aneurysms. *Int. J. Comput. Assist. Radiol. Surg.* **2019**, *14*, 1805–1813. [CrossRef] [PubMed]
16. Rosner, J.; Reddy, V.; Lui, F. *Neuroanatomy, Circle of Willis*, 1st ed.; StatPearls Publishing: St. Petersburg, FL, USA, 2022.
17. Devault, K.; Gremaud, P.A.; Novak, V.; Olufsen, M.S.; Vernières, G.; Peng, Z. Blood Flow in the Circle of Willis: Modelling and Calibration. *Multiscale Model. Simul.* **2008**, *7*, 888–909. [CrossRef] [PubMed]
18. Malm, J.; Birnefeld, J.; Zarrinkoob, L.; Wåhlin, A.; Eklund, A. Hemodynamic Disturbances in Posterior Circulation Stroke: 4D Flow Magnetic Resonance Imaging Added to Computed Tomography Angiography. *Front. Neurosci.* **2021**, *15*, 656769. [CrossRef]
19. Hindenes, L.B.; Håberg, A.K.; Johnsen, L.H.; Mathiesen, E.B.; Robben, D.; Vangberg, T.R. Variations in the circle of willis in a large population sample using 3D TOF angiography: The tromsø study. *PLoS ONE* **2020**, *15*, e0241373. [CrossRef]
20. Chnafa, C.; Brina, O.; Pereira, V.M.; Steinman, D.A. Better Than Nothing: A Rational Approach for Minimizing the Impact of Outflow Strategy on Cerebrovascular Simulations. *Am. J. Neuroradiol.* **2018**, *39*, 337–343. [CrossRef]
21. Castro, M.; Putman, C.; Cebral, J. Computational Fluid Dynamics Modeling of Intracranial Aneurysms: Effects of Parent Artery Segmentation on Intra-Aneurysmal Hemodynamics. *Am. J. Neuroradiol.* **2006**, *27*, 1703–1709.
22. Dennis, K.D.; Kallmes, D.F.; Dragomir-Daescu, D. Cerebral aneurysm blood flow simulations are sensitive to basic solver settings. *J. Biomech.* **2017**, *57*, 46–53. [CrossRef] [PubMed]
23. Britz, G.; Golshani, K.; Ferrell, A.; Zomorodi, A.; Smith, T. A review of the management of posterior communicating artery aneurysms in the modern era. *Surg. Neurol. Int.* **2010**, *1*, 88. [CrossRef] [PubMed]
24. Geuzaine, C.; Remacle, J.F. Gmsh: A 3-D Finite Element Mesh Generator with built-in Pre- and Post-Processing Facilities. *Int. J. Numer. Methods Eng.* **2009**, *79*, 1309–1331. [CrossRef]
25. Hachem, E.; Rivaux, B.; Kloczko, T.; Digonnet, H.; Coupez, T. Stabilized finite element method for incompressible flows with high Reynolds number. *J. Comput. Phys.* **2010**, *229*, 8643–8665. [CrossRef]
26. Coupez, T.; Hachem, E. Solution of high-Reynolds incompressible flow with stabilized finite element and adaptive anisotropic meshing. *Comput. Methods Appl. Mech. Eng.* **2013**, *267*, 65–85. [CrossRef]
27. Meliga, P.; Hachem, E. Time-accurate calculation and bifurcation analysis of the incompressible flow over a square cavity using variational multiscale modeling. *J. Comput. Phys.* **2019**, *376*, 952–972. [CrossRef]
28. Valen-Sendstad, K.; Steinman, D.A. Mind the gap: Impact of computational fluid dynamics solution strategy on prediction of intracranial aneurysm hemodynamics and rupture status indicators. *Am. J. Neuroradiol.* **2014**, *35*, 536–543. [CrossRef] [PubMed]
29. Goetz, A.; Rico, P.J.; Chau, Y.; Sédat, J.; Larcher, A.; Hachem, E. Proposal for Numerical Benchmarking of Fluid-Structure Interaction in Cerebral Aneurysms. *arXiv* **2023**, arXiv:2308.08301.
30. Hodis, S.; Uthamaraj, S.; Smith, A.L.; Dennis, K.D.; Kallmes, D.F.; Dragomir-Daescu, D. Grid convergence errors in hemodynamic solution of patient-specific cerebral aneurysms. *J. Biomech.* **2012**, *45*, 2907–2913. [CrossRef]
31. Abraham, F.; Behr, M.; Heinkenschloss, M. Shape optimization in unsteady blood flow: A numerical study of non-Newtonian effects. *Comput. Methods Biomech. Biomed. Eng.* **2005**, *8*, 201–212. [CrossRef]
32. Xiang, J.; Tremmel, M.; Kolega, J.; Levy, E.I.; Natarajan, S.K.; Meng, H. Newtonian viscosity model could overestimate wall shear stress in intracranial aneurysm domes and underestimate rupture risk. *J. Neurointerv. Surg.* **2012**, *4*, 351–357. [CrossRef] [PubMed]

33. Gambaruto, A.M.; Janela, J.; Moura, A.; Sequeira, A. Sensitivity of hemodynamics in a patient specific cerebral aneurysm to vascular geometry and blood rheology. *Math. Biosci. Eng.* **2011**, *8*, 409–423. [PubMed]
34. Tanaka, H.; Fujita, N.; Enoki, T.; Matsumoto, K.; Watanabe, Y.; Murase, K.; Nakamura, H. Relationship between Variations in the Circle of Willis and Flow Rates in Internal Carotid and Basilar Arteries Determined by Means of Magnetic Resonance Imaging with Semiautomated Lumen Segmentation: Reference Data from 125 Healthy Volunteers. *AJNR Am. J. Neuroradiol.* **2006**, *27*, 1770–1775. [PubMed]
35. Vignon-Clementel, I.E.; Alberto Figueroa, C.; Jansen, K.E.; Taylor, C.A. Outflow boundary conditions for three-dimensional finite element modeling of blood flow and pressure in arteries. *Comput. Methods Appl. Mech. Eng.* **2006**, *195*, 3776–3796. [CrossRef]
36. Janiga, G.; Berg, P.; Beuing, O.; Neugebauer, M.; Gasteiger, R.; Preim, B.; Rose, G.; Skalej, M.; Thevenin, D. Recommendations for accurate numerical blood flow simulations of stented intracranial aneurysms. *Biomed. Tech.* **2013**, *58*, 303–314. [CrossRef]
37. Boussel, L.; Rayz, V.L.; McCulloch, C.; Martin, A.; Acevedo-Bolton, G.; Lawton, M.; Higashida, R.; Smith, W.S.; Young, W.L.; Saloner, D. Aneurysm growth occurs at region of low wall shear stress: Patient-specific correlation of hemodynamics and growth in a longitudinal study. *Stroke* **2008**, *39*, 2997–3002. [CrossRef]
38. Urschel, K.; Tauchi, M.; Achenbach, S.; Dietel, B. Investigation of wall shear stress in cardiovascular research and in clinical practice—From bench to bedside. *Int. J. Mol. Sci.* **2021**, *22*, 5635. [CrossRef]
39. Cebral, J.; Detmer, F.; Chung, B.; Choque-Velasquez, J.; Rezai Jahromi, B.; Lehto, H.; Tulamo, R.; Hernesniemi, J.; Niemela, M.; Yu, A.; et al. Local Hemodynamic Conditions Associated with Focal Changes in the Intracranial Aneurysm Wall. *AJNR Am. J. Neuroradiol.* **2019**, *40*, 510–516.
40. Meng, H.; Tutino, V.M.; Xiang, J.; Siddiqui, A. High WSS or Low WSS? Complex interactions of hemodynamics with intracranial aneurysm initiation, growth, and rupture: Toward a unifying hypothesis. *Am. J. Neuroradiol.* **2014**, *35*, 1254–1262. [CrossRef]
41. Malek, A. Hemodynamic Shear Stress and Its Role in Atherosclerosis. *JAMA* **1999**, *282*, 2035. [CrossRef]
42. Furukawa, K.; Ishida, F.; Tsuji, M.; Miura, Y.; Kishimoto, T.; Shiba, M.; Tanemura, H.; Umeda, Y.; Sano, T.; Yasuda, R.; et al. Hemodynamic characteristics of hyperplastic remodeling lesions in cerebral aneurysms. *PLoS ONE* **2018**, *13*, e0191287. [CrossRef] [PubMed]
43. Mazzi, V.; Gallo, D.; Calò, K.; Najafi, M.; Khan, M.O.; De Nisco, G.; Steinman, D.A.; Morbiducci, U. A Eulerian method to analyze wall shear stress fixed points and manifolds in cardiovascular flows. *Biomech. Model. Mechanobiol.* **2020**, *19*, 1403–1423. [CrossRef] [PubMed]
44. Sheikh, M.A.A.; Shuib, A.S.; Mohyi, M.H.H. A review of hemodynamic parameters in cerebral aneurysm. *Interdiscip. Neurosurg. Adv. Tech. Case Manag.* **2020**, *22*, 100716. [CrossRef]
45. Jongen, J.C.; Franke, C.L.; Ramos, L.M.; Wilmink, J.T.; Van Gijn, J. Direction of Flow in Posterior Communicating Artery on Magnetic Resonance Angiography in Patients with Occipital Lobe Infarcts. *Stroke* **2004**, *35*, 104–108. [CrossRef] [PubMed]
46. Zarrinkoob, L.; Ambarki, K.; Wåhlin, A.; Birgander, R.; Eklund, A.; Malm, J. Blood flow distribution in cerebral arteries. *J. Cereb. Blood Flow Metab.* **2015**, *35*, 648–654. [CrossRef]
47. Mynard, J.P.; Valen-Sendstad, K. A unified method for estimating pressure losses at vascular. *Int. J. Numer. Methods Biomed. Eng.* **2015**, *31*. [CrossRef]
48. Hacein-Bey, L.; Provenzale, J.M. Current Imaging Assessment and Treatment of Intracranial Aneurysms. *Am. J. Roentgenol.* **2011**, *196*, 32–44. [CrossRef]
49. Chen, J.; Li, M.; Zhu, X.; Chen, Y.; Zhang, C.; Shi, W.; Chen, Q.; Wang, Y. Anterior Communicating Artery Aneurysms: Anatomical Considerations and Microsurgical Strategies. *Front. Neurol.* **2020**, *11*, 1020. [CrossRef]
50. Zhang, H.; Fujiwara, N.; Kobayashi, M.; Yamada, S.; Liang, F.; Takagi, S.; Oshima, M. Development of a Numerical Method for Patient-Specific Cerebral Circulation Using 1D–0D Simulation of the Entire Cardiovascular System with SPECT Data. *Ann. Biomed. Eng.* **2016**, *44*, 2351–2363. [CrossRef]
51. Hindenes, L.B.; Ingebrigtsen, T.; Isaksen, J.G.; Håberg, A.K.; Johnsen, L.H.; Herder, M.; Mathiesen, E.B.; Vangberg, T.R. Anatomical variations in the circle of Willis are associated with increased odds of intracranial aneurysms: The Tromsø study. *J. Neurol. Sci.* **2023**, *452*, 120740. [CrossRef]
52. Feng, L.; Mao, H.J.; Zhang, D.D.; Zhu, Y.C.; Han, F. Anatomical variations in the Circle of Willis and the formation and rupture of intracranial aneurysms: A systematic review and meta-analysis. *Front. Neurol.* **2023**, *13*, 1098950. [CrossRef] [PubMed]

fluids MDPI

Article

Hemodynamic Assessment of the Pathological Left Ventricle Function under Rest and Exercise Conditions

Jana Korte [1,2,*], Thomas Rauwolf [1,3], Jan-Niklas Thiel [4], Andreas Mitrasch [3], Paulina Groschopp [1], Michael Neidlin [4], Alexander Schmeißer [1,3], Rüdiger Braun-Dullaeus [1,3] and Philipp Berg [1,5]

1 Research Campus STIMULATE, University of Magdeburg, 39106 Magdeburg, Germany
2 Department of Fluid Dynamics and Technical Flows, University of Magdeburg, 39106 Magdeburg, Germany
3 Department of Cardiology, University Hospital Magdeburg, 39120 Magdeburg, Germany
4 Department of Cardiovascular Engineering, Institute of Applied Medical Engineering, Medical Faculty, RWTH Aachen University, 52074 Aachen, Germany
5 Department of Healthcare Telematics and Medical Engineering, University of Magdeburg, 39106 Magdeburg, Germany
* Correspondence: korte.j@outlook.com

Abstract: Purpose: The analysis of pathological human left ventricular hemodynamics using high-resolved image-based blood flow simulations shows a major potential for examining mitral valve insufficiency (MI) under exercise conditions. Since capturing and simulating the patient-specific movement of the left ventricle (LV) during rest and exercise is challenging, this study aims to propose a workflow to analyze the hemodynamics within the pathologically moving LV. Methods: Patient-specific ultrasound (US) data of ten patients with MI in different stages were captured with three-dimensional real-time echocardiography. US measurements were performed while patients were resting and while doing handgrip exercise (2–4 min work). Patient-specific hemodynamic simulations were carried out based on the captured ventricular wall movement. Velocity and kinetic energy were analyzed for rest and exercise and for the different MI stages. Results: The results reveal a dependency of the kinetic energy over time in the ventricular volume curves. Concerning the comparison between rest and exercise, the left ventricular function reveals lower systolic kinetic energy under exercise (kinetic energy normalized by EDV; mean $\pm$ standard deviation: rest = 0.16 $\pm$ 0.14; exercise = 0.06 $\pm$ 0.05; p-value = 0.04). Comparing patients with non-limiting (MI I) and mild/moderate (MI II/III) MI, lower velocities (mean $\pm$ standard deviation: non-limiting = 0.10 $\pm$ 0.03; mild/moderate = 0.06 $\pm$ 0.02; p-value = 0.01) and lower diastolic kinetic energy (kinetic energy normalized by EDV; mean $\pm$ standard deviation: non-limiting = 0.45 $\pm$ 0.30; mild/moderate = 0.20 $\pm$ 0.19; p-value = 0.03) were found for the latter. Conclusion: With the proposed workflow, the hemodynamics within LVs with MI can be analyzed under rest and exercise. The results reveal the importance of the patient-specific wall movement when analyzing intraventricular hemodynamics. These findings can be further used within patient-specific simulations, based on varying the imaging and segmentation methods.

Keywords: computational fluid dynamics; hemodynamics; kinetic energy; left ventricle; mitral valve insufficiency; moving mesh method

Citation: Korte, J.; Rauwolf, T.; Thiel, J.-N.; Mitrasch, A.; Groschopp, P.; Neidlin, M.; Schmeißer, A.; Braun-Dullaeus, R.; Berg, P. Hemodynamic Assessment of the Pathological Left Ventricle Function under Rest and Exercise Conditions. *Fluids* **2023**, *8*, 71. https://doi.org/10.3390/fluids8020071

Academic Editors: Huidan (Whitney) Yu and Mehrdad Massoudi

Received: 23 January 2023
Revised: 7 February 2023
Accepted: 13 February 2023
Published: 16 February 2023

1. Introduction

Mitral valve insufficiency (MI) is a pathology occurring at the left ventricle of the human heart, causing reflux through the mitral valve back to the atrium [1]. MI occurs with an incidence up to 11.7% for people over 75 years [2] and in different stages (non-limiting—MI I and mild/moderate—MI II/III) [3]. For most patients, the MI stage increases under exercise [4], which has an effect on the development of symptoms specific for each patient [5]. The relationship between rest or exercise conditions and the intraventricular blood flow in patients with MI is still not fully understood [6]. However, the patient-specific blood flow analysis is crucial to obtain knowledge about the disease itself and for the planning of support therapy [7]. This can be realized by performing image-based patient-specific hemodynamic simulations [8,9]. In this regard, the numerical replication of the patient-specific movement of the ventricular wall is crucial to provide realistic flow characteristics [10].

As found by Nquyen et al. [11], magnetic resonance imaging scans can be used as a basis for a patient-specific moving mesh simulation to analyze the intraventricular hemodynamics. Boundary conditions are calculated by the deviation of the inlet and outlet mass flow from the ventricular volume over time [11]. Nevertheless, there is a lack of patient-specific pressure curves at the inflow and outflow tract, affecting the resulting hemodynamics. More recently, Goubergrits et al. and Obermeier et al. [12,13] used CT scans and calculated the movement of the LV walls using scaling based on the volume curve over time. With their approach, the patient-specific volume, including a detailed description of the valves, is considered. However, the patient-specific ventricular wall movement was not taken into account, and radiation exposure argues against the use of additional CT. Bavo et al. [14] developed a promising moving mesh method using ultrasound images from transesophageal echocardiography. The application of this method in clinical routine may, however, be constrained due to its invasive character [15]. Instead, transthoracic echocardiography (TTE) can capture time-resolved images of the ventricular volume in a non-invasive and radiation-free manner [16]. The imaging is even possible while patients perform low impact exercises [17]. Bakkestrøm et al. [18] investigated hemodynamics in patients with MI I and MI II/III, under rest and exercise, using catheterization and imaging modalities (echocardiography and magnetic resonance imaging). Within their study, significant differences in pulmonary capillary wedge and artery pressure were found between patients with MI I and MI II/III; however, highly resolved hemodynamics were not evaluated. Addressing these mentioned limitations, image-based hemodynamic simulations are used to investigate the hemodynamics in LVs of 10 patients with different MI stages, under rest and exercise. TTE is used to capture time-resolved medical images, and a moving mesh method is applied to take the patient-specific ventricular wall movement into account.

Therefore, the aim of this study was to determine potential hemodynamic markers that are able to characterize the heart function depending on the MI stage. In addition, the different impact of rest and exercise on the chosen hemodynamic parameters was identified.

2. Materials and Methods

2.1. Patient-Specific Ultrasound Data and Medical Image Segmentation

The database within this study consisted of the 3D patient-specific ventricular wall geometry of the LV. Data was captured in the Division of Cardiology and Angiology, Department of Internal Medicine, University Hospital Magdeburg, with real-time 3D transthoracic echocardiography using a Siemens ACUSON SC2000 ultrasound medical device (Siemens ACUSON SC2000 Ultrasound System, 2016, SIEMENS Healthineers AG Munich, Germany), with a frequency of 4.5 MHz.

Echocardiographic data was collected as part of the CIRCUS study (Compare Impedance Cardiography with Right Heart Catheterization and UltraSound, German Clinical Trials Register, DRKS00015635). The study was approved by the institutional review board. All patients provided written informed consent.

The data comprises 10 patients with different stages of MI I-III (see Table 1). The time discrete LV wall movement was captured via US as n timesteps of the 3D LV geometry during one cardiac cycle (CC). Each LV movement exhibited $n = 9$ up to $n = 22$ timesteps per CC. Details on each patient's timesteps, as well as the patient data, such as sex, age, size, weight, and end-diastolic volume (EDV), end-systolic volume (ESV), and ejection fraction (EF) can be found in Table 1.

Table 1. Patient-specific parameters for each case 1–10 comprising the stage of MI (I/II/III), sex, age in years, weight in kg, size in cm, ESV (end-systolic volume) in ml, EDV (end-diastolic volume) in ml, EF (ejection fraction), HR (heart rate) in beats per minute (bpm), and the timestep size n of the captured time frames during ultrasound measurements.

Case	MI		Sex	Age	Weight	Height	ESV (mL)	
	rest	exercise			(kg)	(cm)	rest	exercise
1	MI I	MI I	m	75	89	178	82.96	93.18
2	MI I	MI I	m	76	93	175	80.43	64.93
3	MI I	MI I	m	76	76	188	41.18	42.28
4	clipped	clipped	f	78	86	160	51.15	37.13
5	MI I	MI II	f	78	84	180	26.52	25.06
6	MI II	MI II	f	73	79	168	16.46	38.13
7	MI II	MI III	f	67	57	170	15.72	26.14
8	MI III	MI III	m	81	76	176	100.63	93.17
9	MI III	MI III	f	86	65	154	50.98	43.20
10	MI III	MI III	f	78	75	163	30.81	23.92

Case	EDV (mL)		EF (-)		HR (bpm)		Timesteps n	
	rest	exercise	rest	exercise	rest	exercise	rest	exercise
1	132.67	135.50	37.47	31.23	89	110	12	10
2	106.14	91.29	24.22	28.87	71	133	16	14
3	109.02	104.01	62.23	59.35	68	120	14	11
4	106.76	76.12	52.08	51.22	60	89	15	16
5	56.16	62.64	52.78	59.98	77	114	16	16
6	40.90	72.57	59.76	47.46	60	116	19	15
7	47.12	63.57	66.63	58.89	47	124	21	20
8	140.12	132.76	28.19	29.82	67	110	15	23
9	93.65	62.33	45.56	30.69	58	111	20	14
10	85.00	58.30	63.75	58.97	66	117	18	17

The moving LV geometries were captured under rest and exercise, respectively. For exercise measurements, patients performed handgrip exercise for 2–4 min during the data acquisition. Heart rates measured within each patient during data acquisition ranged between 47–89 (66.30 $\pm$ 10.84) during rest and 89–133 (114.40 $\pm$ 10.80) during exercise (see Table 1).

The 3D US data of the LV geometries was segmented with 4-dimensional left ventricle analysis (4D LVA) using the ImageArena 2020 software by TomTec (TOMTEC Imaging Systems, Unterschleißheim, Germany). Initial geometry files (IGF) were provided in Standard Triangle Language (STL). Due to the existing image resolution, the valves could not be segmented from US and were represented as surfaces moving with the ventricular wall (see Figure 1). For each patient (P1–5: first and second column, P6–P10: third and fourth column) the end-systolic state of the moving LV geometry is presented in Figure 1 during rest (first and third column) and exercise (second and fourth column).

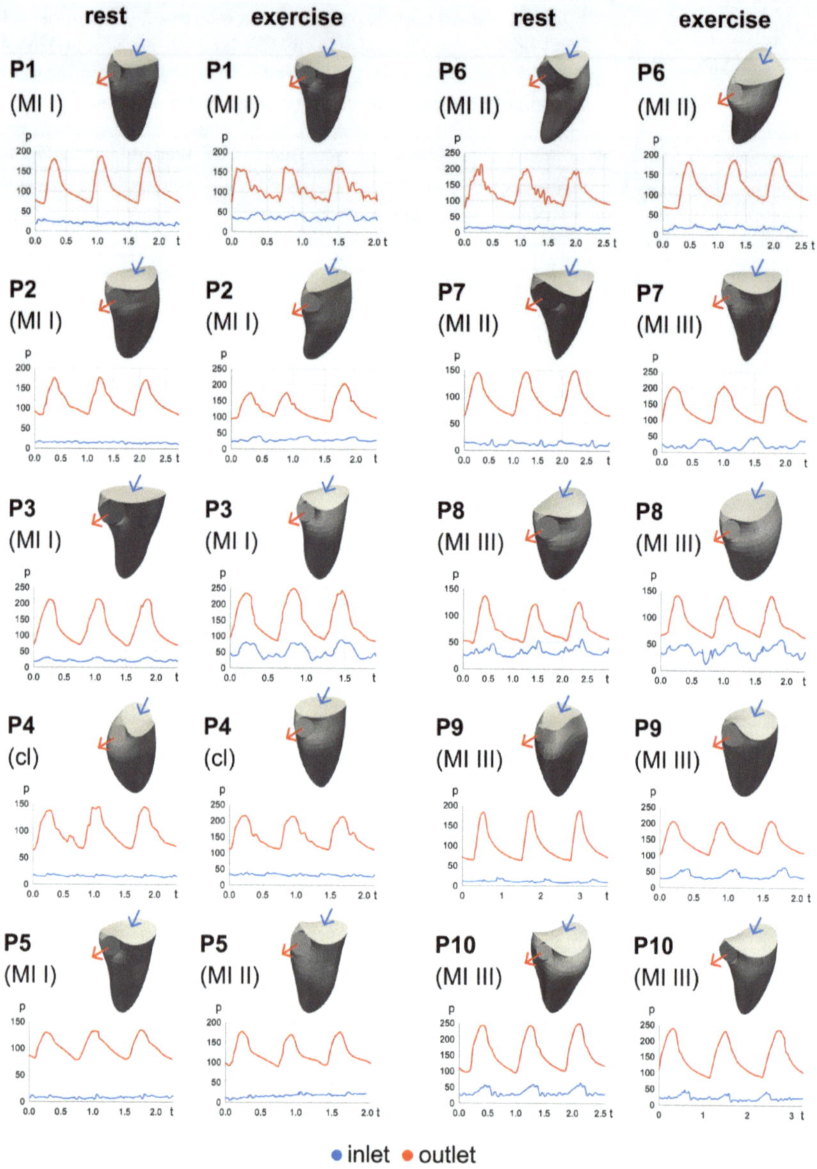

Figure 1. End-systolic LV geometries of all patients (P1–P10) during rest (first and third column) and exercise (second and fourth column) captured with 3D US. Corresponding pressure (p) curves in mmHg over time (t) in seconds at the inlet (blue) and outlet (red) boundaries are shown below each geometry. Pressure curves are shown over 3 CCs.

2.2. Boundary Conditions: Ventricular Wall Movement and Pressure Curves

To implement the patient-specific wall movement as a boundary condition in the simulation, n meshed surfaces (for n timesteps) with corresponding surface nodes were provided. To realize this wall movement, the moving mesh method (https://github.com/mneidlin/movingmesh, accessed on 1 June 2022) proposed by Grünwald et al. was chosen [19]. Concerning the creation of the input data for the moving mesh method, first, the end-systolic timestep ($n = 1$) of the IGF was load into ANSYS Spaceclaim 2021 R2 as STL and was then meshed (meshed geometry file—MGF) in ANSYS Meshing 2021 R2 (Ansys Inc., Canonsburg, PA, USA). A base mesh size of 0.5 mm was set for each patient, resulting in a total number of cells ranging from 1.3 to 3.6 million, depending on the patient-specific ESV. Mesh independency tests were carried out by analyzing the mesh of the ESV geometry of one patient. Details concerning the analyzed parameters with respect to the mesh study can be found in [19].

Second, the MGF surface nodes P_i on the end-systolic geometry (n = 1) were correlated with the triangular faces on the IGF surface nodes (triangular face with edge points P_1 P_2 P_3). Here, for each MGF surface node P_i the location at the IGF triangular face was searched. When found, the correlation between the location P_i and the triangle corner points P_1, P_2, P_3 was calculated. For this, the factors α and β were defined to describe the location of P_i on the triangle faces using the terms of the cross product (CP) and dot product (DP) of two parallel vectors: CP equals zero, and DP equals the product of their lengths. This combination reveals the following formula to calculate α and β:

$$P_i = P_{i2} + \alpha_i * \overrightarrow{P_{i2}P_{i1}} + \beta_i * \overrightarrow{P_{i2}P_{i3}} \tag{1}$$

Since the original STL geometries over the time series ($n_1, n_2, \ldots, n_T$; with T = amount of timesteps) comprise an equal amount of surface nodes, this correlation could then be used to create the exact same mesh for each timestep n over the time series (for the amount of timesteps for each case, see Table 1) [19].

Third, to overcome the time discrete movement, an interpolation was carried out between the P_i to the P_{i+1} meshed points to create a continuous mesh movement of the LV over one CC. A cubic spline interpolation was used with 20 interpolation steps for each LV geometry.

From the velocity curves measured with pulsed waves (PW) and continuous waved (CW) Doppler for each patient using the ultrasound scanner, the pressure curves at the LV inlet (mitral valve) and the LV outflow tract were derived directly within the machine using the Bernoulli's principle [20]:

$$\frac{v^2}{2} + gz + \frac{p}{\rho} = constant \tag{2}$$

with v = velocity, g = gravity acceleration, z = elevation of the point above a reference plane, p = pressure at chosen point, and ρ = fluid density.

The pressure curves over one CC at the inlet and outlet of the LV are presented below the end-systolic geometries for each patient in Figure 1. These derived curves are used as boundary conditions at the inlet and outlet during simulation. Additionally, the velocity curves at the LV inlet were further correlated with the resulting flow from computational fluid dynamics (CFD) (for details, see Appendix A, Figure A1).

2.3. Hemodynamic Moving Mesh Simulation

CFD simulations were performed, solving the Navier–Stokes equations using the finite volume solver ANSYS FLUENT 2021 R2 and the moving mesh method. The governing Navier–Stokes equations of the continuity and momentum are shown within (3) and (4):

$$\frac{\partial}{\partial t}\int_V \rho dV + \int_S \rho\left(\vec{v} - \vec{v_b}\right) * \vec{n} dS = 0 \tag{3}$$

$$\frac{\partial}{\partial t}\int_V \rho\vec{v} dV + \int_S \left(\rho\vec{v}\left(\vec{v} - \vec{v_b}\right) + pl - \vec{\tau}\right) * \vec{n} dS = 0 \tag{4}$$

The moving mesh method is implemented throughout user-defined functions written in the programming language C and compiled within the solver [19]. During each timestep of the simulation, the current mesh was loaded before the fluid solver calculations, with 30 iterations per timestep. Thus, the timestep size depends on the mesh and the cycle duration. Valves were simulated as walls, when closed. Concerning the MI stage, the orifice was assumed as a circular part in the middle of the inlet surface area, which remains a pressure-inlet during systole (~10% of the surface area for MI I and ~20% of the surface area for MI II/III). Blood was assumed to be Newtonian, with a dynamic viscosity of 0.004 Pa × s and incompressible with a density of 1055 kg/m^3. A pressure-based coupled solver and a k-ω SST model was used, with an absolute criteria of 10^{-3} for k and ω [21,22]. Concerning the convergence, residuals for continuity were set to 3×10^{-4} and to 10^{-3} for all velocity components, respectively. Since Grünwald et al. [19] determined that flow results reveal cycle independence after the second CC, the simulations ran for three cycles, and the calculated parameters during the third cycle were evaluated.

2.4. Hemodynamic Analysis

Simulation results of the intraventricular blood flow were analyzed in ANSYS EnSight 2021 R2 and MATLAB R2022a (The Mathworks Inc, Natick, MA, USA). Parameters taken into account were the intraventricular velocity (VEL), vorticity (ω), and kinetic energy (KE) [23]. Vorticity is defined with the components $\zeta_x, \zeta_y, \zeta_z$ (using velocity components u, v, w and directions x, y, z):

$$\zeta_x = \frac{\partial_w}{\partial_y} - \frac{\partial_v}{\partial_z}, \zeta_y = \frac{\partial_u}{\partial_z} - \frac{\partial_w}{\partial_x}, \zeta_z = \frac{\partial_v}{\partial_x} - \frac{\partial_u}{\partial_y} \quad (5)$$

Kinetic energy is defined as:

$$\mathrm{KE} = \frac{1}{2} * \rho \left(\sqrt{u^2 + v^2 + w^2}\right)^2, \quad (6)$$

Mean values describe the spatial mean inside the ventricle averaged over one CC. Vortex core lines were estimated from the 3D flow field's velocity gradient tensor using ANSYS EnSight, and the associated vortex core line length (VCL) was calculated during late systole. Statistical tests were performed comparing rest vs. exercise and MI I vs. MI II/III, using a two-sided Student's t-test with a significance level of $p = 0.05$.

3. Results

Within this study, the blood flow in patient-specific moving LV geometries was analyzed numerically for 10 patients during rest and exercise. In this section, the resulting chosen hemodynamic parameters, as well as the comparison between rest and exercise and the different MI stages (I and II/II), are presented.

3.1. Intraventricular Kinetic Energy and Volume Curves

The moving mesh method allows for the establishment of a continuous movement of the LV geometries during simulation. In Figure 2, the resulting continuous volume curves of each patient during rest (P1–P5, first, and P6–P10, third column) and exercise (P1–P5, second, and P6–P10, fourth column) are shown in green, along with the resulting kinetic energy curves normalized by EDV (red curves).

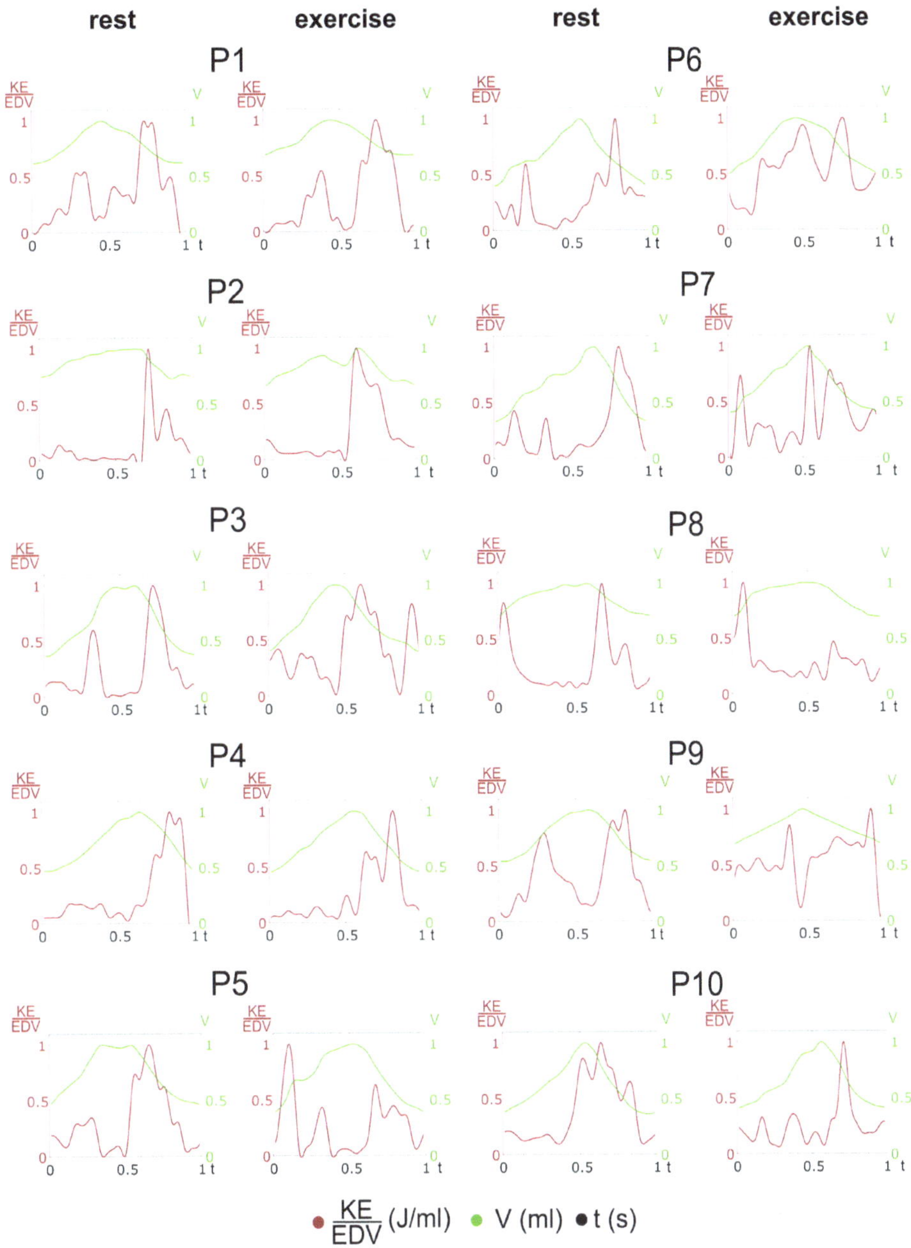

Figure 2. Presentation of the volume (green) and KE/EDV (red) curves of each patient during rest and exercise. The graphs (KE/EDV, V, and t) are normalized according to their maximum for better demonstration of the course. For non-normalized mean values, see Tables 2 and 3 and Figure 3.

Table 2. Comparison of hemodynamic parameters averaged over all patients between rest and exercise. Mean values and standard deviation (std) are shown for each parameter (VCL—vortex core line length, ω—mean vorticity, KE/EDV—mean kinetic energy normalized by end-diastolic volume, VEL—mean velocity).

	Mean Rest	Std Rest	Mean Ex	Std Ex	*p*-Value
VCL (m)	0.20	0.14	0.27	0.16	0.36
ω (1/s)	16.49	7.24	18.66	8.42	0.56
KE/EDV (J/mL)	0.09	0.06	0.10	0.08	0.73
VEL (m/s)	0.08	0.03	0.08	0.03	0.60
max dias KE/EDV (J/mL)	0.32	0.22	0.33	0.34	0.94
max sys KE/EDV (J/mL)	0.16	0.14	0.06	0.05	0.04

Table 3. Comparison of mean KE/EDV and mean VEL between P1–5 (MI I) and P6–10 (MI II/III), independent of rest or exercise state.

	Mean P1–5	Std P1–5	Mean P6–10	Std P6–10	*p*-Value
	(ex + rest)	(ex + rest)	(ex + rest)	(ex + rest)	
KE/EDV (J/mL)	0.12	0.07	0.07	0.05	0.08
VEL (m/s)	0.10	0.03	0.06	0.02	0.01
max dias KE/EDV (J/mL)	0.45	0.30	0.20	0.19	0.03
max sys KE/EDV (J/mL)	0.11	0.11	0.10	0.11	0.80

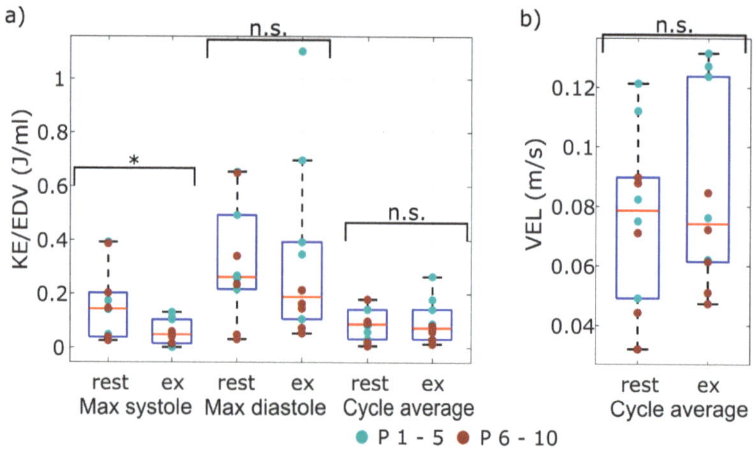

Figure 3. Boxplots with scatters for max (during diastole and systole) and mean KE/EDV (**a**) and mean velocity (**b**) for all patients. Patient data for P1–5 (MI I) is presented in green and for P6–10 (MI II/III), with red scatter points. Differences are signed with "*" if significant and with "n.s." if not significant.

For each patient, the curves are normalized by their maximum value to present the progression of the graph and the difference in scale, respectively. The graphs show the dependency of the resulting KE from the ventricular wall movement. For each patient, the decrease in volume results in a peak within the KE graph. As the volume increases, the KE graphs reveal smaller peaks as well. The movement of the LV geometries strongly differs between rest and exercise, not only in volume (see Table 1), but also as indicated: EDV rises during exercise for P1, P5–7; EF rises for P2, P5, and P8, but also in the course of the graph (see Figure 2). The volume curves of P3re (MI I), P4re (cl), P5re (MI I), P8re (MI III), and P9re (MI III) reveal a decrease during diastole before the peak volume is reached and the volume increases again. For P7re (MI II), this decrease appears earlier during the filling phase. This decrease is also presented in the graphs of P2ex (MI I), P7ex (MI III), and P10ex (MI III).

The KE curves (red curves) show peaks during systole, except for P5ex (MI II) and P8ex (MI III). A peak during diastole also occurs for each patient during rest and exercise, except for P2 (MI I), P4 (cl), and P10 (MI III). Concerning the peak waves of the KE graphs, differences between rest and exercise occur only for P8 when looking at the peak formation within the course of the graph. For each patient, this peak is higher during systole than during diastole. Concerning the scale, KE reveals lower values within patients with MI II/III (P7–10) than within P2–5 (MI I). Higher KE is visible, especially for patients with MI I (P1–5). Only P7 (MI II/III) and P10 (MI III) reveal high KE during systole. For each patient at rest, despite P1 (MI I), KE is higher during systole than during diastole. Within exercising patient results, KE is lower during systole for P1–5 (MI I) and higher for P6–10 (MI II/III).

A qualitative visualization of the KE distribution and the vortex formation (using isosurfaces of the Q-criterion Q = 200 $1/s^2$) within one patient (P10), over 10 timesteps per CC during rest and exercise can be found in the Appendix A, Figure A2. Videos showing the continuously moving LV of patient 6 during rest and exercise, with the qualitative KE distribution and vortex formation, can be found in the Supplementary Material (Video S1 and S2).

3.2. *Hemodynamic Intraventricular Average and Peak Parameters*

Mean values (mean over all patients) of VCL (end-systolic state), ω (spatial and temporal average), KE/EDV (spatial and temporal average), VEL (spatial and temporal average), max sys KE/EDV (peak during systole), and max dias EDV (peak during diastole) are shown in Table 2. Details on the patient-specific hemodynamic parameters can be found in Appendix A, Table A1.

The mean values of VCL are lower during rest for P1–P10, but do not show statistical significance (p-value = 0.36). Max sys KE/EDV is significantly lower during exercise (p-value = 0.04). Mean values reveal differences between rest and exercise for KE, ω, VEL, and max dias KE/EDV as well, but no significance was found. To analyze the values specific for each patient, the scatterplots of kinetic energy (max and mean values) and velocity (spatial and temporal mean) are shown in Figure 3. Values of patients with MI I are marked with green dots, and values of patients with MI II/III are marked with red dots, respectively. Specifically for the mean velocity, the values of P1–5 (MI I) are higher than those of P6–10 (MI II/III) for rest and exercise.

In Table 3, the comparison between patients with MI I and with MI II/III is shown. The mean values during rest and exercise for both cohorts reveal differences for KE/EDV (p-value = 0.08) and max sys KE/EDV (p-value = 0.80). The results show significant differences for VEL (p-value = 0.01) and max dias KE/EDV (p-value = 0.03).

4. Discussion

In this study, the LV function of patients with MI during rest and exercise conditions was analyzed by specifically taking the patient-specific wall motion into account. This was carried out by acquiring US scanning and PW/CW-Doppler measurements during rest and exercise and performing image-based blood flow simulations of the moving LV at both states. With US, a medical imaging technology is chosen, providing major clinical applicability [24]. Due to negligible risks for the patients, echocardiography is useful for MI with exercise, as analyzed by Voilliot et al. [25]. In contrast to CT or MRI imaging, US is a low-cost method, which is easily accessible for clinicians [26,27].

Based on the US measurements and the interpolation method, a continuous wall movement could be established within the simulations. The results show the dependency of the kinetic energy on the volume (recall Figure 2). This stands in agreement with the findings of Al-Wakeel et al. [28], who showed that KE and volume can be related in left ventricular hemodynamics in patients with MI.

The comparison of resulting intraventricular hemodynamics between rest and exercise revealed differences in the max systolic KE (normalized by EDV), with a lower value occurring during exercise ($p = 0.04$). This observation is in accordance to Grewal et al. [29], who identified that the exercise capacity for patients with a mild heart dysfunction is lower than for hearts with normal function. Thus, systolic KE might be a marker for cardiac function in patients with MI.

Additionally, differences in ω and in vortex formation during late diastole by looking at the VCL for rest vs. exercise was observed (recall Table 2). Higher ω and VCL occur during exercise, but when comparing the different MI stages, higher ω occurs during rest in patients with MI II/III. This is in line with ref. [30], who investigated several vortices in LVs in patients with MI compared to less left ventricular vortex formation in probands. However, these differences were not statistically significant, and further statements concerning the comparison between rest and exercise could be made.

When looking at the comparison between patients with MI I and MI II/III, velocity and max diastolic KE (normalized by EDV) particularly showed higher results for patients with MI I than with MI II/III (significant for velocity with a p-value = 0.03 and for max diastolic KE/EDV with a p-value = 0.01). This goes in line with the resulting EF (see Table 1), since the EF for patients with MI II/III was lower during exercise than during rest, which stands in agreement with [29] as well. This leads to the assumption that the LV function is lower under exercise for patients with heart dysfunction.

This study has several limitations. First, the underlying segmentation is a limitation of this study. Since the implementation of the moving mesh method allows for the simulation of the patient-specific ventricular wall movement, the effect of this movement on the intraventricular hemodynamics can be analyzed. Within this study, the high impact of the movement is shown, as can be seen in Figure 2. Kinetic energy curves depend on the volume change of the LVs. Since the TomTec software is used to automatically segment the patient data, the ventricular walls are unrealistically smooth and therefore, biased.

Second, the valves are simplified and represented as surface areas, leading to the assumption of a circular orifice area during systole of patients with MI II/III. The effect of the valve on the flow is not reflected in this analysis. Nevertheless, the patient-specific wall movement is captured and used in the simulation, and the underlying segmentation method was equal for each patient. This leads to an equally biased analysis of the flow parameters, which can be used as a basis for further investigations. There is still the possibility in the provided method to use improved segmentations, including valves. The overall (wall and valves) patient-specific LV movement could then be used as a boundary condition in the moving mesh simulation, presenting reliable results for the intraventricular blood flow.

Finally, substantial differences of inflow velocities at the mitral valve were observed for all patients. For patients with MI I, the results exhibit a very low degree of correlation to the experimental data. Still, the inlet velocity values of patients with MI II/III correlate to the CW Doppler results (R = 0.69, $p = 0.02$; see Appendix A, Figure A1). The difference might be caused by the modeling approach of the inflow section into the moving domain with MI II/III. Simulations of patients with MI I (P1–P4 and P5re) were performed with the valve areas as closed surfaces during the filling (AV closed) and ejection (MV closed) phases. Patients with MI II/III (P5ex, P6–P10) were modeled with an orifice at the inlet surface area during the ejection phase. Since the blood is assumed to be incompressible, the LV wall movement has the strongest effect on the incoming velocity curves using this approach. When performing simulations on patients with MI, the inlet surface area is never fully closed, and the pressure boundary curves have a higher effect on the inlet VEL. This limitation can be traced back to the lacking segmentation, since it is assumed that with a more precise segmentation (including segmentation of the valves), hemodynamics could appear to be more realistic.

Despite these limitations, the approach in this study provides a useful tool to investigate cardiac function on a patient-specific basis using non-invasive imaging data. The results reveal differences in cardiac function of MI patients between rest and exercise and underline the variations for different MI stages. The implemented workflow for the data processing and hemodynamic simulations allows for the analysis of the movement of the patient-specific human ventricular wall and its impact onto the intraventricular flow. Advantages of this numerical analysis are the highly resolved results of the flow field and the non-invasive patient-specific data acquisition via US. The latter is especially important for patients with existing heart failure, since invasive treatment can have fatal consequences [31].

Future work will include the improvement of the segmentation to obtain a more realistically shaped ventricle. Here, the comparison to higher resolved medical imaging techniques (e.g., MRI) will be taken into account. This includes the implementation of the LV valves, either by improved segmentation or through manual reconstruction. Moreover, a larger amount of patient data can increase the clinical significance of the results. This also includes the comparison to a proband cohort, without heart failure as a control. Additionally, the validation of the intraventricular blood flow will be part of the future work, although different approaches are possible. For instance, Xu et al. [32] showed the validation of CFD velocity curves using particle image velocimetry (PIV) results after implementing a moving mesh method. Eriksson et al. [33] carried out an analysis of the hemodynamics with medical imaging (cardiac magnetic resonance). Combining imaging techniques would not only serve as a possible validation, but also as a basis for the method to be improved in regards to segmentation and temporal resolution.

5. Conclusions

In this study, an analysis of LV functions for patients with MI during rest and exercise using patient-specific wall movement is carried out. By using patient-specific US data and implementing the captured wall movement into the simulation, patient-specific results for hemodynamic parameters were calculated. The results reveal significant differences between rest and exercise for max systolic KE (p-value = 0.04). Differences are found between patients with MI I and MI II/III for mean VEL and max diastolic KE, respectively. The KE curves over a cardiac cycle reveal a dependency on the volume change for each patient, which leads to the conclusion that a precise segmentation of the ventricular wall movement is crucial for deriving patient-specific blood flow parameters. The feasibility of the proposed workflow is shown within this study, and it can be implemented onto a larger cohort of patients.

Supplementary Materials: The following supporting information can be downloaded at: https://www.mdpi.com/article/10.3390/fluids8020071/s1, Video S1: Ventricle vortex formation and kinetic energy distribution in P6 during rest; Video S2: Ventricle vortex formation and kinetic energy distribution in P6 during exercise.

Author Contributions: Conceptualization, J.K., T.R. and A.S.; methodology, J.K.; software, J.K., T.R., M.N. and J.-N.T.; validation, J.K.; formal analysis, J.K.; investigation, J.K.; resources, J.K., J.-N.T., M.N. and P.B.; data curation, J.K., T.R., A.S., A.M., P.B. and R.B.-D.; writing—original draft preparation, J.K.; writing—review and editing, J.K., T.R., A.M., J.-N.T., P.B. and M.N.; visualization, J.K. and P.G.; supervision, P.B.; project administration, J.K.; funding acquisition, P.B. All authors have read and agreed to the published version of the manuscript.

Funding: This study was funded by the German Federal Ministry of Education and Research within the Research Campus STIMULATE (grant no. 13GW0473A) and the German Research Foundation (SPP2311, project number: 465189657). The CIRCUS study (Comparison of Impedance Cardiography with Right Heart Catheterization and Ultrasound, German Clinical Trials Register, DRKS00015635) was supported by the Research Network, "Autonomy in Old Age" (EFRE IP* 1a—EU Structural Funds).

Institutional Review Board Statement: The study was conducted in accordance with the Declaration of Helsinki, and approved by the Institutional Review Board of University of Magdeburg and University hospital Magdeburg (protocol code: DRKS00015635, 22 May 2017).

Informed Consent Statement: Informed consent was obtained from all subjects involved in the study.

Data Availability Statement: The data presented in this study are available on request from the corresponding author. The data are not publicly available.

Conflicts of Interest: The authors declare that they have no conflict of interest.

Appendix A

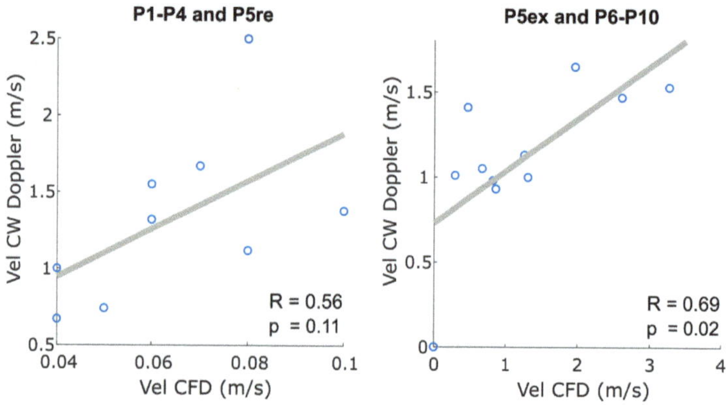

Figure A1. Correlation of max values at the inlet boundary measured with CW Doppler and calculated with computational fluid dynamic (CFD) simulations for each patient during rest and exercise.

Table A1. Patient-specific hemodynamic parameters: Vortex core length during late systole and all remaining parameters, spatially and temporally (over one CC) averaged.

Case	VCL (m)		Mean Vorticity (1/s)		Mean KE (J)		Mean VEL (m/s)	
	re	ex	re	ex	re	ex	re	ex
1	0.01	0.39	13.46	9.57	11.07	4.39	0.11	0.06
2	0.46	0.69	10.23	35.23	3.41	16.34	0.05	0.12
3	0	0.07	22.61	24.04	15.45	14.86	0.12	0.13
4	0.17	0.18	10.74	15.02	6.30	6.79	0.07	0.08
5	0.27	0.26	21.70	28.69	8.04	16.56	0.08	0.13
Mean	0.18	0.32	15.75	22.51	8.85	11.79	0.09	0.10
Std	0.17	0.21	5.35	9.23	4.13	5.14	0.03	0.03
6	0.10	0.18	29.34	22.09	7.34	5.93	0.09	0.08
7	0.16	0.21	17.70	16.65	4.64	4.27	0.07	0.07
8	0.28	0.23	6.81	10.34	0.96	1.92	0.03	0.05
9	0.29	0.19	8.50	7.58	1.96	1.94	0.04	0.05
10	0.30	0.27	23.82	17.40	7.81	3.51	0.09	0.06
Mean	0.23	0.22	17.23	14.81	4.54	3.51	0.06	0.06
Std	0.08	0.03	8.66	5.20	2.76	1.51	0.02	0.01

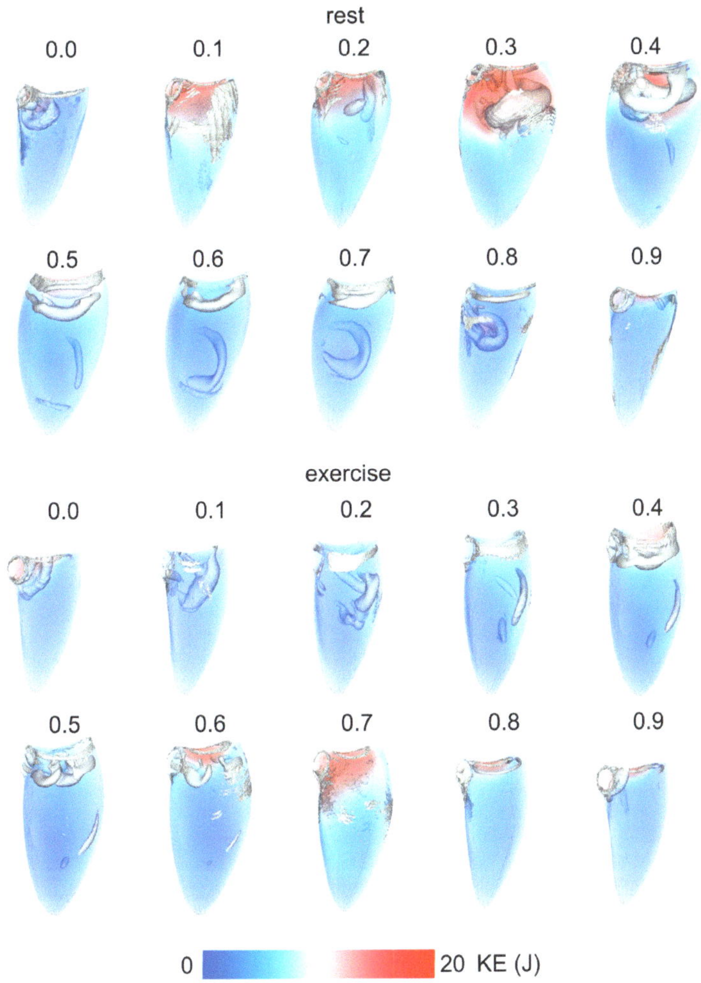

Figure A2. Ventricle vortex formation (isosurfaces Q-criterion Q = 200 $1/s^2$, shown in white) and kinetic energy distribution (shown with volume rendering) in P10 for 10 LV instances over one CC. Instances are normalized by CC period, and are shown above each visualization.

References

1. Faletra, F.; La Marchesina, U.; Bragato, R.; Grimaldi, A. Insufficienza mitralica. *Ital. Heart J. Suppl.* **2002**, *3*, 486–494. [PubMed]
2. Nkomo, V.T.; Gardin, J.M.; Skelton, T.N.; Gottdiener, J.S.; Scott, C.G.; Enriquez-Sarano, M. Burden of valvular heart diseases: A population-based study. *Lancet* **2006**, *368*, 1005–1011. [CrossRef] [PubMed]
3. Nagai, T.; Anzai, T. NYHA functional classification and AHA/ACC Stages for heart failure management. *Nihon rinsho. Jpn. J. Clin. Med.* **2016**, *74* (Suppl. 6), 340–344.
4. Sugimoto, T.; Bandera, F.; Generati, G.; Alfonzetti, E.; Barletta, M.; Losito, M.; Labate, V.; Rovida, M.; Caracciolo, M.; Pappone, C.; et al. Left Atrial Dynamics During Exercise in Mitral Regurgitation of Primary and Secondary Origin: Pathophysiological Insights by Exercise Echocardiography Combined With Gas Exchange Analysis. *JACC Cardiovasc. Imaging* **2020**, *13*, 25–40. [CrossRef]
5. Izem, O.; Maufrais, C.; Obert, P.; Rupp, T.; Schuster, I.; Nottin, S. Kinetics of Left Ventricular Mechanics during Transition from Rest to Exercise. *Med. Sci. Sports Exerc.* **2019**, *51*, 1838–1844. [CrossRef] [PubMed]
6. Zoghbi, W.A.; Adams, D.; Bonow, R.O.; Enriquez-Sarano, M.; Foster, E.; Grayburn, P.A.; Hahn, R.T.; Han, Y.; Hung, J.; Lang, R.M.; et al. Recommendations for Noninvasive Evaluation of Native Valvular Regurgitation: A Report from the American Society of Echocardiography Developed in Collaboration with the Society for Cardiovascular Magnetic Resonance. *J. Am. Soc. Echocardiogr.* **2017**, *30*, 303–371. [CrossRef] [PubMed]

7. Vakil, K.; Roukoz, H.; Sarraf, M.; Krishnan, B.; Reisman, M.; Levy, W.C.; Adabag, S. Safety and efficacy of the MitraClip® system for severe mitral regurgitation: A systematic review. *Catheter. Cardiovasc. Interv.* **2014**, *84*, 129–136. [CrossRef] [PubMed]
8. Borazjani, I.; Westerdale, J.; McMahon, E.M.; Rajaraman, P.K.; Heys, J.J.; Belohlavek, M. Left Ventricular Flow Analysis: Recent Advances in Numerical Methods and Applications in Cardiac Ultrasound. *Comput. Math. Methods Med.* **2013**, *2013*, 395081. [CrossRef]
9. Collia, D.; Zovatto, L.; Tonti, G.; Pedrizzetti, G. Comparative Analysis of Right Ventricle Fluid Dynamics. *Front. Bioeng. Biotechnol.* **2021**, *9*, 667408. [CrossRef]
10. Larsson, D.; Spuhler, J.H.; Petersson, S.; Nordenfur, T.; Colarieti-Tosti, M.; Hoffman, J.; Winter, R.; Larsson, M. Patient-Specific Left Ventricular Flow Simulations From Transthoracic Echocardiography: Robustness Evaluation and Validation Against Ultrasound Doppler and Magnetic Resonance Imaging. *IEEE Trans. Med. Imaging* **2017**, *36*, 2261–2275. [CrossRef]
11. Nguyen, V.-T.; Wibowo, S.N.; Leow, Y.A.; Nguyen, H.-H.; Liang, Z.; Leo, H.L. A Patient-Specific Computational Fluid Dynamic Model for Hemodynamic Analysis of Left Ventricle Diastolic Dysfunctions. *Cardiovasc. Eng. Technol.* **2015**, *6*, 412–429. [CrossRef]
12. Goubergrits, L.; Vellguth, K.; Obermeier, L.; Schlief, A.; Tautz, L.; Bruening, J.; Lamecker, H.; Szengel, A.; Nemchyna, O.; Knosalla, C.; et al. CT-Based Analysis of Left Ventricular Hemodynamics Using Statistical Shape Modeling and Computational Fluid Dynamics. *Front. Cardiovasc. Med.* **2022**, *9*, 901902. [CrossRef] [PubMed]
13. Obermeier, L.; Vellguth, K.; Schlief, A.; Tautz, L.; Bruening, J.; Knosalla, C.; Kuehne, T.; Solowjowa, N.; Goubergrits, L. CT-Based Simulation of Left Ventricular Hemodynamics: A Pilot Study in Mitral Regurgitation and Left Ventricle Aneurysm Patients. *Front. Cardiovasc. Med.* **2022**, *9*, 828556. [CrossRef] [PubMed]
14. Bavo, A.; Pouch, A.; Degroote, J.; Vierendeels, J.; Gorman, J.; Gorman, R.; Segers, P. Patient-specific CFD models for intraventricular flow analysis from 3D ultrasound imaging: Comparison of three clinical cases. *J. Biomech.* **2017**, *50*, 144–150. [CrossRef] [PubMed]
15. Hilberath, J.N.; Oakes, D.A.; Shernan, S.K.; Bulwer, B.E.; D'Ambra, M.N.; Eltzschig, H.K. Safety of Transesophageal Echocardiography. *J. Am. Soc. Echocardiogr.* **2010**, *23*, 1115–1127. [CrossRef] [PubMed]
16. Matulevicius, S.A.; Rohatgi, A.; Das, S.R.; Price, A.L.; DeLuna, A.; Reimold, S.C. Appropriate Use and Clinical Impact of Transthoracic Echocardiography. *JAMA Intern. Med.* **2013**, *173*, 1600–1607. [CrossRef]
17. Marino, P.N.; Zanaboni, J.; Degiovanni, A.; Sartori, C.; Patti, G.; Fraser, A.G. Left atrial conduit flow rate at baseline and during exercise: An index of impaired relaxation in HFpEF patients. *ESC Hear. Fail.* **2021**, *8*, 4334–4342. [CrossRef]
18. Bakkestrøm, R.; Banke, A.; Christensen, N.L.; Pecini, R.; Irmukhamedov, A.; Andersen, M.; Borlaug, B.A.; Møller, J.E. Hemodynamic Characteristics in Significant Symptomatic and Asymptomatic Primary Mitral Valve Regurgitation at Rest and During Exercise. *Circ. Cardiovasc. Imaging* **2018**, *11*, e007171. [CrossRef]
19. Grünwald, A.; Korte, J.; Wilmanns, N.; Winkler, C.; Linden, K.; Herberg, U.; Groß-Hardt, S.; Steinseifer, U.; Neidlin, M. Intraventricular Flow Simulations in Singular Right Ventricles Reveal Deteriorated Washout and Low Vortex Formation. *Cardiovasc. Eng. Technol.* **2021**, *13*, 495–503. [CrossRef]
20. Qin, R.; Duan, C. The principle and applications of Bernoulli equation. *J. Physics: Conf. Ser.* **2017**, *916*, 012038. [CrossRef]
21. Zhang, J.; Zhang, P.; Fraser, K.; Griffith, B.P.; Wu, Z.J. Comparison and Experimental Validation of Fluid Dynamic Numerical Models for a Clinical Ventricular Assist Device. *Artif. Organs* **2013**, *37*, 380–389. [CrossRef] [PubMed]
22. Ghodrati, M.; Khienwad, T.; Maurer, A.; Moscato, F.; Zonta, F.; Schima, H.; Aigner, P. Validation of numerically simulated ventricular flow patterns during left ventricular assist device support. *Int. J. Artif. Organs* **2021**, *44*, 30–38. [CrossRef] [PubMed]
23. Mut, F.; Löhner, R.; Chien, A.; Tateshima, S.; Viñuela, F.; Putman, C.; Cebral, J.R. Computational hemodynamics framework for the analysis of cerebral aneurysms. *Int. J. Numer. Methods Biomed. Eng.* **2011**, *27*, 822–839. [CrossRef] [PubMed]
24. Cheitlin, M.D.; Armstrong, W.F.; Aurigemma, G.P.; A Beller, G.; Bierman, F.Z.; Davis, J.L.; Douglas, P.S.; Faxon, D.P.; Gillam, L.D.; Kimball, T.R.; et al. ACC/AHA/ASE 2003 guideline update for the clinical application of echocardiography: Summary article: A report of the American college of cardiology/American heart association task force on practice guidelines (ACC/AHA/ASE committee to update the 1997 guidelines for the clinical application of echocardiography). *J. Am. Coll. Cardiol.* **2003**, *42*, 954–970.
25. Voilliot, D.; Lancellotti, P. Exercise Testing and Stress Imaging in Mitral Valve Disease. *Curr. Treat. Options Cardiovasc. Med.* **2017**, *19*, 17. [CrossRef]
26. Marwick, T.H.; Anderson, T.; Williams, M.; Haluska, B.; Melin, J.A.; Pashkow, F.; Thomas, J.D. Exercise echocardiography is an accurate and cost-efficient technique for detection of coronary artery disease in women. *J. Am. Coll. Cardiol.* **1995**, *26*, 335–341. [CrossRef]
27. Pearlman, A.S.; Ryan, T.; Picard, M.H.; Douglas, P.S. Evolving Trends in the Use of Echocardiography: A Study of Medicare Beneficiaries. *J. Am. Coll. Cardiol.* **2007**, *49*, 2283–2291. [CrossRef]
28. Al-Wakeel, N.; Ms, J.F.F.; Amiri, A.; Siniawski, H.; Goubergrits, L.; Berger, F.; Kuehne, T. Hemodynamic and energetic aspects of the left ventricle in patients with mitral regurgitation before and after mitral valve surgery. *J. Magn. Reson. Imaging* **2015**, *42*, 1705–1712. [CrossRef]
29. Grewal, J.; McCully, R.B.; Kane, G.C.; Lam, C.; Pellikka, P.A. Left ventricular function and exercise capacity. *JAMA* **2009**, *301*, 286–294. [CrossRef]
30. Stöhr, E.J.; González-Alonso, J.; Bezodis, I.N.; Shave, R. Left ventricular energetics: New insight into the plasticity of regional contributions at rest and during exercise. *Am. J. Physiol. Circ. Physiol.* **2014**, *306*, H225–H232. [CrossRef]
31. Vahanian, A.; Iung, B. Mitral regurgitation. Timing of surgery or interventional treatment. *Herz* **2016**, *41*, 3–9. [CrossRef] [PubMed]

32. Xu, F.; Kenjereš, S. Numerical simulations of flow patterns in the human left ventricle model with a novel dynamic mesh morphing approach based on radial basis function. *Comput. Biol. Med.* **2021**, *130*, 104184. [CrossRef] [PubMed]
33. Eriksson, J.; Carlhäll, C.-J.; Dyverfeldt, P.; Engvall, J.; Bolger, A.F.; Ebbers, T. Semi-automatic quantification of 4D left ventricular blood flow. *J. Cardiovasc. Magn. Reson.* **2010**, *12*, 9. [CrossRef]

fluids MDPI

Article

Modelling The Hemodynamics of Coronary Ischemia

Abdulaziz Al Baraikan [1,2], Krzysztof Czechowicz [1], Paul D. Morris [1], Ian Halliday [1,*], Rebecca C. Gosling [1], Julian P. Gunn [1], Andrew J. Narracott [1], Gareth Williams [1] Pankaj Garg [1,3,4], Maciej Malawski [5], Frans van de Vosse [6], Angela Lungu [7], Dan Rafiroiu [7] and David Rodney Hose [1]

1 Department of Infection, Immunity and Cardiovascular Disease, The Medical School, Sheffield S10 2RX, UK
2 College of Applied Medical Sciences, King Saud bin Abdulaziz University for Health Sciences, Riyadh 14611, Saudi Arabia
3 Norwich Medical School, University of East Anglia, Norfolk NR9 3LX, UK
4 Norfolk and Norwich University Hospital NHS Foundation Trust, Norfolk NR4 7UY, UK
5 Sano Centre for Computational Medicine, 30-072 Krakow, Poland
6 Biomedical Engineering Department, Eindhoven University of Technology, 5612AE Eindhoven, The Netherlands
7 Department of Electrotechnics and Measurements, Universitatea Tehnica Cluj-Napoca, 400114 Cluj, Romania
* Correspondence: i.halliday@sheffield.ac.uk; Tel.: +44-0114-222-5522

Abstract: Acting upon clinical patient data, acquired in the pathway of percutaneous intervention, we deploy hierarchical, multi-stage, data-handling protocols and interacting low- and high-order mathematical models (chamber elastance, state-space system and CFD models), to establish and then validate a framework to quantify the burden of ischaemia. Our core tool is a compartmental, zero-dimensional model of the coupled circulation with four heart chambers, systemic and pulmonary circulations and an optimally adapted windkessel model of the coronary arteries that reflects the diastolic dominance of coronary flow. We guide the parallel development of protocols and models by appealing to foundational physiological principles of cardiac energetics and a parameterisation (stenotic Bernoulli resistance and micro-vascular resistance) of patients' coronary flow. We validate our process first with results which substantiate our protocols and, second, we demonstrate good correspondence between model operation and patient data. We conclude that our core model is capable of representing (patho)physiological states and discuss how it can potentially be deployed, on clinical data, to provide a quantitative assessment of the impact, on the individual, of coronary artery disease.

Keywords: haemodynamics; compartmental models; parameter identification; fluid dynamics

Citation: Baraikan , A.A.; Czechowicz, K.; Morris, P.D.; Halliday, I.; Gosling, R.C.; Gunn, J.P.; Narracott, A.J.; Williams, G.; Garg, P.; Malawski, M.; et al. Modelling The Hemodynamics of Coronary Ischemia. *Fluids* **2023**, *8*, 159. https://doi.org/10.3390/fluids8050159

Academic Editors: Huidan (Whitney) Yu and D. Andrew S. Rees

Received: 7 February 2023
Revised: 6 April 2023
Accepted: 27 April 2023
Published: 17 May 2023

1. Introduction

Myocardial function is dependent on sufficient coronary blood flow (CBF), the rate of which is matched closely to the fluctuating metabolic requirements of the heart. CBF is compromised in ischaemic heart disease (IHD), the commonest cause of death in the world. The most frequent cause of IHD is coronary artery disease (CAD), whereby the epicardial coronary arteries become stenosed or occluded, thus restricting CBF, initially under exercise conditions, but eventually at rest; see Section 3.2.2. Although there are data correlating global myocardial ischaemic burden with clinical outcomes, there are little data linking it with myocardial energetics, function or cardiac output. Here we demonstrate a multi-compartment, zero-dimensional (0D) model [1–4] of the coronary and systemic circulation, based on clinical data, with the potential to address myocardial energetics. Specifically, we postulate that a zero-dimensional (0D) model, adapted to readily available, patient-specific data, can accurately simulate coronary (patho)physiology, with the potential to personalise [5] and predict an individual patient's global ischaemic burden and associate this with LV energetics and cardiac output. Our aims are to:

1. Devise and implement a 0D model and robust data handling protocols able to replicate in silico haemodynamic dysfunction;
2. Make optimal use of appropriate haemodynamic data, collected in the clinical PCI pathway;
3. Demonstrate our process' ability to capture patient-specific (patho)physiological states in silico.

2. Background

CFD modelling can provide a detailed assessment of local, usually single-vessel coronary haemodynamics which may be patient-specific [6]. It has also allowed for less or even non-invasive diagnostic tools, such as virtual fractional flow reserve (vFFR) [7], and in the novel form of the lattice Boltzmann simulation [8], it may be extended to describe flow within multiple small vessels [9]. However, CFD alone cannot provide a description of the perfusion of the entire myocardium necessary to characterise global ischaemic burden. To quantify the effects of multi-vessel disease upon total myocardial blood flow, one must set the diseased vessel(s) within the context of the whole coronary tree. This requires a methodology which can interact with CFD *and* wider physiological data.

Windkessel models are not new [10,11], neither are low-order representations of cardiac function [12–14] and coupled pseudo-mechanical system models, which have both been utilised here have been available, open source, for years [15]. We are concerned with computational medicine however, where a parsimonious model design should be constrained by data sources, with suitable interpretation protocols in mind. A composite of low-order models comprising a mechanical cardio-vascular 0D model of left heart function, valvular function and systemic circulation has been shown to have the capacity to ingest CFD-derived data and prognose the systemic effects of aortic and mitral valve disease [16]. Czechowicz et al. took CFD-derived valve properties, ingested them into a 0D model and formulated hypotheses connecting valve physiology, personalised cardiac energetics and valve function, to develop a clinical decision support tool. Here, we attempt to extend that methodology to a wider range of clinical data and disease states which are more difficult to characterise. This requires:

1. A coronary network topology conforming to the clinical pattern of atherosclerosis;
2. Modelling that reflects diastolic dominance of myocardial flow;
3. The ability to incorporate all available clinical patient data;
4. Sufficient computational simplicity for clinical deployment.

These demands make it necessary to simplify the task from the outset. To reduce the burden of model personalisation, we parameterise its right circulation parameters to normal values [17]. (This decision is supported by our input parameter sensitivity analysis in Appendix B, which shows that the coronary, left heart and systemic circulation input parameters are most influential). Further, the fact that resting coronary blood flow (250 mL min^{-1}; 0.8 mL min^{-1} g^{-1} of heart muscle) constitutes about 5% of cardiac output [18] allows us to assume that coronary flow does not directly influence the systemic circulation. It also allows us to approach our problem by adjusting, with the relevant coronary flow data a posteriori, a model in which systemic and ventricular outputs are tuned first. Further, we appeal, hierarchically, to the key physiological principles of LV performance and the systemic circulation to construct each patient's baseline and only then incorporate detailed CFD-derived information related to the coronaries' microvascular resistance (MVR), etc., while retaining a minimum number of assumptions necessary to impose physiological principles. We use cardiac energetics and the concepts encapsulated in Wiggers' diagram [19] to refine our low-order model of LV contractility and compliance; then, we consider ventricular–aortic (VA) coupling (which characterises the interaction between the myocardial contractile function and LV afterload and defines cardiovascular performance and efficiency [20]), and finally, the role of passive ventricular filling and valves. Only once an accurate model of general cardiac function is established will coronary flow be considered.

3. Methods and Materials

Broadly, we first describe our model (Section 3.1), then its data (Section 3.2). However, since model operation and data are related, some model details *bleed* into Section 3.2.

3.1. Methods

Personalisation cost function hyper-surfaces are intricate in the presence of non-linearities. Similar to valve functions, the cost function varies at very different rates with displacement in different parameter directions. Then, the task of locating the global minimum becomes computationally expensive. We address this difficulty with the multi-step protocol outlined in Table 1. The six steps declared there buffer clinical data into our tool, using low-order haemodynamic models, a closed loop system model and finally 3D CFD.

3.1.1. Low-Order Models

To express heart chamber biomechanics and valve characteristics appropriately, we use the double Hill elastance model [12,14] and the valves' pressure-flow characteristics. See Appendix A.2 for details; see also Czechowicz et al. [16], who give a fuller account of valve characteristic functions. Our passive systemic and pulmonary circulation windkessels, with proximal inertance (L), viscous dissipation (R) and multiple compliance (C), are also well documented [21]. Our pulmonary and systemic windkessels are designated *LRCRC*, having two capacitances to separate arterial and venous compliance. Note, each coronary branch is characterised by a compliance with its back plate pressure set to that in the relevant ventricle. Note also that elastance is not used solely to activate our system model. We will also use it, together with the LV PV loop [22], to clarify the relationship between our LV pressure and volume time-series data in Section 3.2.1.

3.1.2. System Model

Figure 1 shows, in relation to our CFD models, our assembly of windkessels, elastance functions and valves. It is designated a compartmental, electrical analogue or 0D model. It has four heart chambers as well as systemic arterial and venous, pulmonary arterial and venous, and several coronary arterial compartments. The overhead of four heart chambers is offset by enhanced performance [17]. We set the right circulation parameters to population average values. We compute N sampled time-series outputs $y_n(t_i), n = 1, .., N$, derived from internal states, $x_n(t_i), n = 1, .., N$ with the sample times t_i drawn from the converged heart cycle. Table 2 declares our compartment numbering convention and Table 3 our names, conventions, equivalent subscripting and input parameter base values. Note, micro-vascular resistance (MVR) is distributed relative to arterial compliance, with a proximal-to-distal ratio controlled by input parameter α.

It is typical to express the dynamics of the system in Figure 1 in coupled ordinary differential equations (ODEs), or differential algebraic equations (DAEs [23]), typically with compartment pressures serving as internal states. In state-space form [24], using relative time, our model is expressed as

$$\frac{d\underline{x}}{d\tau} = \underline{f}(\underline{x}(\tau); \underline{\Theta}), \quad \underline{y}(\tau) = \underline{h}(\underline{x}(\tau); \underline{\Theta}),$$

Vector function $\underline{h}(*)$, which maps between internal states and outputs, is often the identity. Outputs $\underline{y}(\tau)$ can, of course, be sampled to generate a discrete output metrics vector. $\underline{\Theta}$ denotes the system model's full input parameter set and

$$\underline{x}(\tau) = (p_1(\tau), p_2(\tau), .., p_{10}(\tau))^T.$$

is our instantaneous state vector. We use $\underline{\Theta}$ ($\underline{\theta}$) to denote the full set (a subset) of input parameters. Particular examples of the system expressed in equations in Section 3.1.2 are given in Appendix A. Deriving equations in Section 3.1.2 uses recursive applications

of simple haemodynamic sub-models, to formulate, in closed form, all the necessary relationships between compartmental pressures, volumes and flows (see Appendix A). A simple but lengthy process of eliminating flow and volume variables results in a DAE system [23] which is readily transformed into state-space form, by differentiating equations where necessary. Typically, the system is solved using a time-marching numerical solution, e.g., a fourth-order Runge–Kutta algorithm. A time step of 1×10^{-3} was chosen for the numerical solution. It will occasionally be helpful to replace the numerical compartment index (subscript) with an abbreviation of its name, e.g., $p_1(t) = p_{LV}(t)$, with LV denoting left ventricle, etc.; see Tables 2 and 3. For example model output data see Figure 2, which shows flow in a coronary artery illustrating a key feature- to capture diastolic dominance in coronary flow, our coronary vessels' compliance is biased by the LV or RV pressure, as appropriate.

Table 1. Stages in the personalisation process and their rationale. Physiological effects (column 2) are sequentially ingested into the workflow, with the objective declared in column 3, culminating, at step 5, in a patient-specific in silico model of the resting patient state.

Step	Processing Tool	Physiology	Objective	Comment
0	statistical analysis	LV function	ensemble average, downsample time-series	synchronised to ECG1 R-R interval
1	elastance model	LV function	derive LV elastance; ingest into 0D model	characterised by double Hill LV parameters
2	0D system model	LV–aortic coupling in systole	primary tune of 0D model systemic circulation windkessel parameters	large sub-space scanned by GA tuning
3	0D system model	LV–atrial coupling and diastolic filling	ancillary tuning of 0D left atrial elastance and mitral valve resistances	uses gradient descent tuning.
4	CFD analysis	coronary flows	ingest into 0D model computed R_{sten}, MVR	use software VirtuHEART^{TM}
5	0D system model	system-coronary	in silico representation of the patient at rest	

Table 2. Indexing and naming conventions for system model variables. Free subscript * is to be replaced by the appropriate symbol. Flows generally are trans-compartment; pressures and volumes apply to compartments directly.

Subscript		
	Name	
Number	$p_*(t), v_*(t)$	$q_*(t)$
1	LV	ao val
2	RA	tri val
3	RV	pul val
4	LA	mit val
5	sys art	sys prox
6	pul art	pul prox
7	sys ven	sys dist
8	pul ven	pul dist
9	cor art	cor prox
10	cor ven	cor dist

Table 3. System model input parameter nomenclature and status. The 0D model chamber parameters are classified by compartment. The status of the parameter is either fixed (at population average value) or tuned, from an initial value declared above. In the latter, the step in which the parameter is tuned is identified below. Note, the coronary LAD circulation parameters 74, ..., 80 and the coronary RCA circulation parameters 81, ..., 87 follow the naming convention of those declared above for the left circumflex artery, with the text "LAD" and "RCA" replacing the text "Cx" and identical initialisations.

Compartment	Index	Symbol	Description	Unit	Status	Base Value
NA	1	HR	Heart rate	bpm	Allocated	
NA	2	mcfp	Mean circulatory filling pressure	mmHg	tuned, step 3	7
Left Ventricle (LV)	3	$E_{LV,min}$	LV min. elastance ("compliance")	mmHg/mL	assigned, step2	0.08
	4	$E_{LV,max}$	LV max. elastance ("contractility")	mmHg/mL	assigned, step3	3
	5	$n_{LV,1}$	LV contraction rate constant	-	tuned, step 2	1.32
	6	$n_{LV,2}$	LV relaxation rate constant	-	tuned, step 2	27.4
	7	$\tau_{LV,1}$	LV systolic time constant	-	tuned, step 2	0.269
	8	$\tau_{LV,2}$	LV diastolic time constant	-	tuned, step 2	0.452
	9	$V_{LV,0}$	LV volume offset	mL	fixed	10
	10	$T_{LV,shift}$	LV fractional time shift	-	fixed	0
Left Atrium (LA)	11	$E_{LA,min}$	LA max. elastance ("compliance")	mmHg/mL	tuned, step 4	0.08
	12	$E_{LA,max}$	LA min. elastance ("contractility")	mmHg/mL	tuned, step 4	0.17
	13	$n_{LA,1}$	LA contraction rate constant	-	fixed	1.32
	14	$n_{LA,2}$	LA relaxation time constant	-	fixed	13.1
	15	$\tau_{LA,1}$	LA systolic time	-	tuned, step 4	0.11
	16	$\tau_{LA,2}$	LA diastolic time	-	tuned, step4	0.18
	17	$V_{LA,0}$	LA volume offset	mL	fixed	3
	18	$T_{LA,shift}$	LA fractional time shift	-	fixed	0.85
Right Ventricle (RV)	19	$E_{RV,min}$	RV min. elastance ("compliance")	mmHg/mL	fixed	0.04
	20	$E_{RV,max}$	RV max. elastance ("contractility")	mmHg/mL	fixed	0.6
	21	$n_{RV,1}$	RV contraction rate constant	-	fixed	1.32
	22	$n_{RV,2}$	RV relaxation rate constant	-	fixed	27.4
	23	$\tau_{RV,1}$	RV systolic time	-	fixed	0.269
	24	$\tau_{RV,2}$	RV diastolic time	-	fixed	0.452
	25	$V_{RV,0}$	RV volume offset	mL	fixed	55
	26	$T_{RV,shift}$	RV fractional time shift	-l	fixed	0
Right Atrium (RA)	27	$E_{RA,min}$	RA max. elastance ("compliance")	mmHg/mL	fixed	0.04
	28	$E_{RA,max}$	RA min. elastance ("contractility")	mmHg/mL	fixed	0.15
	29	$n_{RA,1}$	RA contraction rate constant	-	fixed	1.32
	30	$n_{RA,2}$	RA relaxation rate constant	-	fixed	13.1
	31	$\tau_{RA,1}$	RA systolic time	-	fixed	0.11
	32	$\tau_{RA,2}$	RA diastolic time	-	fixed	0.18
	33	$V_{RA,0}$	RA volume offset	mL	fixed	17
	34	$T_{RA,shift}$	RA fractional time shift	-	fixed	0.85
Systemic Circulation LR CRC windkessel	35	$R_{sysart,dist}$	Systemic Resistance Distal	mmHg·s/mL	tuned, step 3	0.033
	36	$R_{sysart,prox}$	Systemic Resistance Proximal	mmHg·s/mL	tuned, step 3	0.9
	37	$C_{sysart,prox}$	Systemic Arterial Proximal Capacitance	mL/mmHg	tuned, step 3	1
	38	$V_{sysart,prox,0}$	Systemic Arterial Proximal Unstressed Volume	mL	fixed	720
	39	$C_{sysart,dist}$	Systemic Arterial Distal Capacitance	mL/mmHg	fixed	11
	40	$V_{sysart,dist,0}$	Systemic Arterial Distal Unstressed Volume	mL	fixed	3000
	41	$L_{sysart,prox}$	Systemic Arterial Inertance (Proximal)	mmHg·s^2/mL	tuned, step 3	0.005

Table 3. *Cont.*

Compartment	Index	Symbol	Description	Unit	Status	Base Value
Pulmonary Circulation LR CRC windkessel	42	$R_{pulart,dist}$	Pulmonary Resistance Distal	mmHg·s/mL	fixed	0.01
	43	$R_{pulart,prox}$	Pulmonary Resistance Proximal	mmHg·s/mL	fixed	0.05
	44	$C_{pulart,prox}$	Pulmonary Arterial Proximal Capacitance	mL/mmHg	fixed	10
	45	$V_{pulart,prox,0}$	Pulmonary Arterial Proximal Unstressed Volume	mL	fixed	180
	46	$C_{pulart,dist}$	Pulmonary Arterial Distal Capacitance	mL/mmHg	fixed	15
	47	$V_{pulart,dist,0}$	Pulmonary Arterial Distal Unstressed Volume	mL	fixed	720
	48	$L_{pulart,prox}$	Pulmonary Arterial Inertance (Proximal)	mmHg·s^2/mL	fixed	0.0017
Aortic Valve	49	$a_{2,aortic}$	Aortic Valve quadratic coefficient	mmHg·s^2/mL^{2}mL	fixed	1.60×10^{-5}
	50	$a_{1,aortic}$	Aortic Valve linear coefficient	mmHg·s/mL	fixed	0
	51	$a_{2,aortic,regurg}$	Aortic Valve regurgitant quadratic coefficient	mmHg·s^2/mL^{2}mL	fixed	0
	52	$a_{1,aortic,regurg}$	Aortic Valve regurgitant linear coefficient	mmHg·s/mL	fixed	0
Mitral Valve	53	$a_{2,mitval}$	Mitral Valve quadratic coefficient	mmHg·s^2/mL^{2}mL	tuned, step 4	6.25×10^{-6}
	54	$a_{1,mitval}$	Mitral Valve linear coefficient	mmHg·s/mL	tuned, step 4	0
	55	$a_{2,mitval,regurg}$	Mitral Valve regurgitant quadratic coefficient	mmHg·s^2/mL^{2}mL	fixed	0
	56	$a_{1,mitval,regurg}$	Mitral Valve regurgitant linear coefficient	mmHg·s/mL	fixed	0
Pulmonary Valve	57	$a_{2,pulval}$	Pulmonary Valve quadratic coefficient	mmHg·s^2/mL^{2}mL	fixed	8.16×10^{-6}
	58	$a_{1,pulval}$	Pulmonary Valve linear coefficient	mmHg·s/mL	fixed	0
	59	$a_{2,pulval,regurg}$	Pulmonary Valve regurgitant quadratic coefficient	mmHg·s^2/mL^{2}mL	fixed	0
	60	$a_{1,pulval,regurg}$	Pulmonary Valve regurgitant linear coefficient	mmHg·s/mL	fixed	0
Tricuspid Valve	61	$a_{2,trival}$	Tricuspid Valve quadratic coefficient	mmHg·s^2/mL^{2}mL	fixed	6.25×10^{-6}
	62	$a_{1,trival}$	Tricuspid Valve linear coefficient	mmHg·s/mL	fixed	0
	63	$a_{2,trival,regurg}$	Tricuspid Valve regurgitant quadratic coefficient	mmHg·s^2/mL^{2}mL	fixed	0
	64	$a_{1,trival,regurg}$	Tricuspid Valve regurgitant linear coefficient	mmHg·s/mL	fixed	0
Coronary Left Main	65	$a_{2,LMart,regurg}$	Left main stenosis, quadratic coefficient	mmHg·s^2/mL^2	assigned, step 5	0
	66	$a_{1,LMart,regurg}$	Left main stenosis, linear coefficient	mmHg·s/mL	assigned, step 5	0
Coronary Circumflex RCRC windkessel	67	$a_{2,Cx,art}$	Circumflex, quadratic coefficient	mmHg·s^2/mL^{2}mL	assigned, step 5	0
	68	$a_{1,Cx,art}$	Circumflex, linear coefficient	mmHg·s/mL	assigned, step 5	0
	69	$R_{Cx,MVR}$	Circumflex, RCR total resistance	mmHg·s/mL	assigned, step 5	35
	70	C_{Cx}	Circumflex, RCR capacitance	mL/mmHg	tuned, step 5	0.25
	71	α_{Cx}	Circumflex, RCR fraction proximal resistance	-	tuned, step 5	0.9
	72	$V_{Cx,0}$	Circumflex unstressed volume	mL	tuned, step 5	8
	73	β_{Cx}	Circumflex capacitance back pressure fraction	-	fixed	0.5

3.1.3. Input Parameters, Local Sensitivity and Orthogonality Analysis

Of our 87 input parameters, a majority take population average values and are not tuned; see Table 3. The interrogated input parameter sub-space is spanned by parameters of cardiac energetics, coronary perfusion, LV and left atrial elastance, the mitral valve, the coronary and systemic windkessels and mean circulatory filling pressure. The underlying full local sensitivity and input parameter orthogonality analysis is shown in Appendix B, along with a reduced analysis, based upon the most relevant inputs and outputs.

3.1.4. Model Personalisation or Input Parameter Identification

Personalisation involves inverse operation, guided by a cost function, together with an appropriate multi-variate functional minimisation algorithm. Our personalisation process is multi-stage. It is summarised in the inter-relating Tables 1, 4 and 5. For all stages, the minimised cost function is a scalar weighted composite of: (i) the L^2 norm of the difference between model-predicted and target data interpolated onto the same times, t_i, and (ii) a *derived* metric, Y_n

$$f(\underline{\Theta}) = \sum_{j=1}^{M}\left(w_j^{(1)}\left\{\sum_{i=1}^{N}\left(y_j(\underline{x}(t_i;\underline{\Theta})) - y_j^{target}(t_i)\right)^2\right\} + w_j^{(2)}Y_n^2\right). \quad (1)$$

For example, a derived metric might be systolic systemic arterial (compartment 5) pressure

$$Y_5 = \max_{t_i}(p_5(t_i;\underline{\Theta})) - \max_{t}\left(p_5^{target}(t)\right),$$

with $w_j^{(1)}$ and $w_j^{(2)}$ denoting the weights. We defer further discussion until Section 3.2.

Having summarised our processing and its rationale and declared our notation, we may now conclude this sub-section by defining the processing in key steps 3 and 4 of our protocol, which require inverse operation, or tuning, of the 0D system model. In Table 6 we state the model input parameters which are tuned and the corresponding patient data that is used to perform the tuning in each of the steps 3 and 4.

Table 4. Patient-specific data. Data are classified according to physical and mathematical significance, how the data are used, patient state and the 0D model compartment(s) to which they apply. Table 5 provides more information on the sub-models used to introduce these data into our 0D system model. The free subscript * in column 6 denotes one of the resolved coronary arteries, LMS (left main stem), LCA (left coronary artery), LCX (left circumflex) and RCA.

Data Source	Compartment	Role	Unit	Data Form	Notation	State			Pathway Stage	
						Rest	Exercise	Hyperemia	Pre PCI	Post PCI
six-minute walk test	all (general)	cardiac output proxy	∼	statistics	∼	yes	yes	no	yes	yes
activity monitoring	all (general)	cardiac output proxy	∼	statistics	∼	yes	yes	no	yes	yes
CMR scan	left ventricle	volume	ml	time-series	$v_{LV}(t)$	yes	no	no	yes	no
coronary angiograms	coronaries	flow	mL/s	discrete	$< Q_* >_t$	yes	no	yes	yes	yes
coronary angiograms	coronaries	MVR	mmHg·s/mL	discrete	$R_{\mu LCA}$ etc.	yes	no	yes	yes	yes
PCI pressure catheter	left ventricle	pressure	mmHg	time-series	$p_{LV}(t)$	yes	no	yes	yes	yes
PCI pressure catheter	systemmic	pressure	mmHg	time-series	$p_{AO}(t)$	yes	no	yes	yes	no
PCI pressure catheter	coronaries	pressure	mmHg	time-series	$p_*(t)$ etc.	yes	no	yes	yes	yes

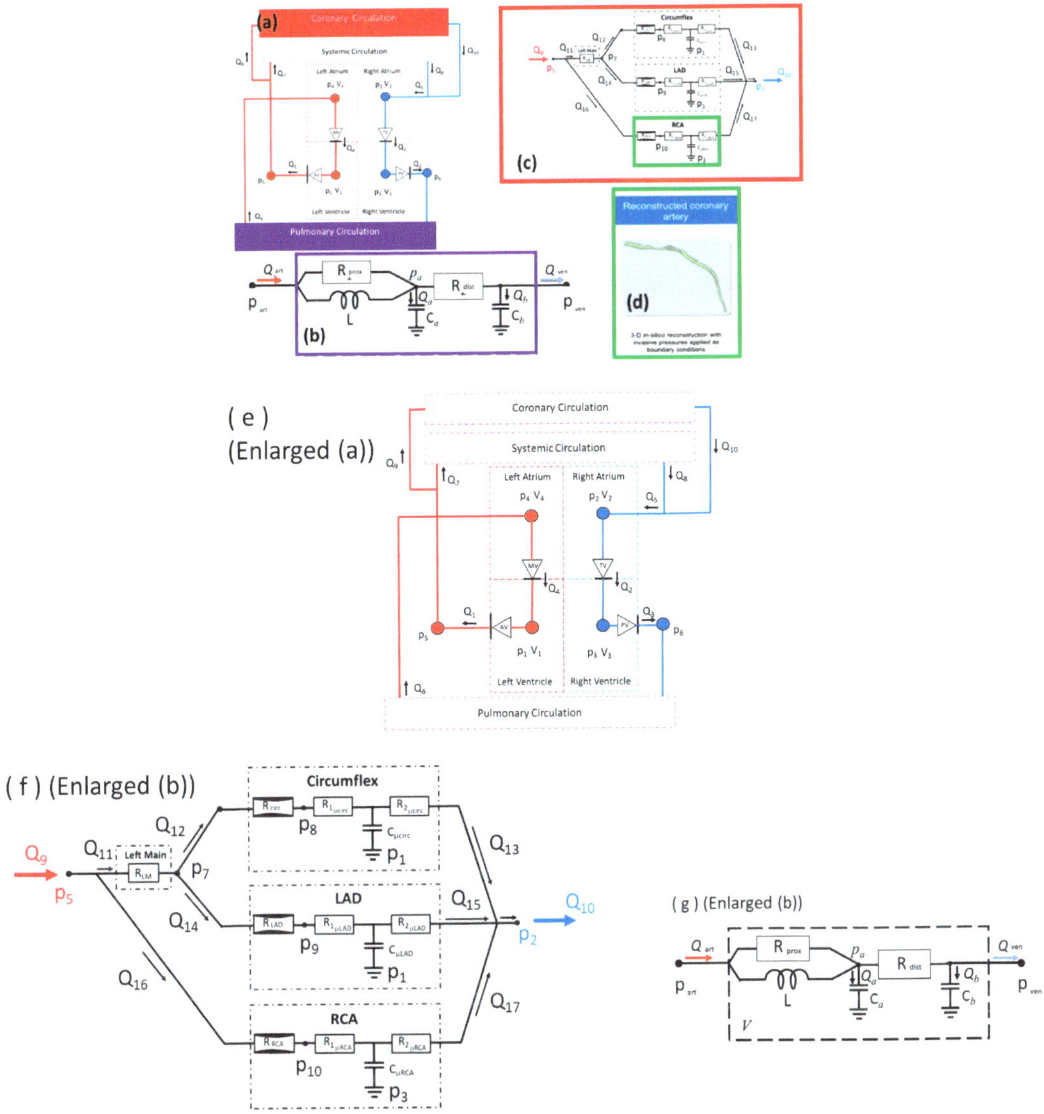

Figure 1. Interacting 0D system and CFD models. Clockwise from the top. First panel uses colour coding to expose the interaction between (**a**) core heart module; (**b**) systemic and LRCRC windkessels; (**c**) coronary module; and (**d**) 3D steady CFD workflow used to parameterise the stenotic resistance and MVR. (**e**) An enlargement of (**a**), the heart module. Important pressure, flow and volume variables are displayed. Valves are all diodes. Chamber elastances are represented by double Hill elastance functions. The subscripting of the compartment state variables $(p_n, v_n, q_n), n = 1, \ldots, 10$ is that which is used throughout, with the convention that compartment, inlet pressure, volume and outlet flow are all subscripted by the compartment number. (**f**) An enlargement of (**c**), the coronary windkessel set. There is a RCR windkessel for each of the three principal vessels, designated LCx, LAD and RCA. The back plates of the vessel compliances are all coupled to the relevant ventricular pressure. (**g**) An enlargement of (**b**), the systemic and pulmonary circulation windkessels.

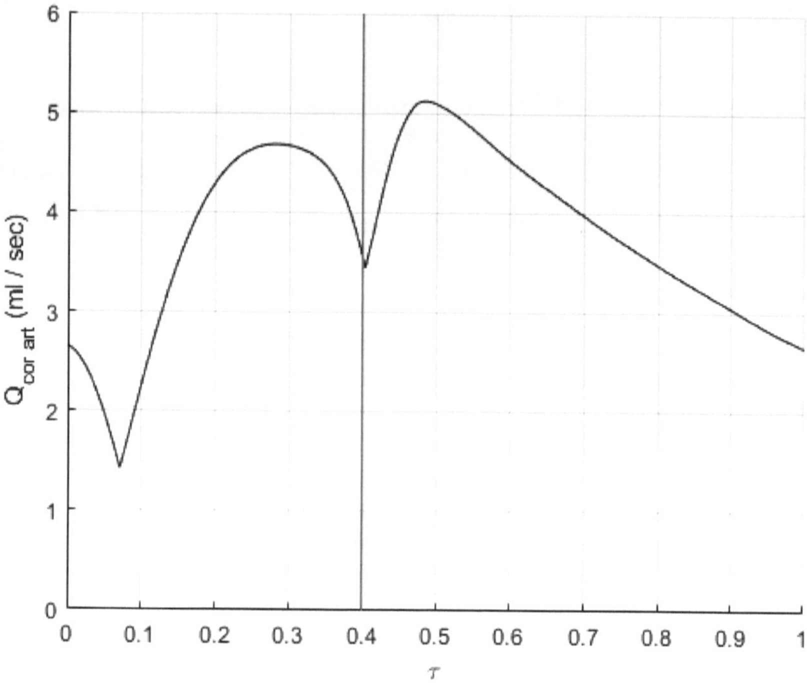

Figure 2. Diastolic dominance of coronary artery flow in the system model. It is essential to capture the diastolic dominance of flow in the coronaries. Here, the vertical red line indicates the end of systole. These illustrative data show that computed flow peaks in the diastolic phase.

Table 5. Sub-models used to pre-process patient data. We provide further information on how patient data outlined in Table 4 are prepared. Here, $< . >_{EA}$ denotes an ensemble average, taken over equivalent intervals of dimensionless time τ. This table shows the interaction between the various data sources and re-processing, to allow them to be ingested into the 0D compartmental system model (column 3). All pressure data were downsampled onto the CMR phase times; see Equation (2).

Source Data	Pre-Processing	Ingests into 0D System Model as
$< v_{LV}(\tau) >_{EA}$ $< p_{LV}(\tau) >_{EA}$	fit to double Hill elastance function	personal elastance function parameters
$< v_{LV}(\tau) >_{EA}$	extract extrema	personal LV volume extrema
$< p_{LV}(\tau) >_{EA}$	extract extrema	personal LV pressure extrema
$< p_{AO}(\tau) >_{EA}$		personal systemmic (aortic) pressure time-series
coronary angiograms $< p_{LMS}(\tau) >_{EA}$ etc.	steady-state CFD simulations	personal MVRs for treated vessels
coronary angiograms $< p_{LMS}(\tau) >_{EA}$ etc.	steady-state CFD simulation	personal stenotic Bernoulli resistance quadratic and linear coefficients in treated vessels
6mwt monitoring	clinical interpretation	validation data

Table 6. Supplementary information for Table 1. Additional information on personalisation steps 3, 4, declared in Table 1. The search target data and input parameter sub-space are identified here. Note that search targets accumulate between steps 3 and 4.

Step	Personalisation Search Definition	
	Input Parameters Adjusted	**Output Metrics Targeted**
3	mcfp, L, R_{prox}, R_{dist}, C_{prox}	$\max_\tau(v_{LV}(\tau)), \min_\tau(v_{LV}(\tau)), \max_\tau(p_{LV}(\tau)), \min_\tau(p_{LV}(\tau)), \{p_{AO}(\tau_i); i = 1, .., 29\}$
4	$a_{1,mit}, a_{2,mit}, n_1^{LA}, n_2^{LA}, \tau_1^{LA}, \tau_2^{LA}, E_{LA,max}, E_{LA,min}$	to the above, add $\{v_{LV}(\tau_i); i = 1, .., 29\}$

3.1.5. 3D CFD Models

We proceed to consider the CFD tools and modelling used in our study. Patient-specific stenotic Bernoulli resistance coefficients and MVRs are provided in our core model using two validated CFD based methods, (i) VirtuHEART^{TM} (University of Sheffield, Sheffield, United Kingdom) and (ii) VirtuQ^{TM} (University of Sheffield, Sheffield, United Kingdom) [6,7]. In the domain Ω, these tools solve the incompressible Navier–Stokes and continuity equations

$$\begin{aligned} (\underline{v}(\underline{r}) \cdot \underline{\nabla})\underline{v}(\underline{r}) &= -\frac{1}{\rho}\nabla p(\underline{r}) + \nu\nabla^2\underline{v}(\underline{r}), \\ \underline{\nabla} \cdot \underline{v}(\underline{r}) &= 0, \end{aligned} \tag{2}$$

for steady, i.e., time-average flow; all symbols have their usual meaning. An accepted approximate value of Reynolds' number $Re \approx 600$ for the coronary arteries, which indicates transitional flow. Note that we assume blood is a Newtonian fluid. Patient-specific boundary conditions are derived as follows. A closed coronary artery geometry (Figure 1d) is decomposed into a luminal boundary, $\partial\Omega_1$, an entrance face, $\partial\Omega_2$, and an exit face, $\partial\Omega_3$. Using epipolar geometry and two compatible angiogram views, taken at the time of PCI, we construct $\partial\Omega_1$, generate a mesh and apply a Dirichlet, no-slip condition

$$\underline{v}(\underline{r}) = 0, \quad \underline{r} \in \partial\Omega_1.$$

We remark that a model of vessel sequestration using Murray's law [25,26] to assign a flux across the luminal boundary based upon local geometry is available [7,27]; however, zero leak was specified here. Lesion proximal and distal time-averaged pressures are also recorded at PCI. Solutions were computed using Fluent^{TM} (Ansys Corporation, Canonsburg, PA, United States of America), with boundaries $\partial\Omega_2$ and $\partial\Omega_3$ processed in each of two ways:

1. Following VirtuQ methodology, apply measured proximal and distal average pressure on $\partial\Omega_2$, $\partial\Omega_3$, respectively,

   $$p(\underline{r}) = \langle p_{*,prox}(\tau)\rangle_\tau, \quad \underline{r} \in \partial\Omega_2; \quad p(\underline{r}) = \langle p_{*,dist}(\tau)\rangle_\tau, \quad \underline{r} \in \partial\Omega_3,$$

   we compute the steady-state solution and deduce vessel flow

   $$q_* = \int\int_{\partial\Omega_3} \underline{v}(\underline{r}) \cdot d\underline{A};$$

   then, we use Ohm's law and $\langle p_{*,dist}(\tau)\rangle_\tau$ to compute a patient-specific MVR.
2. Following VirtuHeart methodology, assume an MVR based on patient characteristics [6]; compute vessel inlet and outlet areas

   $$A_{in} = \int\int_{\partial\Omega_2} dA, \quad A_{out} = \int\int_{\partial\Omega_3} dA,$$

set two physiologically plausible inlet flows ($q_* = 1, 3$ mL/s), find uniform inlet and outlet streamwise velocities

$$v_s^{in}(\underline{r}) = \frac{q_*}{A_{in}}, \quad \underline{r} \in \partial\Omega_2; \quad v_s^{out}(\underline{r}) = \frac{q_*}{A_{in}}, \quad \underline{r} \in \partial\Omega_3,$$

compute the two corresponding pressure drops, δp_*^1, δp_*^3 and hence infer the two Bernoulli resistance coefficients of the stenosis.

3.2. Materials

We analysed anonymised clinical datasets from 40 CAD patients with angina symptoms. The elective PCI workflow culminates in re-vascularisation of one or more of the coronaries in a majority of cases. The decision to stent is guided by the average pressure drop across a lesion. This is measured using a pressure wire (PRESSUREWIRE™X Abbott, downloading software Coroflow™, Coroventis Uppsala, Sweden) with the patient in a hyperaemic state of pharmacological stress induced by adenosine (a strong vasodilator); from this, one computes

$$FFR = \frac{p_{distal}}{p_{proximal}}, \tag{3}$$

with re-vascularisation indicated for $FFR < 0.8$. The clinical pathway generates PCI pressure data, cardiac MRI of the LV, six-minute walk tests (6mwt) before and after PCI and home activity monitoring (see below); these data fall into two categories

1. Measured data (such as pressure and volume, expressed as time-series);
2. Exercise and physical activity data (like patient 6mwt (exercise) and home monitoring (activity) measures).

Table 4 identifies data available for every patient, for Table 5, the particular form in which pre-processed data modalities, declared in Table 4, are ingested into the system model, and Table 1, a commensurate sequence of handling protocols.

Our personalisation process steps $0, \ldots, 5$ (Table 1) introduce physiological effects sequentially, considering

1. Functional form (qualitative data trends should accord with accepted physiology);
2. Modality (pressure is measured directly, using a manometer; CMR volumes, in contradistinction, involve reciprocal space reconstruction).

Our raw data take the form of time-series at two sampling rates. Pressure data were all sampled for 120 s, using a sampling rate of 100 Hz. CMR volumes were acquired at a nominal sampling rate of 30 images per heart cycle. Pressure and volume can be co-registered to the start of ventricular contraction using the RR interval of the concurrent ECG1 traces, also recorded. Pressure and CMR data were acquired at different stages in the patient pathway (Table 4), with patients often in different physiological states. For all pressure signals, the sampled wave-train had its low frequency contribution removed and was decomposed, using time-domain analysis, into single beat pressure excursions, identified from the concurrent ECG 1 R-R interval, and then ectopic beats were filtered, and the remaining cycle pressure samples were downsampled onto the 30 dimensionless times defined as follows

$$\tau_n = \frac{n}{30}\tau, \quad n \in \mathbb{N}, \quad 0 \leq n \leq 29; \quad \tau = \frac{t}{T}, \quad T = \frac{60}{HR}, \tag{4}$$

corresponding to CMR data sampling times. This set was then ensemble averaged, with the sample standard deviations in each dimensionless time channel providing an error. Example aortic and LV pressure data are shown in Figure 3. CMR-derived LV volume data are also effectively ensemble averaged (acquisition involves a reciprocal space reconstruction of data acquired at a given cycle phase, over several consecutive beats). Our CMR total LV volume measurements are attributed to 30 equispaced dimensionless times. Each was computed by segmenting the chamber, in a set of between 9 and 12 cross-sections, with the base and apical frames frequently discarded, owing to error. Frames were positioned along the long axis of the approximately cylindrical LV. Our segmentation and subsequent volume computation were performed using the MASS^{TM} software (Leiden University, Version 2018 EXP, Leiden University Medical Centre, Leiden, The Netherlands). Figure 3, right panel, shows a typical CMR volume time-series.

Downsampled pressure and volume are acquired relative to a common clock provided by the ECG 1 RR interval which can potentially synchronise them, to yield a personal double Hill elastance function, $E_{LV}(t)$, etc. However, there are demonstrable inconsistencies, such as heart rate (HR) miss-match. Pressure data have a certain HR, and CMR data another. We will demonstrate that careful pressure and volume time-series co-registration is essential, and we will develop a minimal rigid shift which adjusts only relative phase.

3.2.1. Physical Data Processing

Chamber elastance functions (Appendix A.2) are low-order models developed from studied physical data. Elastance may be defined as follows

$$E_*(\tau) \equiv \frac{p_*(\tau)}{(v_*(\tau) - v_0)}, \tag{5}$$

where subscript $*$ now denotes the chamber and v_0 its unstressed volume. We take the clinical approximation, $v_0 = 0$. Clearly in Equation (5), $E_*(\tau)$ is sensitive to the relative phase of $p_*(\tau)$ and $v_*(\tau)$, but the importance of co-registration is perhaps better illustrated by another metric—the PV loop. The latter is a closed, two-dimensional parametric, Cartesian curve

$$(p_{LV}(\tau), v_{LV}(\tau)), \quad \tau \in [0,1].$$

Figure 4, top panel, shows a personal PV loop in which the LV pressure and volumes are correctly co-registered. The ventricular–aortic coupling parameter (blue line) and the wasted mechanical work, as well as the useful mechanical work (loop area), can be deduced. In contradistinction, Figure 4, bottom panel, shows the same patient's PV loop using the putative common clock of the RR interval, to co-register the two time-series signals (red data). This *primitive* co-registration is implausible: it has no identifiable iso-volumetric phases, and it implies a high level of valvular regurgitation and irregular aortic and mitral valve actuation pressures. The difference between Figure 4, top and bottom panels. argues for a principled co-registration other that the EGG 1 RR signal. Reserving for subsequent work a more elaborate co-registration, possibly involving signal dilations, we chose rigidly to delay the LV volume samples $\{v_{LV}(\tau_n), 0 \leq n \in \mathbb{N}, n \leq 29\}$ relative to a fixed pressure sample set $\{p_{LV}(\tau_n), n \in \mathbb{N}, n \leq 29\}$, by an integral number of dimensionless time steps, $\delta\tau = (\tau_n - \tau_{(n-1)})$. We observed a cohort-average shift of two CMR phases. Patient-specific synchronisation was accomplished by seeking the instant at which the aortic and mitral valves open, in the separate pressure and volume time-series, using the multiple criteria declared in the fourth column of Table 7.

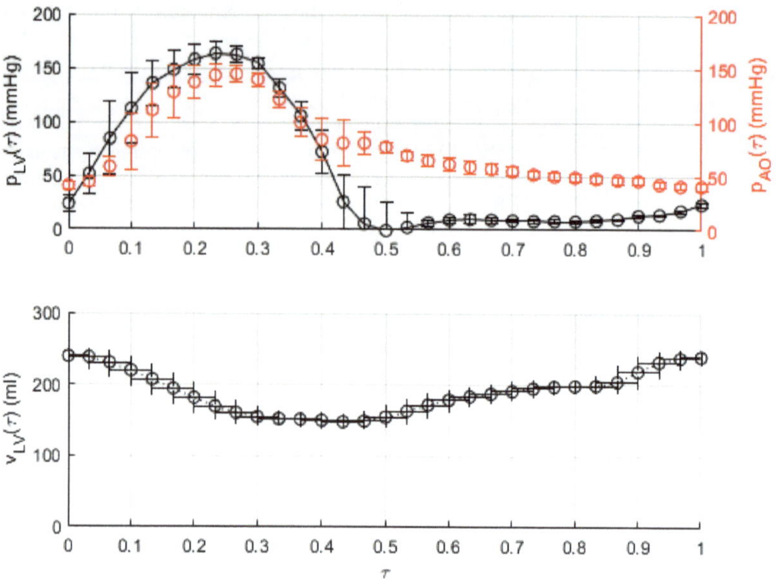

Figure 3. Raw time-series data. Top panel. Downsampled aortic (right ordinate) and left ventricular (left ordinate) pressure variation. These pressure time-series (original sampling rate 120 Hz) were acquired with a concurrent ECG time-series (ECG 1 lead) in the catheterisation laboratory while the patient was undergoing elective PCI. The independent signals were co-registered, using the common background clock of the RR interval of the accompanying ECG 1 recordings. The error bar on these data is derived from the standard deviation of the downsampled pressure time-series sample bins. Bottom panel. LV volume variation. These time-series data are expressed in 30 equispaced CMR phases times. The horizontal error bar corresponds to two CMR phases because the principal source of error in the CMR signals was deemed to originate in the pressure–volume co-registration, which was corrected by a rigid shift, relative to the pressure signals, of about two CMR phases. The data in both panels have time period $\frac{60}{HR}$ secs, but the patient HR and systolic time typically vary between the data collections.

When times T^p_{ao} and T^v_{ao} are not simultaneous, a plausible co-registration is imposed by postulating a shift, or delay in the data $\{v_{LV}(\tau_i), i = 0, .., 29\}$ of $\delta^{ao}_t = (T^p_{ao} - T^v_{ao})$ CMR phases. Similarly, $\delta^{mit}_t = (T^p_{mit} - T^v_{mit})$ was evaluated. The final rigid shift of CMR volume time-series data is

$$S \approx \frac{(\delta^{ao}_t + \delta^{mit}_t)}{2}, \quad S \in \mathbb{N}, \quad 0 \leq \quad S \leq 29.$$

Here, the symbol $\approx$ is used to denote a rounded value. The result of applying the rigid shift defined in Table 7 to our volume time-series data transforms the PV loop; see Figure 4, bottom panel. Our processing paradigm may be summarised as: *constrain the LV pressure and volume time-series to conform to accepted PV loop physiology*.

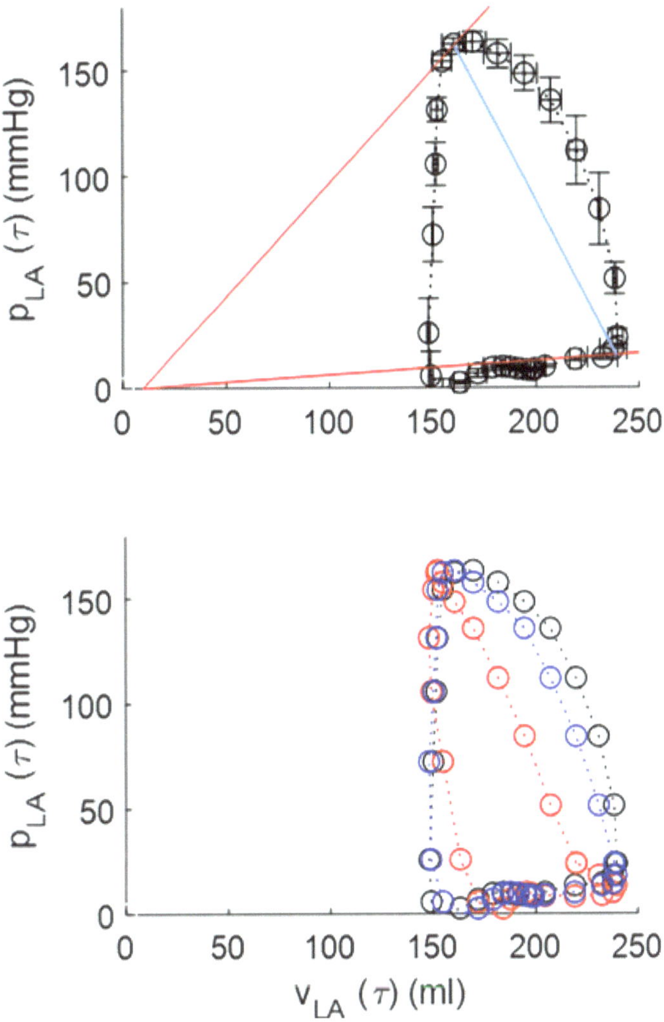

Figure 4. Patient-specific PV loops. Top panel. The closed curve, or loop, derives from the raw data in Figure 3, appropriately co-registered. Ventricular–arterial coupling [20] is measured by the modulus of the gradient of the blue solid line. Isovolumetric compression (right) and relaxation (left) segments are apparent. The volume enclosed within the PV loop is the useful mechanical work, applied to ejected fluid. Wasted work is calculated from the area bounded by the red lines and the left-hand side of the PV loop. Bottom panel. One PV loop based up the same data as shown in the top panel, but using different co-registrations of pressure and volume time-series data. The primitive co-registration (red) has no rigid shift applied to the volume data; the blue dataset shows the result of a rigid shift of one CMR phase, and the black data show the result of a shift of two CMR phases (the value of the shift obtained by synchronising the valve timing points). These data demonstrate the clear need for careful co-registration of the two signals.

With an acceptable co-registration of pressure and volume data, we derive the LV double Hill elastance function parameters. $E_{LV,max}$ (widely termed "contractility") and $E_{LV,min}$ ("compliance") are assigned directly

$$E_{LV,max} = \max_i \left(\frac{p_{LV}(t_i)}{v_{LV}(t_i)} \right), \quad E_{LV,min} = \min_i \left(\frac{p_{LV}(t_i)}{v_{LV}(t_i)} \right). \tag{6}$$

Table 7. Identification of mitral and aortic valve opening times, in LV pressure and volume time-series data. Each valve opening instant was identified by multiple criteria (column 3). Discrete time-series data temporal derivatives were evaluated by the class of finite difference approximation stated in column 4. Column 5 defines the notation for the identified cycle time fraction.

Valve	Time-Series	Valve Opening Signature	Difference	Notation
Aortic	pressure	$p_{LV} \gtrsim 80$ mmHg	central	T^p_{ao}
		$\max_t\left(\frac{dp_{LV}}{dt}\right)$		
	volume	$\frac{dv_{LV}}{dt} \lesssim 5$ mL/s	forward	T^v_{ao}
		close to $\max_t(v_{LV})$		
Mitral	pressure	$\frac{dp_{LV}}{dt} < 0$	central	T^p_{mit}
		$\frac{d^2p_{LV}}{dt^2} \gtrsim 0$		
		$p_{LV} \approx 10$ mmHg		
	volume	$\left(\frac{dv_{LV}}{dt}\right) > 0$	forward	T^v_{mit}
		$\left(\frac{d^2v_{LV}}{dt^2}\right) > 0$		

The remaining activation function parameters accrue by seeking the extremum of a multi-variate function, which is assumed to be differentiable

$$f(n_1, n_2, \tau_1, \tau_2) = \sum_{i=0}^{29} (p_{LV}(\tau_i) - E(\tau_i; \underline{\theta}) v_{LV}(\tau_i)), \quad \underline{\theta} = (n_1, n_2, \tau_1, \tau_2)^T \subset \underline{\Theta}. \tag{7}$$

Note, n_1, n_2, τ_1, τ_2 are variables (parameters) on the left (right)-hand side of Equation (7). Minimisation was performed using a gradient descent, initialised to the population average [12]. Figure 5 below shows a derived elastance function.

In this way, our mechanical time-series signals are downsampled onto a common rate of $\frac{60}{30HR}$ Hz, suitably co-registered, and then used to deduce a personalised $E_{LV}(\tau)$ for each patient. These data will be recycled as 0D system model personalisation targets, both as discrete time-series and derived metrics.

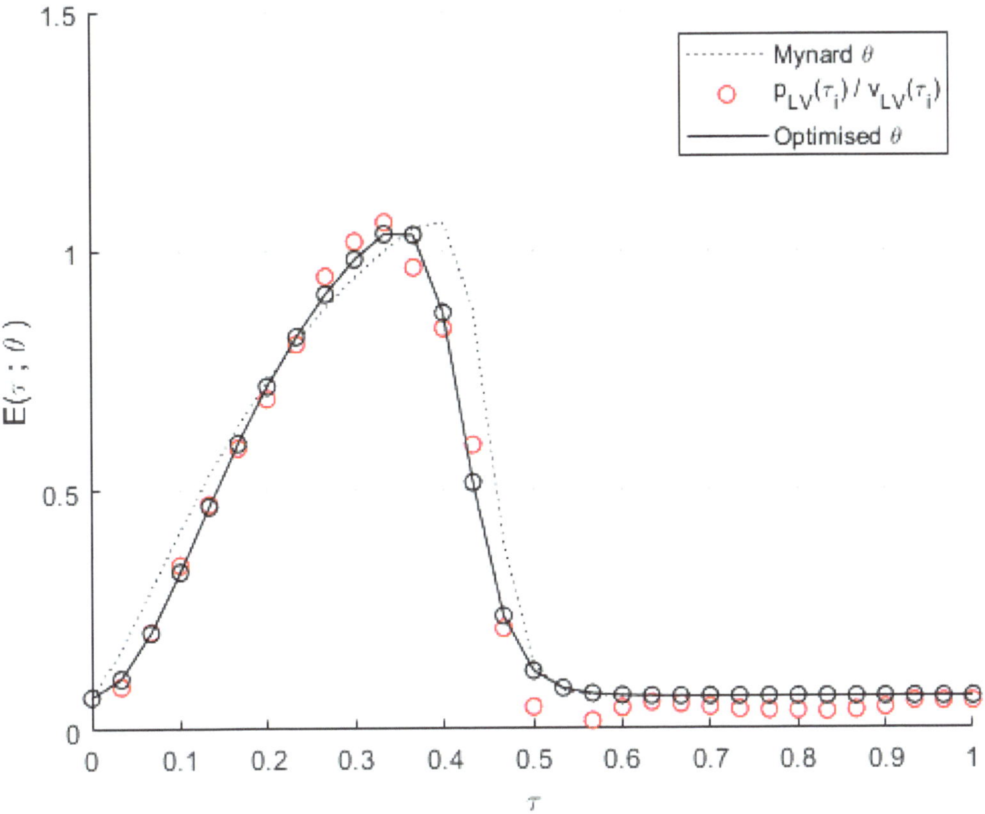

Figure 5. A personalised elastance function. The solid line corresponds to the optimum elastance function, obtained from suitably co-registered coordinates $(p_{LV}(\tau_i), v_{LV}(\tau_i))$, $i = 0, \ldots, 29$ using the cost function in Equation (7), minimised using a gradient descent simplex method. Broken line: our common initialisation, using Mynard's population average [12] parameterisation, θ. Open red circles: raw data, with the appropriate shift applied. Black circles and line: optimised fit on the shifted data minimising the residual declared in Equation (7).

3.2.2. Exercise/Monitoring Data Processing

CAD and cardiorespiratory fitness are linked. Endothelial nitric oxide synthase (eNOS) prevents platelet aggregation and white cell adhesion and inhibits vascular smooth muscle cell proliferation [28], and regular physical exercise enhances bioavailability of eNOS and regeneration of the vascular endothelium [29]. Conversely, inactivity is associated with pathological processes preceding atherosclerosis and CAD [30]. As a result, a dose–response relationship exists between physical activity and the risk of CAD [31], with exercise capacity predictive of mortality, myocardial infarction and risk of re-vascularisation in patients with established CAD [32]; see Figure 6. Assessing patients' activity and exercise capacity is key in the management and assessment of CAD. For completeness, therefore, we summarise, in two categories, the study exercise and activity data which will, in future work, quantify the change in patient exercise tolerance attending measured PCI-related physiological changes.

1. Home activity monitoring. Fitbit™ Charge 4 wrist watches (Healthy Metrics Research Inc. San Francisco, CA, USA) and a smartphone were used. Data monitoring started at recruitment and extended up to six months post PCI. All data were uploaded

to the Fitbit website simultaneously through a smartphone application and then exported as coarse-grained time-series (mechanical energy consumption, distance walked, minutes spent sedentary, light, fairly, and great activity). Concurrent HR data were also exported as downsampled HR data, which were simultaneously updated throughout the day as time-averaged values every 10–15 min.

2. Formal six-minute walk tests (6MWT). These were performed at baseline before the PCI (ideally one day) with a repeat assessment after three and six months. In these standard clinical assessments, the data collected were cuff blood pressure, HR and distance walked.

We remark that exercise training in CAD is superior to PCI in improving event-free survival and exercise capacity and is at a much lower cost [33], and the total distance walked is a proxy for cumulative CO.

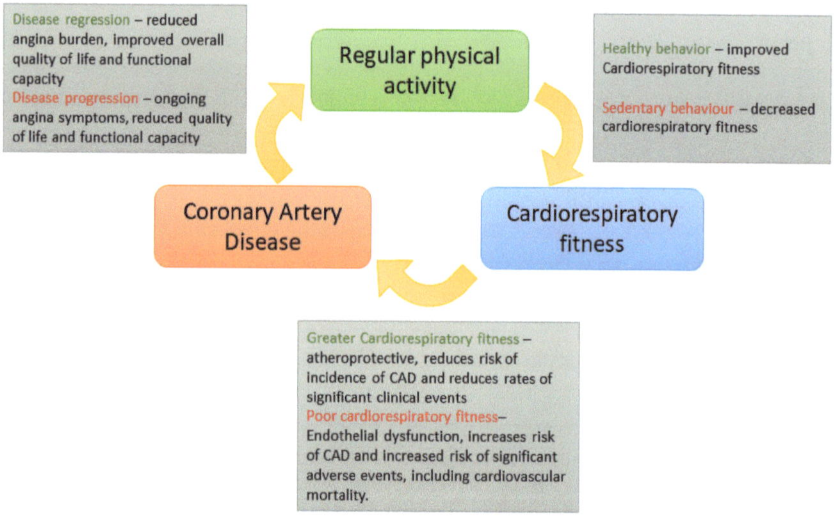

Figure 6. Schematic representation of the relationship between exercise, coronary artery disease progression and cardio-pulmonary fitness and the underlying physiology. The exercise and activity data gathered in this study will eventually expose quantitative links between physical activity, cardio-respiratory fitness and coronary artery disease.

4. Results

Three randomly chosen patients designated $A, \ldots, C$ were used to validate our personalisation process, by demonstrating an accurate in silico description of patients' pre-PCI rest state. Since changes in arterial geometry accompanying re-vascularisation are recorded, PCI-modified MVR and stenotic Bernoulli resistance coefficients will accrue with our approach. Thus, data such as those presented here will eventually be available on the post-operative patient; see Section 5. Results fall into two categories: (i) data on our methodology as summarised in Tables 1 and 4 and (ii) the emergent patient representations.

4.1. Elastance Function Evaluation

Downsampling and ensemble averaging of patient LV and aortic pressure time-series datasets were performed straightforwardly. We proceed to protocol step 2 (a personal $E_{LV}(\tau)$)), where we encounter the sensitivity of $E_{LV}(\tau)$, to the co-registration of pressure and volume data. See Table 8 and Figure 7, which describe patient B, who required the largest shift. The PV loop illustrates the importance of co-registration, in Figure 4, where the PV loop shape progressively acquires accepted physiological features of iso-volumetric compression and relaxation phases. The corresponding sensitivity of $E_{LV}(\tau)$, in Figure 7, is quantified by the parameterisations in Table 8. We defer further comment to Section 5.

Figure 8 shows the personalised $E_{LV}(\tau)$ (left) and PV loop (right) for patients A, …, C. The elastance data take correctly co-registered LV pressure and volumes and show the initial (red) and optimised (black) fit. The PV loops in the right-hand column show the importance of co-registration case by case, with the PV loop obtained without co-registration shown in red and that with an appropriate co-registration in black.

Table 8. Sensitivity of LV double Hill elastance function parameterisation. For a single patient B, fitted double Hill elastance function shape parameters are declared for a range of co-registration shifts. We deem a shift of 5 in the last column to be the correct co-registration for this particular patient. The corresponding elastance function sequence is plotted in Figure 7.

	Shift	1	2	3	4	5
double Hill LV elastance parameter	n_1	1.91	1.83	1.78	1.68	1.52
	n_2	16.46	17.93	18.06	17.52	17.04
	τ_1	0.28	0.32	0.38	0.47	0.73
	τ_2	0.47	0.48	0.49	0.49	0.49
	$E_{LV,min}$	0.04	0.04	0.04	0.04	0.04
	$E_{LV,max}$	2.95	2.86	2.68	2.48	2.27

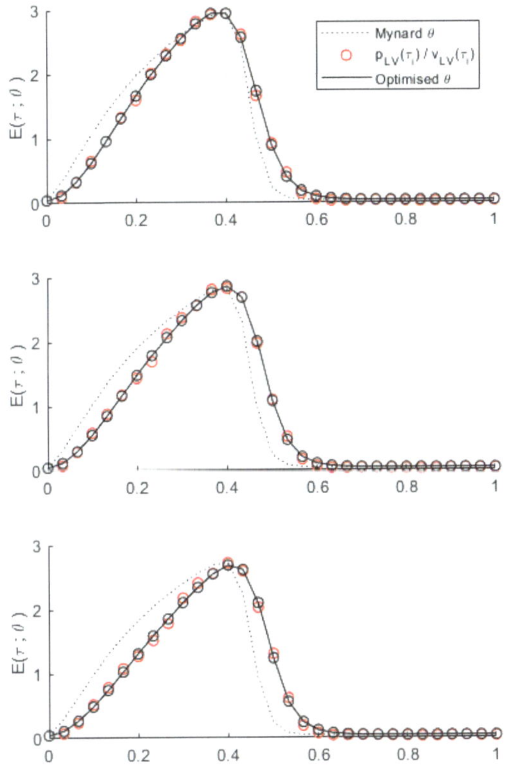

Figure 7. *Cont.*

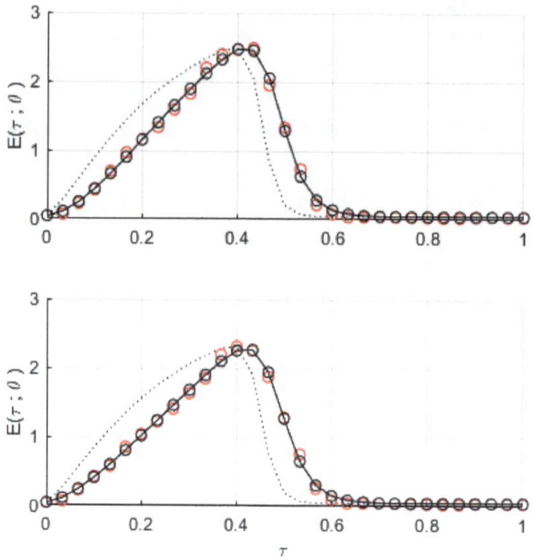

Figure 7. Sensitivity of a patient elastance function to co-registration shift. Data correspond to the double Hill elastance function parameters in Table 8. The shift increases top to bottom with the bottom panel corresponding to the expected shift of 5 phases, deduced using the co-registration declared in Table 9. The broken line shows our common initialisation using Mynard's population average [12] parameterisation. Open red circles show the raw data with the appropriate shift applied. The black circles and the line show the optimised fit on the shifted data from a gradient descent simplex method applied to Equation (7).

Table 9. Patient-specific LV double Hill elastance function parameters. These parameterisations result from fitting the product $v_{LV}(\tau_i) \times E(\tau_i; \underline{\theta})$ to the measured, downsampled pressure $p_{LV}(\tau_i)$, using the cost function defined in Equation (7) and a gradient descent method. The applied co-registration rigid shift, or relative delay, applied to the volume data is declared in the first row.

Patient ID		**A**	**B**	**C**
Co-Registration Rigid Shift		**1**	**5**	**2**
double Hill LV elastance parameter	τ_1	0.29	0.54	0.82
	τ_2	0.42	0.5	0.42
	n_1	1.58	1.56	1.26
	n_2	17.53	17.96	19.95
	$E_{LV,max}$	1.1	2.22	2.17
	$E_{LV,min}$	0.05	0.03	0.03

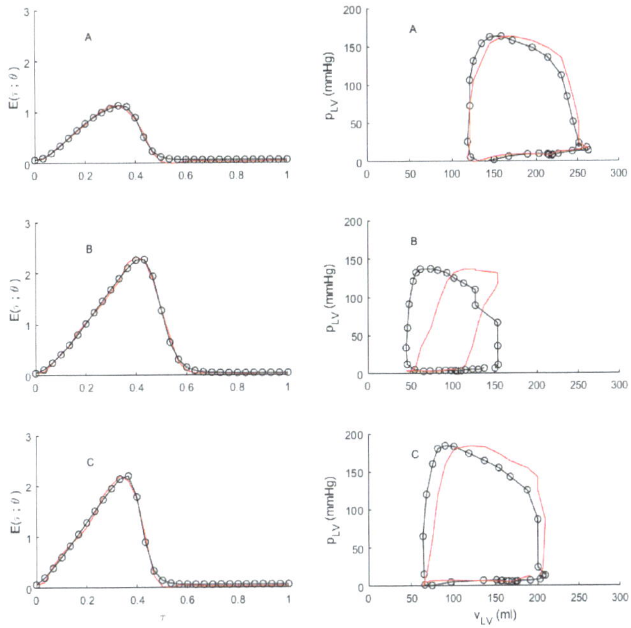

Figure 8. Patient-specific PV loops and elastance functions, for patients A, . . . , C. Left column. Patient-specific elastance functions. Red lines correspond to LV pressure and volume time-series data, which are correctly co-registered according to the criteria declared in Table 7 but which is not optimised. Black circles, with black lines to guide the eye, correspond to an optimised, double Hill elastance function parameterisation, obtained by minimising the residual declared in Equation (7), targeting the 30 downsampled LV pressure time samples, using the 30 LV volume time samples. Right column. The corresponding patient-specific PV loops. These data were processed as described in the text, with the personal co-registration shifts identified in Table 8. The red line shows the result which would be obtained without any co-registration. The black circles, with black lines to guide the eye, are the co-registered data.

4.2. *Tuning Protocol Evaluation*

Steps 0, 1, and 2 culminate in a personal $E_{LV}(\tau)$ which is input into the system 0D model. The latter then provides time-series outputs for LV and aortic pressures and LV volume at the end of step 3; see Figure 9. The cost function used to acquire these data, Equation (1), equally weights its five targets. Step 3 uses a genetic algorithm (GA) to localise the global cost function minimum, by adjusting the systemic windkessel. In general, the LV diastolic phase exhibits the largest discrepancy between model outputs and targets. This suggests parameters, such as mitral valve resistances and left atrial (LA) elastance, as secondary targets for step 4, which uses a less expensive gradient descent search, on the assumption that previous GA tuning will first have located the correct region. The initial values of the input parameters of the mitral valve Bernoulli resistance and $E_{LA}(\tau)$ are specified in Tables 1 and 5. Step 4 retains the targets of step 3 but adds the LV volume time-series. That step 4 favours the diastolic phase, particularly for cases A and C, is apparent from Figure 9, which shows outputs at the end of step 3 in the left column and outputs from the end of step 4 in the right column.

Having assigned systemic and left-heart energetic parameters, it remains to tune flow in patients' coronary circulations. The coronaries sequester a small fraction of cardiac output [18]. It is reasonable to suppose that this does not perturb already identified systemic and left heart parameters. The tuning of coronaries in step 5 relies upon CFD data to assign the MVR and stenotic Bernoulli resistance coefficients. The latter rely on data declared in Tables 10 and 11 and are derived as described in Section 4.3.

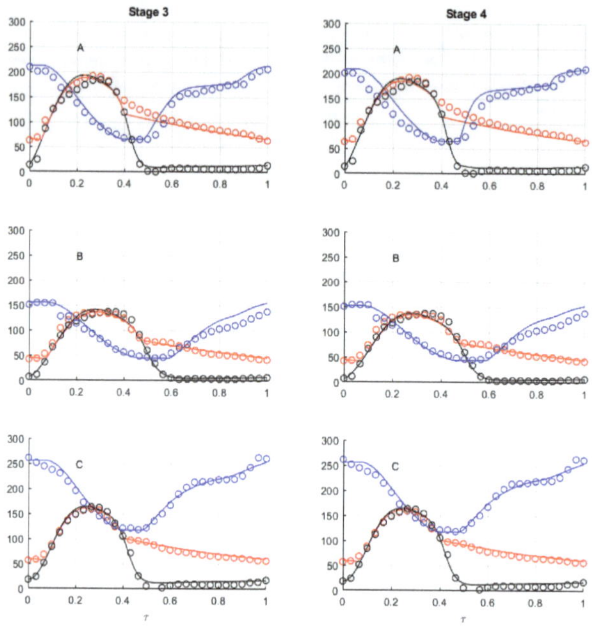

Figure 9. Systemic parameter tuning for patients A, B, C. Key 0D system model outputs (solid lines) are shown alongside the corresponding ensemble-averaged and downsampled data (open circles). Black line is LV pressure (mmHg), red line is aortic pressure (mmHg), and blue line is LV volume (mL). All data that were acquired after the appropriate personal elastance function assigned in step 2 had been ingested. Left column, step 3. The search used the GA guided by the cost function in Equation (1), with equal weights applied to patients' discrete LV pressure and volume extrema and the aortic time-series data. These data result from tuning the following subset of the 0D model input parameters $\{L, R_{prox}, C_{prox}, R_{dist}\}$ of the systemic circulation. Right column, step 4. The search used a gradient descent method based on the cost function, Equation (1), with equal weights applied to patients' discrete LV pressure, volume extrema, aortic time-series data and now LV volume data. These data result from tuning the following subset of the 0D model input parameters $\{a_{mit}, b_{mit}, E_{LA,min}, E_{LA,max}, \tau_{1,mit}, \tau_{2,mit}\}$.

4.3. CFD Data Evaluation

Patients' stenotic Bernoulli resistance coefficients were derived, as discussed above, from VIRTUheart^{TM} (University of Sheffield, Sheffield, UK). VirtuQ^{TM} (University of Sheffield, Sheffield, UK) was used to compute the corresponding MVRs. Both tools assume steady flow. Furthermore, VIRTUheart approximates pressure distal to the MVR as zero. The first assumption is raised within our *transient* 0D system model. Moreover, the 0D system model, with its four hear chambers, develops an unconstrained pressure distal to the MVR. The stenotic Bernoulli resistance coefficients ingested into the system 0D model (Table 10) are deemed to be unaffected by these differences. However, the MVR yielded by our CFD tool requires the correction in Equation (8)

$$MVR \rightarrow \left(\frac{\langle p_d(\tau) \rangle_\tau - \langle p_{RA}(\tau) \rangle_\tau}{\langle p_d(\tau) \rangle_\tau} \right) MVR. \tag{8}$$

Tables 10 and 11 declare the patient-specific CFD data inputted into the system 0D model. There, we encounter that sparsity and incompleteness which are typical of clinical data- none of our selected patients have three vessel disease and others have total coronary occlusion (TCO). The latte precludes all treatment but which can still be mapped to the coronary artery module in Figure 1 however, by setting large stenotic resistance coefficients. The latter are used to determine the pressure drop, δP_*, across a lesion from the flow, using Equation (9)

$$\delta P_* = R_{sten*,a} q_*^2 + R_{sten*,b} q_*. \tag{9}$$

Above, $*$ identifies the particular vessel.

Table 10. Coronary artery average pressure data. Time average pressure for lesion-proximal, aortic sinus, and lesion-distal locations. Study data are recorded for treated arteries only. Pressures were obtained using a pressure transducer by catheterisation during the PCI process.

Patient	Vessel					
	LAD		LCx		RCA	
	p_{prox}	p_{dist}	p_{prox}	p_{dist}	p_{prox}	p_{dist}
UNIT	mmHg	mmHg	mmHg	mmHg	mmHg	mmHg
A	88.60	74.50	~	~	~	~
B	103.71	99.17	103.95	101.69	~	~
C	121.74	115.73	~	~	127.23	125.89

Table 11. CFD-derived data for the patient coronary arteries. Stenotic Bernoulli resistance coefficients and MVRs for patients A,B,C.

Patient	Metric	Unit	Vessel		
			LAD	LCx	RCA
A	MVR	mmHg·s/mL	1.50	~	~
	$R_{sten,a}$	mmHg·s^2/mL^2	28.45	~	~
	$R_{sten,b}$	mmHg·s/mL	28.58	~	~
B	MVR	mmHg·s/mL	0.77	1.10	~
	$R_{sten,a}$	mmHg·s^2/mL^2	16.56	10.22	~
	$R_{sten,b}$	mmHg·s/mL	22.34	14.33	~
C	MVR	mmHg·s/mL	1.75	~	1.01
	$R_{sten,a}$	mmHg·s^2/mL^2	22.34	~	16.67
	$R_{sten,b}$	mmHg·s/mL	12.33	~	14.11

5. Discussion

We have characterised the pre-PCI patient state, but equivalent data exist for post-PCI patient states; thus, the approach described here may be used to examine post-intervention physiology. Our methodology brings within scope for the first time, a means to relate qualitative changes in physiology to changes in, e.g., exercise tolerance. The importance of a careful co-registration of our LV pressure and volume time-series data is underscored by the data in Figure 7 and the corresponding double Hill parameterisations, declared in Table 8. These data show LV contractility ($E_{LV,max}$) and systolic activation function parameters τ_1, n_1 to be the most sensitive elastance parameters to co-registration shift. These parameters respectively quantify LV contractility and the vigour of systolic onset and are very important indicators of cardiac health.

Step 2 of our protocol is seen to be robust and necessary. It represents a successful means of dealing with nonsimultaneous pressure and volume time-series data. From the data of Figure 8, it is apparent that a range of elastance functions and PV loops are successfully addressed. The change in the elastance function as correct co-registration is progressively applied demonstrates a significant degree of sensitivity in the elastance function. This underscores the importance of step 2 in our protocol. From Figure 8, it is also apparent that the PV loop of some cases is rather more sensitive than others to our rigid shift co-registration of pressure and volume time-series data.

From Figure 8, we see that the optimisation process enshrined in Equation (7) makes relatively little difference to the final shape of the elastance function, certainly compared with that produced by the volume and pressure time-series co-registrations.

6. Conclusions

We have verified the ability of appropriate protocols, PCI-acquired data and a suitable 0D system model, robustly interacting with 3D CFD models to recover a quantitative in silico representation of three individual patients' physiological and pre-PCI cardiovascular state. This depiction relies on parsing diverse clinical data, compiled during the elective PCI clinical pathway. Specifically, our hierarchical, multi-stage protocol is shown to identify prioritised input model parameters of cardiac energetics and coronary perfusion, within a 0D (compartmental) four-chamber system model with data-adapted coronary artery topology. Using this, we reached verifiable in silico representations of significant LV aortic and coronary artery volume, flow and pressure time-series. This process of input parameter identification is known in the clinical context as model personalisation. It is a key determinant of clinical utility, which urges a parsimonious model or (as here) the prioritisation of a subset of model input parameters, chosen with the supporting sensitivity and orthogonality analysis, provided in the appendices. The present study requires accurate representation of coronary flow states which are dominated by diastolic dynamics. The coupling of the coronary arteries' compliances (capacitances) to the appropriate chamber pressure (see Figure 1) apparently recovers diastolic dominance in coronary flow.

Time-series data play a defining role. To utilise them optimally, it was necessary to reconcile data collected at different stages of the clinical pathway with patients at different points in their physiological envelopes. This is deemed to be our largest source of error. The time-series data in question are non-invasive left ventricular volume, derived from CMR, and invasive LV and aortic pressures measured during PCI. Our approach involves downsampling pressure to accord with MRI phases and devising a suitable rigid phase shift, based upon LV valve timings, to overcome an identified co-registration error; results were assessed by the qualitative credibility of a physiological properties inferred from the LV PV loop. The commensurate LV elastance function is an input into the system 0D model personalisation process. It is shown to be sensitive to the co-registration of time-series data. Possibly, the most consequential, purely clinical outcome of this work is that we have been able to obtain credible results for patient PV loops and left-ventricular elastance functions by using a straightforward co-registration of non-simultaneous invasive pressure and non-invasive volume data when guided by rudimentary and therefore robust physiological principles.

Despite sources of error, it is apparent that there remains sufficient depth in the clinical pathway dataset to build appropriate digital twins, and we present as evidence data for three patients in the pre-PCI state. A route to the complementary, post-intervention description of the patient now lies open. One simply needs to apply the current workflow to extant data which are of equivalent form to that considered here. Therefore, using our methodology, one can expect to develop both pre- and post-PCI representation of patients from which to measure physiological changes attending PCI. Section 3.2.2 outlines a range of relevant activity measures which might be used meaningfully to contextualise this change in global ischaemic burden. Put another way, the act of correlating these model-derived statistics of the study data has the potential to reveal relationships between coronary physiology, cardiac energetics and physical activity, as well as providing clinical decision support. The model represents a means not only to establish a connection between global coronary perfusion and cardiac energetics, but to quantify it.

Author Contributions: A.A.B. and G.W., P.G. collected, curated and prepared all data. D.R.H., D.R. and F.v.d.V. devised the original model elements. K.C., A.L., I.H., A.J.N., M.M., F.v.d.V. and D.R.H. devised, implemented and adapted the model elements and designed the data-handling protocols. Clinicians P.D.M., R.C.G., P.G. and J.P.G. performed the PCI interventions from which all data are harvested. J.P.G. designed and led the study. All authors have read and agreed to the published version of the manuscript.

Funding: This publication is supported by the European Union's Horizon 2020 research and innovation programme under grant agreement Sano No. 857533 and by the Sano project carried out within the International Research Agendas Programme of the Foundation for Polish Science, co-financed by the European Union under the European Regional Development Fund.

Institutional Review Board Statement: The study was conducted in accordance with the Declaration of Helsinki, and approved by the UK's NHS Health Research Authority, sponsored by Sheffield Teaching Hospitals NHS Foundation Trust, (IRAS project ID 272069, REC reference 20/NS/0063, approved 23 March 2020).

Informed Consent Statement: Informed consent was obtained from all subjects involved in the study. Written informed consent has been obtained from the patients to publish this paper.

Data Availability Statement: Study data are not publicly available due to ethical restrictions, which apply to personal patient data. The data presented in this study are available on request from the corresponding author, Ian Halliday.

Conflicts of Interest: There are no conflict of interest.

Appendix A. System Model

Appendix A.1. System Model Dynamical Equations

It is possible to formulate the 0D model in terms of ordinary differential equations or differential algebraic equations [23]. Compartments are specified by their time-dependant dynamic pressure p (mmHg), inlet flow q (mL/s) and volume v (mL). The equations relating to the passive compartmental state variables all take the form

$$\frac{dv_i}{dt} = q_i - q_{i+1}, \quad \frac{dp_i}{dt} = \frac{1}{C_i}(q_i - q_{i+1}), \quad q_i = \frac{p_i - p_{i+1}}{R_i}. \tag{A1}$$

Above, the subscripts $(i-1), i, (i+1)$ respectively represent the proximal, present and distal system compartments, v_i(mL) denotes the circulating (stressed) volume and C_i (mL/mmHg) and R_i (mmHgs/mL) denote compartmental compliance and the Ohmic, or Bernoulli, resistance between compartments i, $(i+1)$. We turn to the active system compartments, described by so-called activation functions, and valve characteristics.

Appendix A.2. Sub-Models—Chamber Elastances and Valve Characteristic Functions

The double Hill elastance used here [12,14] is that in most widespread use within the clinical and physiological communities. The same essential model is used to express the biomechanics of all four heart chambers. Shi proposed an alternative model, with more compact support, which is arguably more intuitive [34], having similar properties. Neither takes into account stress stiffening in systole. Arts et al. [35] and Bovendeerd et al. [36] developed a model based on a more formal approach which accounts for stress-stiffening of the chamber. The double Hill function for it is expressed in terms of an activation or shape function, which is a product of monotonically increasing and decreasing expressions, as follows:

$$e(\tau; n_1, n_2, \tau_1, \tau_2) = \frac{x}{(1+x)(1+y)}, \quad x = \left(\frac{\tau}{\tau_1}\right)^{n_1}, \quad y = \left(\frac{\tau}{\tau_2}\right)^{n_2}.$$

A model elastance is then assigned

$$E(\tau; \underline{\theta}) = E_{max}\left(\frac{e(\tau; n_1, n_2, \tau_1, \tau_2)}{\max_\tau(e(\tau; n_1, n_2, \tau_1, \tau_2))}\right) + E_{min}.$$

Above, $\underline{\theta}$ denotes the set elastance parameters $\{E_{max}, E_{min}, n_1, n_2, \tau_1, \tau_2\}$. E_{max} and E_{min} specify chamber contractility and compliance, respectively. Parameter n_2 (the relaxation rate constant) is commonly set to large values to produce a sharp cut-off in $E(\tau;\underline{\theta})$ at the systolic to diastolic boundary. The double Hill parameter, τ_2, (the diastolic time constant) is the principal determinant of the systolic to diastolic ratio. Amplitude and shape parameters differ between all four chambers. Moreover, atrial elastance functions have adjustable phase, relative to ventricular elastances.

Valve characteristic functions express the relationship between flow through a valve, Q_{valve}, and pressure difference, δp, between separated compartments. Valves used in this work are diodes, having quadratic (Bernoulli) flow-pressure characteristics under forward bias with no *leakage* flow under reverse bias

$$q_*(\tau) = \begin{cases} a_{1,*}\delta p(\tau)^2 + a_{2,*}\delta p(\tau) & \delta p(\tau), > 0, \\ 0 & \delta p(\tau) \leq 0. \end{cases}$$

Above, the free subscript $*$ denotes one of our four valve flows.

Appendix B. Sensitivity and Orthogonality Analysis

Personalisation is a parameter identifiability analysis [5] applied to a chosen subset of input parameters which are often hypothesised to serve as clinical biomarkers and to a set of output metrics which correspond to available clinical data. The foundations of input parameter identifiability are sensitivity and input parameter orthogonality analysis. Using such, one can turn to formal methods (see, e.g., Li et al. [37]) which help to determine subsets of model input parameters which are optimal for personalisation [38]. Relative sensitivity and input parameter orthogonality analyses, derived from the operation of our system 0D model, are presented here in two forms. First, we consider an full analysis, based on the whole set of our system model's input parameters with an impractically large range of discrete outputs. Then we present the same analyses, with the list of input parameters restricted, to reflect those which we use in our patient personalisation process. For this subset, we also restrict the range of discrete model outputs to be representative of our study's patient data.

The relative sensitivity of the mth discrete output, X_m, on the nth discrete input, θ_n, is measured by a numerical approximation to the normalised partial derivative

$$S_{nm} = \frac{\theta_{n,0}}{X_m(\underline{\Theta}_0)} \left[\frac{\partial X_m}{\partial \theta_n}\right]_{\underline{\Theta}_0}, \quad 1 \leq m \leq M, \quad 1 \leq n \leq N. \tag{A2}$$

Above, the model base state, $\underline{\Theta}_0$, corresponds to our normal patient, where N (M) is the number of model inputs (outputs) and S_{nm} is a sensitivity matrix element. Noting the system input index, n, which identifies rows of $\mathbf{S}$, we define a relative sensitivity row vector, characteristic of the nth input

$$\underline{S}_n = (S_{n1}, S_{n2}, ..., S_{nM}), \quad 1 \leq n \leq N. \tag{A3}$$

From such sensitivity vectors, it is possible to measure the linear independence of the action across all outputs, of two chosen input parameters using a simple, intuitive orthogonality measure, which is a straightforward generalisation of the scalar product of two vectors

$$d_{nn'} = \sin\left(\cos^{-1}\left(\frac{\underline{S}_n \cdot \underline{S}_{n'}}{\sqrt{\underline{S}_n \cdot \underline{S}_n}\sqrt{\underline{S}_{n'} \cdot \underline{S}_{n'}}}\right)\right), \quad 1 \leq n, n' \leq N. \tag{A4}$$

Above, for two input parameters having identical action on all outputs, we find $d_{nn'} = 0$. Put another way, if $d_{nn'} = 0$, then incrementing θ_n and $\theta_{n'}$ will move all the system outputs in the same direction. Under such circumstances, while θ_n and $\theta_{n'}$ might each have considerable influence, they do not act upon them differentially. Optimal choices of input parameters for personalisation should clearly be chosen on the basis of influence *and* orthogonality. Clearly, $d_{nn'} = d_{n'n}$, and it is convenient to survey the orthogonality of the model input parameters by presenting the square, symmetric matrix $\boldsymbol{d}$, as a heatmap.

Figure A1 shows the full sensitivity matrix, based upon all inputs and a large range of discrete outputs. Of course, no practical study, however extensive, will have access to the range of outputs shown here. For the sake of completeness, however, we also present the corresponding input parameter orthogonality matrix and the statistics of that dataset in Figures A2 and A3, respectively. These figures are to be interpreted using Table A1.

Of much greater practical significance, certainly for the present study, is the partial sensitivity matrix, now based upon that subset of model inputs which are considered in this study alongside a set of outputs which are representative of the data we access. These data are presented in Figures A4–A6. They suggest that the systemic circulation parameters influence coronary flow. This, of course, is consistent with our assumptions: the recognised principal determinant of LV afterload is the systemic arterial windkessel parameterisation and a reduction in (say) systemic vascular resistance will increase flow to the system, and thereby, the fixed fraction which is sequestered to the coronaries increases. Conversely, changes in the parameterisation of the coronaries have little effect on the systemic metrics. The block diagonal structure in Figure A6 reinforces the conclusion that there is only weak coupling between the coronaries and the systemic and cardiac parameters of our model.

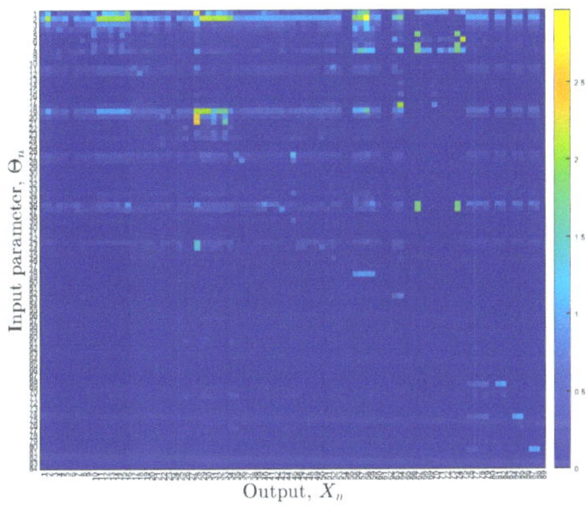

Figure A1. Full relative sensitivity matrix of the model, at base state. The sensitivity matrix is represented here as a heatmap. These data represent the fullest assessment of the studied four-chamber model and are presented for the sake of completeness. Input parameters and outputs are identified by their numerical subscripts, identified in Table A1. Many of the sensitivities are of restricted utility as they are based upon unobservable outputs. A restricted sensitivity analysis is presented in Figure A4 where the range of model inputs and outputs is restricted, in line with data available in the present study. Here, and in a majority of complex models, the heatmap is dominated by a few large sensitivities.

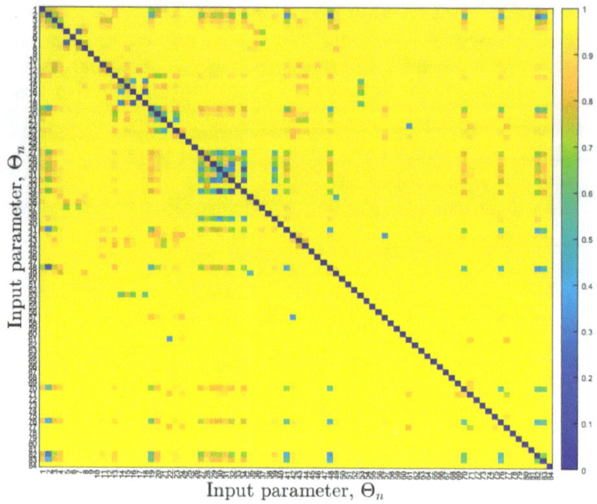

Figure A2. Full input parameter orthogonality analysis of the model sensitivity vectors. This heatmap is a representation of the matrix $\boldsymbol{d}$ (Equation (A4)). Input parameters are identified by their numerical subscripts identified in Table A1. Data are based upon sensitivity vectors, derived from the full model sensitivity matrix, in Figure A1. A more meaningful input parameter orthogonality analysis is presented in Figure A4, where the range of model inputs and outputs is restricted, in line with the data available in the present study.

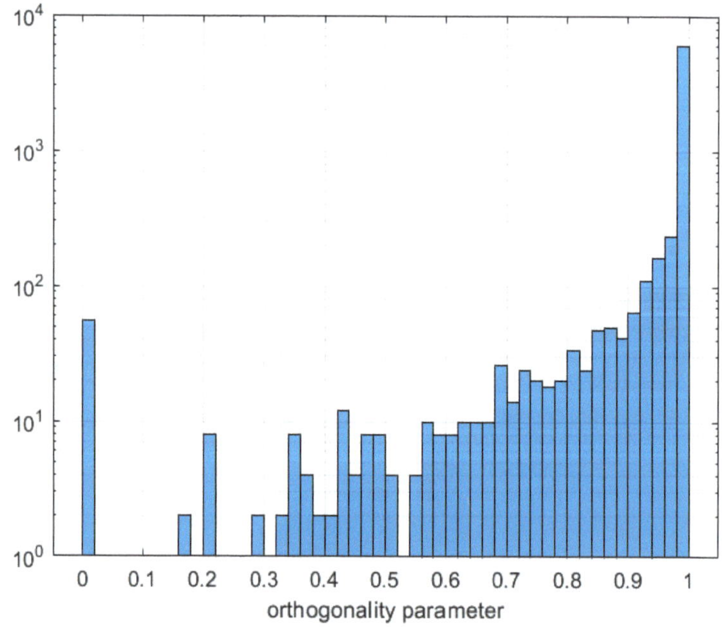

Figure A3. Statistics of the full input parameter orthogonality analysis. The statistical distribution of the matrix elements d_{nm} are shown in heatmap form in Figure A2. The distribution of these data tends to suggest that with a comprehensive set of outputs, a majority of model input parameters are identifiable.

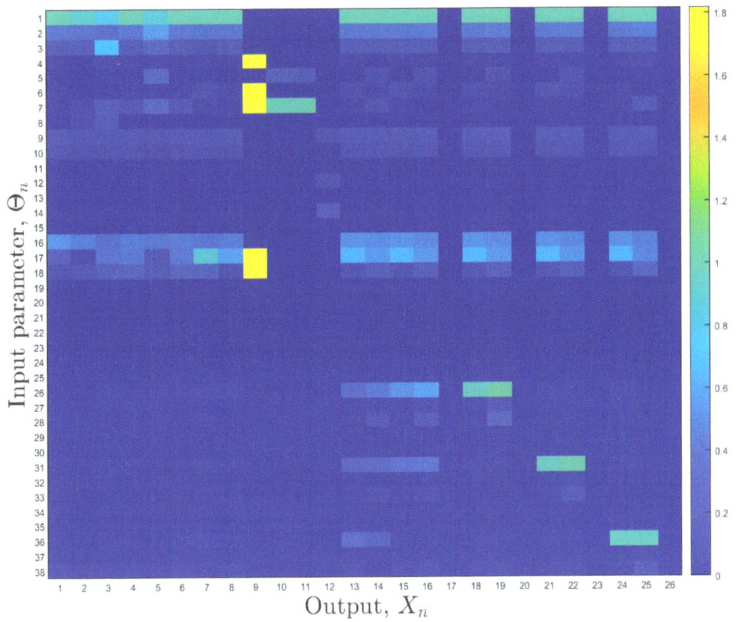

Figure A4. Restricted relative sensitivity matrix of the model, at base state. Relative to the data in Figure A1, here we show the sensitivity matrix based upon model input parameters which are emphasised within this study, combined with outputs which are acquired within the patient clinical pathway. Input parameters and outputs are identified by the numerical subscripts identified in Table A1.

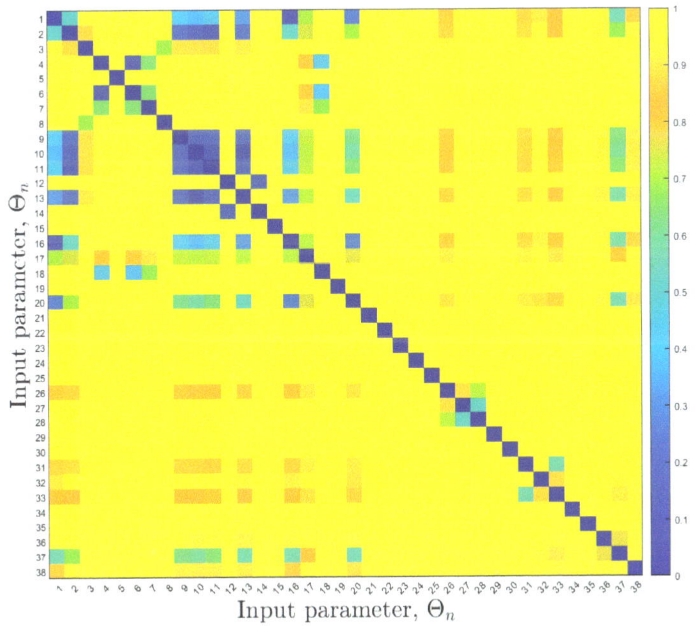

Figure A5. Restricted input parameter orthogonality analysis. These data represent the subset of model input parameters used in this study. The output parameters which are embedded in these data are those effectively declared in the data of figure A4. Input parameters are identified by the subscripts declared in Table A1.

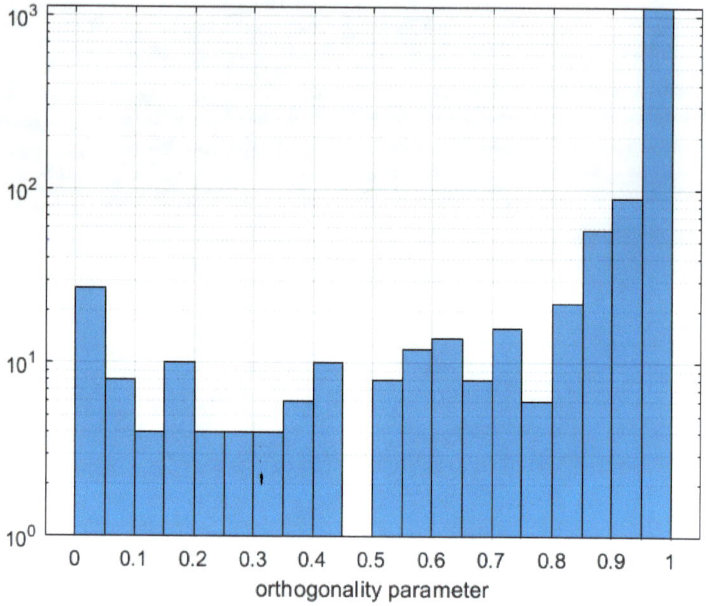

Figure A6. Statistics of the restricted input parameter orthogonality analysis. The statistical distribution of the matrix elements d_{nm} are shown in heatmap form in Figure A5.

Table A1. Key to full and restricted sensitivity and orthogonality analyses in Figures A1, A2, A4 and A5. First column, index, and second and third columns are restricted analysis, fourth and fifth columns are full analysis. Θ_n denotes an input parameter; X_n denotes an output parameter.

n	Θ_n	X_n	Θ_n	X_n
1	Heart Rate	Cardiac Output	Heart Rate	Cardiac Output
2	Mean circ. filling press.	LV End Diastolic Volume	Mean circ. filling press.	Useful Cardiac Power
3	ELVmin	LV End Systolic Volume	ELVmin	LV End Diastolic Volume
4	ELVmax	LV Maximum Pressure	ELVmax	LV End Systolic Volume
5	n1LV	LV Minimum Pressure	n1LV	LV Stroke Volume
6	n2LV	Systemic Artery Systolic Press.	n2LV	LV Ejection Fraction
7	Tau1fLV	Systemic Artery Diastolic Pres.	Tau1fLV	LV Maximum Pressure
8	Tau2fLV	Systemic Artery True MAP	Tau2fLV	LV Minimum Pressure
9	LV fractional time shift	AV Opens	LV volume offset	LV End Diastolic Pressure
10	ELAmin	AV Closes	LV fractional time shift	LV dp/dt (peak)
11	ELAmax	MV Opens	ELAmin	LV Stroke Work per Beat
12	n1LA	MV Closes	ELAmax	LV Stroke Power (Mean)
13	n2LA	Mean Coronary Flow	n1LA	LV Stroke Power (Peak)
14	Tau1fLA	Maximum Coronary Flow	n2LA	LV Myocardial Power (mean)
15	Tau2fLA	Left Main Flow (mean)	Tau1fLA	LV Myocardial Power (peak)
16	LA fractional time shift	Left Main Flow (peak)	Tau2fLA	LV Wasted Myocardial Power (Suga)
17	Systemic Resistance (Prox.)	Left Main 'FFR' (<Pd> / <Pa>)	LA volume offset	LA Maximum Volume
18	Systemic Resistance Distal	Cx Flow (mean)	LA fractional time shift	LA Minimum Volume
19	Systemic Art. Cap (Prox.)	Cx Flow (peak)	ERVmin	LA Maximum Pressure

Table A1. *Cont.*

n	Θ_n	X_n	Θ_n	X_n
20	Systemic Art. Inertance (Prox.)	Cx 'FFR' (<Pd> / <Pa>)	ERVmax	LA Minimum Pressure
21	Mitral Valve quadratic coeff.	LAD Flow (mean)	n1RV	RV End Diastolic Volume
22	Mitral Valve linear coeff.	LAD Flow (peak)	n2RV	RV End Systolic Volume
23	Left main, quadratic coeff.	LAD 'FFR' (<Pd>/<Pa>)	Tau1fRV	RV Stroke Volume
24	Left main, linear coeff.	RCA Flow (mean)	Tau2fRV	RV Ejection Fraction
25	Cx, quadratic coeff.	RCA Flow (peak)	RV volume offset	RV Maximum Pressure
26	Cx, linear coeff.	RCA 'FFR' (<Pd> / <Pa>)	RV fractional time shift	RV Minimum Pressure
27	Cx, RCR total res.	~	ERAmin	RV End Diastolic Pressure
28	Cx, RCR cap.	~	ERAmax	RV dp/dt (peak)
29	Cx Cap. back press. fraction	~	n1RA	RV Stroke Work per Beat
30	LAD, quadratic coeff.	~	n2RA	RV Stroke Power Expenditure
31	LAD, linear coeff.	~	Tau1fRA	RV Stroke Power (Peak)
32	LAD, RCR total resistance	~	Tau2fRA	RV Myocardial Power (mean)
33	LAD, RCR cap,	~	RA volume offset	RV Myocardial Power (peak)
34	LAD Cap. back press. fraction	~	RA fractional time shift	RV Wasted Myocardial Power (Suga)
35	RCA, quadratic coeff.	~	Systemic Resistance Prox.	RA End Diastolic Volume
36	RCA, linear coeff.	~	Systemic Resistance Dist.	RA End Systolic Volume
37	RCA, RCR total resistance	~	Systemic Arterial Prox. Cap.	RA Maximum Pressure
38	RCA, RCR capacitance	~	Systemic Arterial Prox. Unstressed Vol.	RA Minimum Pressure
39	~	~	Systemic Arterial Distal Cap.	Systemic Artery Systolic Pressure
40	~	~	Systemic Arterial Distal Unstressed Vol.	Systemic Artery Diastolic Pressure
41	~	~	Systemic Arterial Inertance (Prox.l)	Systemic Artery Nominal MAP
42	~	~	Pulmonary Resistance Prox.	Systemic Artery True MAP
43	~	~	Pulmonary Resistance Distal	Systemic Artery Pulse Pressure
44	~	~	Pulmonary Arterial Proximal Cap.	Systemic Vein Mean Pressure
45	~	~	Pul. Art. Prox. Unstressed Vol.	Systemic Vein Pulse Pressure
46	~	~	Pul. Art. Distal Cap.	Pulmonary Artery Systolic Pressure
47	~	~	Pulmonary Arterial Distal Unstressed Vol.	Pulmonary Artery Diastolic Pressure
48	~	~	Pulmonary Arterial Inertance (Prox.)	Pulmonary Artery Nominal MAP
49	~	~	Aortic Valve quadratic coeff.	Pulmonary Artery True MAP
50	~	~	Aortic Valve linear coeff.	Pulmonary Artery Pulse Pressure
51	~	~	Aortic Valve regurgitant quadratic coeff.	Pulmonary Vein Mean Pressure
52	~	~	Aortic Valve regurgitant linear coeff.	Pulmonary Vein Pulse Pressure
53	~	~	Mitral Valve quadratic coeff.	AV Forward Flow [ml/beat]
54	~	~	Mitral Valve linear coeff.	AV Regurgitation [ml/beat]
55	~	~	Mitral Valve regurgitant quadratic coeff.	AV Regurgitant Fraction [-]
56	~	~	Mitral Valve regurgitant linear coeff.	AV Maximum Pressure Drop

Table A1. *Cont.*

n	Θ_n	X_n	Θ_n	X_n
57	~	~	PulmonaryValve quadratic coeff.	AV Mean Pressure Drop
58	~	~	Pulmonary Valve linear coeff.	(Stroke Power Lost):(AV Resistance)
59	~	~	Pulmonary Valve regurgitant quad. coeff.	(Stroke Power Lost):(AV Resistance)
60	~	~	Pulmonary Valve regurgitant linear coeff.	MV Forward Flow [ml/beat]
61	~	~	Tricuspid Valve quadratic coeff.	MV Regurgitation [ml/beat]
62	~	~	Tricuspid Valve linear coeff.	MV Regurgitant Fraction [-]
63	~	~	Tricuspid Valve regurgitant quad. coeff.	MV Maximum Pressure Drop
64	~	~	Tricuspid Valve regurgitant linear coeff.	MV Mean Pressure Drop
65	~	~	Left main, quadratic coeff.	Stroke Power Lost to MV Regurgitation
66	~	~	Left main, linear coeff.	(Stroke Power Lost):(MV Regurgitation)
67	~	~	Cx. quad. coefficient	AV Opens
68	~	~	Cx. linear coefficient	AV Closes
69	~	~	Cx. RCR total resistance	MV Opens
70	~	~	Cx. RCR cap.	MV Closes
71	~	~	Cx. RCR fraction proximal resistance	Period of Systole
72	~	~	Cx. cap. back pressure fraction	Period of Diastole
73	~	~	LAD, quadratic coefficient	Proportion of Heart Period in Systole
74	~	~	LAD, linear coefficient	Period of Isovolumetric Contraction
75	~	~	LAD, RCR total resistance	Period of Isovolumetric Relaxation
76	~	~	LAD, RCR capacitance	Mean Coronary Flow
77	~	~	LAD, RCR fraction proximal resistance	Maximum Coronary Flow
78	~	~	LAD cap. back pressure fraction	Left Main Flow (mean)
79	~	~	RCA, quadratic coefficient	Left Main Flow (peak)
80	~	~	RCA, linear coefficient	Left Main 'FFR' (<Pd> / <Pa>)
81	~	~	RCA, RCR total resistance	Circumflex Flow (mean)
82	~	~	RCA, RCR capacitance	Circumflex Flow (peak)
83	~	~	RCA, RCR fraction proximal resistance	Circumflex 'FFR' (<Pd>/ <Pa>)
84	~	~	LAD cap, back pressure fraction	LAD Flow (mean)
85	~	~	~	LAD Flow (peak)
86	~	~	~	LAD 'FFR' (<Pd>/ <Pa>)
87	~	~	~	RCA Flow (mean)
88	~	~	~	RCA Flow (peak)
88	~	~	~	RCA 'FFR' (<Pd> / <Pa>)

References

1. Shi, Y.; Lawford, P.; Hose, D.R, Review of Zero-D and 1-D Models of Blood Flow in the Cardiovascular System. *BioMed. Eng.* **2011**, *10*, 33. [CrossRef] [PubMed]
2. Hose, D.R.; Lawford, P.V.; Huberts, W.; Hellevik, L.R.; Omholt, S.; van de Vosse, W. Cardiovascular models for personalised medicine: Where now and where next? *Med. Eng. Phys.* **2019**, *72*, 38. [CrossRef] [PubMed]
3. Shi, Y. *Lumped-Parameter Cardiovascular Model with Windkessel Afterload*; CellML: Auckland, New Zealand , 2007. Available online: http://models.cellml.org/exposure/ea64608ab564ee085bef7cde3ed1731e (accessed on 1 May 2023).
4. Korakianitis, T.; Shi, Y. A concentrated parameter model for the human cardiovascular system including heart valve dynamics and atrioventricular interaction. *Med. Eng. Phys.* **2006**, *28*, 613–628. [CrossRef]
5. Saltelli, A.; Ratto, M.; Andres, T.; Campolongo, F.; Cariboni, J.; Gatelli, D.; Saisana, M.; Tarantole, S. *Global Sensitivity Analysis: The Primer*; John Wiley and Sons: Hoboken, NJ, USA, 2008 .

6. Morris, P.D.; Ryan, D.; Morton, A.C.; Lycett, R.; Lawford, P.V.; Hose, D.R.; Gunn, J.P. Virtual Fractional Flow Reserve From Coronary Angiography: Modeling the Significance of Coronary Lesions: Results From the VIRTU-1 (VIRTUal Fractional Flow Reserve From Coronary Angiography). *JACC Cardiovasc. Interv.* **2013**, *6*, 149–157. [CrossRef]
7. Morris, P.D.; Gosling, R.; Zwierzak, I.; Evans, H.; Aubiniere-Robb, L.; Czechowicz, K.; Evans, P.C.; Hose, D.R.; Lawford, P.V.; Narracott, A.J. A novel method for measuring absolute coronary blood flow and microvascular resistance in patients with ischaemic heart disease. *Cardiovasc. Res.* **2021**, *117*, 1567–1577. [CrossRef] [PubMed]
8. Succi, S. *The Lattice Boltzmann Equation for Fluid Dynamics and Beyond*; Clarendon Press: Oxford, UK, 2001.
9. Groen, D.; Richardson, R.; Coy, R.; Schiller, U.D.; Chandrashekar, H.; Robertson, F.; Coveney, P.V. Validation of Patient-Specific Cerebral Blood Flow Simulation Using Transcranial Doppler Measurements. *Front. Physiol.* **2018**, *9*, 1–12. [CrossRef] [PubMed]
10. Westerhof, N.; Lankhaar, J.W.; Westerhof, B.E. The arterial windkessel. *Med. Biol. Eng. Comput.* **2009**, *47*, 131–141. [CrossRef]
11. Sagawa, K.R.; Lie, J. Schaefer, Translation of Otto Frank's paper "Die Grundform des arteriellen Pulses" Zeitschrift fur Biologie 37: 483–526 (1899) [Translation of Otto Frank's paper, The basic shape of the arterial pulse]. *J. Mol. Cell. Cardiol.* **1990**, *22*, 253–277. [CrossRef]
12. Mynard, J.P.; Davidson, M.R.; Penny, D.J.; Smolich, J.J. A simple, versatile valve model for use in lumped parameter and one-dimensional cardiovascular models. *Numer. Methods Biomed. Eng.***2012**, *28*, 626–641. [CrossRef]
13. Walley, K.R. Left ventricular function: Time-varying elastance and left ventricular aortic coupling. *Crit. Care* **2016**, 20, 626–641 . [CrossRef]
14. Seemann, F.; Arvidsson, P.; Nordlund, D.; Kopic, S.; Carlsson, M.; Arheden, H.; Heiberg, E.; Noninvasive quantification of pressure-volume loops from brachial pressure and cardiovascular magnetic resonance. *Circ. Cardiovasc.* **2019**, *12*, e008493. [CrossRef] [PubMed]
15. Available online: https://models.physiomeproject.org/welcome (accessed on 1 May 2023).
16. Czechowicz, K.; Archer, G.; Franz, J.; Nowakowski, P.; Narracott, A.J.; Hose, D.R. *Modelling the Aortic and Mitral valve replacement using personalised 0D and 3D Computational Models of Left Heart Circulation*; To be submitted to Fluids; Multi-Disciplinary Publishing Institute: Basel, Switzerland, 2023.
17. Hann, C.E.; Chase, J.G.; Desaive, T.; Froissart, C.; Revie, J.; Stevenson, D.; Lambermont, B.; Ghuysen, A.; Kohl, P.; Shaw, G.M.; Unique parameter identification for cardiac diagnosis in critical care using minimal data sets. *Comput. Methods Programs Biomed.* **2010**, *29*, 75. [CrossRef]
18. Ramanathan, T.; Skinner, H. Coronary blood flow. *Contin. Educ. Anaesth. Crit. Care Pain* **2005**, *5*, 61–64. [CrossRef]
19. Wiggers, C. *Modern Aspects of the Circulation in Health and Disease*; Lea and Febiger: Philadelphia, PA, USA, 1923.
20. Monge Garcia, M.I.; and Santos, A. Understanding ventriculo-arterial coupling. *Ann. Transl. Med.* **2020**, *8*, 795–805. [CrossRef] [PubMed]
21. Stergiopulos, N.; Westerhof, B.E.; Westerhof, N. Total arterial inertance as the fourth element of the windkessel model. *Am. J. Physiol.* **1999**, *276*, 81–88. [CrossRef] [PubMed]
22. Bastos, M.B.; Burkhoff, D.; Maly, J.; Daemen, J.; den Ui, C.A.L, Ameloot, K.; Lenzen, M.; Mahfoud, F.; Zijlstra, F.; Schreuder, J.J.; et al. Invasive left ventricle pressure-volume analysis: Overview and practical clinical implications. *Eur. Heart J.* **2020**, *21*, 1286–1297. [CrossRef]
23. Kunkel, P.; Mehrmann, V.L. *Differential-Algebraic Equations: Analysis and Numerical Solution*; European Mathematical Society: Zurich, Switzerland, 2006; ISBN 978-3-03719-017-3.
24. Jacobs, O.L.R. *Introduction to Control Theory*; Oxford University Press: Oxford, UK , 1984; ISBN 0-19-856148-2.
25. Murray, C.D.; The Physiological Principle of Minimum Work: I. The Vascular System and the Cost of Blood Volume. *Proc. Natl. Acad. Sci. USA* **1923**, *12*, 207–214. [CrossRef]
26. Huo, Y.; Kassab, G.S.; Intraspecific scaling laws of vascular trees. *J. R. Soc. Interface* **2012**, *9*, 190–200. [CrossRef]
27. Gosling, R.C.; Study, J.; Morris, P.D.; Fossan, F.E.; Hellevik, L.R.; Lawford, P.; Hose, D.R.; Gunn, J.P.; Effect of side branch flow upon physiological indices in coronary artery disease, *J. Biomech.* **2020**, *103*, 109698. [CrossRef]
28. Foerstermann, U.; Muenzel, T. Endothelial Nitric Oxide Synthase in Vascular Disease From Marvel to Menace. *Circulation* **2006**, *113*, 1708–1714. [CrossRef]
29. Linke, A.; Erbs, S.; Hambrecht, R. Effects of exercise training upon endothelial function in patients with cardiovascular disease. *Front Biosci.* **2008**, 13, 424–432. [CrossRef]
30. Laufs, U.; Wassmann, S.; Czech, T.; Muenzel, T.; Eisenhauer, M.; Boehm, M.; Nickenig, G. Physical inactivity increases oxidative stress, endothelial dysfunction, and atherosclerosis, Arteriosclerosis. *Thromb. Vasc. Biol.* **2005**, *25*, 809–814. [CrossRef]
31. Sattelmair, J.; Pertman, J.; Ding, E.L.; Kohl, H.W.; Haskell, W.; and Lee, I.M.; Dose response between physical activity and risk of coronary heart disease: A meta-analysis. *Circulation* **2011**, *124*, 789–795. [CrossRef]
32. Hung, R.K.; Al-Mallah, M.H.; McEvoy, J.W.; Whelton, S.P.; Blumenthal, R.S.; Nasir, K.; Schairer, J.R.; Brawner, C.; Alam, M.; Keteyian, S.J.; et al. Prognostic value of exercise capacity in patients with coronary artery disease: The FIT (Henry Ford ExercIse Testing) project. *Mayo Clin. Proc.* **2014**, *89*, 1644–1654. [CrossRef] [PubMed]
33. Hambrecht, R.; Walther, C.; Moebius-Winkler, S.; Gielen, S.; Linke, A.; Conradi, K.; Erbs, S.; Kluge, R.; Kendziorra, K.; Sabri, O.; et al. Percutaneous coronary angioplasty compared with exercise training in patients with stable coronary artery disease: A randomized trial. *Circulation* **2004**, *109*, 371–378. [CrossRef]

34. Shi, Y.; Korakianitis, T. Numerical simulation of cardiovascular dynamics with left heart failure and in-series pulsatile ventricular assist device. *Artif. Organs* **2006**, *30*, 929–948. [CrossRef]
35. Arts, T.; Bovendeerd, P.H.M.; Prinzen, F.W.; Reneman, Relation between left ventricular cavity pressure and volume and systolic fiber stress and strain in the wall. *Biophys. J.* **1991**, *59*, 93–102. [CrossRef]
36. Bovendeerd, P.H.; Borsje, P.; Arts, T.; van de Vosse, F.N. Dependence of intramyocardial pressure and coronary flow on ventricular loading and contractility, a model study. *Ann. Biomed. Eng.* **2006**, *34*, 133–1145. [CrossRef] [PubMed]
37. Li, R.; Henson, M.A.; Kurtz, M.J. Selection of model parameters for off-line parameter estimation. *IEEE Trans. Control. Syst. Technol.* **2004**, *12*, 402–412. [CrossRef]
38. Bjordalsbakke, N.I.; Sturdy, J.T.; Hose, D.R.; Hellevik, L.R. Parameter estimation for closed loop lumped parameter models in the systemic circulation using synthetic data. *Math. Biosci.* **2002**, *343*, 108731. [CrossRef] [PubMed]

fluids

MDPI

Article

Repaired Tetralogy of Fallot Pressure Assessment: Insights from 4D-Flow Pressure Mapping

Safia Ihsan Ali [1,2], David Patton [3,4], Kimberley A. Myers [3] and Julio Garcia [2,4,5,6,*]

1 Department of Biomedical Engineering, University of Calgary, Calgary, AB T2N 1N4, Canada; safia.ihsanali@ucalgary.ca
2 Stephenson Cardiac Imaging Centre, University of Calgary, Calgary, AB T2N 1N4, Canada
3 Department of Pediatrics, Cumming School of Medicine, University of Calgary, Calgary, AB T2N 1N4, Canada; david.patton@albertahealthservices.ca (D.P.); kim.myers@albertahealthservices.ca (K.A.M.)
4 Department of Cardiac Sciences, Cumming School of Medicine, University of Calgary, Calgary, AB T2N 1N4, Canada
5 Libin Cardiovascular Institute, University of Calgary, Calgary, AB T2N 1N4, Canada
6 Department of Radiology, Cumming School of Medicine, University of Calgary, Calgary, AB T2N 1N4, Canada
* Correspondence: julio.garciaflores@ucalgary.ca

Abstract: Tetralogy of Fallot (TOF) is the most prevalent cyanotic congenital heart defect (CHD) that alters normal blood flow through the heart and accounts for 10% of all CHD. Pulmonary stenosis and regurgitation are common in adults who have undergone TOF repair (rTOF) and can impact the load on the right ventricle, blood flow pressure, and pulmonary hemodynamics. Pressure mapping, obtained through 4D-flow magnetic resonance imaging (4D-flow MRI), has been applied to identify abnormal heart hemodynamics in CHD. Hence, the aim of this research was to compare pressure drop and relative pressures between patients with repaired TOF (rTOF) and healthy volunteers. An in vitro validation was performed, followed by an in vivo validation. We hypothesized that pressure drop is a more stable pressure mapping method than relative pressures to detect altered hemodynamics. A total of 36 subjects, 18 rTOF patients and 18 controls underwent cardiac MRI scans and 4D-flow MRI. Pressure drops and relative pressures in the MPA were higher in rTOF patients compared to the controls ($p < 0.05$). Following the in vitro validation, pressure drops proved to be a more stable pressure mapping method than relative pressures, as the flow loses its laminarity and becomes more turbulent. In conclusion, this study demonstrated that flow hemodynamics in rTOF can exhibit altered pressure maps. Pressure mapping can help provide further insight into rTOF patients' hemodynamics to improve patient care and clinical decisions.

Keywords: repaired tetralogy of Fallot; magnetic resonance imaging; 4D-flow MRI; heart hemodynamics; pressure mapping

Citation: Ihsan Ali, S.; Patton, D.; Myers, K.A.; Garcia, J. Repaired Tetralogy of Fallot Pressure Assessment: Insights from 4D-Flow Pressure Mapping. *Fluids* **2023**, *8*, 196. https://doi.org/10.3390/fluids8070196

Academic Editors: Huidan (Whitney) Yu and D. Andrew S. Rees

Received: 30 April 2023
Revised: 26 June 2023
Accepted: 27 June 2023
Published: 29 June 2023

1. Introduction

The cardiovascular system keeps blood flowing efficiently to achieve laminar flow through the vessels and the chambers of the heart. In numerous cases of congenital heart disease, particularly in individuals with surgically repaired tetralogy of Fallot (rTOF), there is evidence of nonlaminar flow in the right ventricle (RV), which includes transitional and turbulent flow [1,2]. TOF is the most prevalent cyanotic congenital heart defect (CHD) that alters normal blood flow through the heart and accounts for 10% of all CHD. TOF is characterized by a ventricular septal defect (VSD), aortic override, RVOT obstruction, and right ventricular hypertrophy. Surgical repair, which includes VSD closure and RVOT reconstruction, is typically performed in infancy [3,4]. The primary repair encompasses the closure of the VSD, removal of the obstructive infundibular muscle, and alleviation of the pulmonary stenosis. Patch reconstructions are typically used to repair a narrowed pulmonary artery [5]. Despite having improved clinical outcomes, patients with rTOF require clinical and imaging follow-up to evaluate for post-surgical problems. The most common complications in patients with rTOF are pulmonary regurgitation (PR) and/or residual or reoccurring pulmonary stenosis (PS). PR, when severe, causes right ventricular dilatation, dysfunction, and arrhythmia. To prevent irreversible right ventricular failure, pulmonary valve replacement is then necessary [6].

Echocardiography and cardiac magnetic resonance (CMR) are the imaging modalities of choice for the follow-up of TOF patients. Both modalities can assess a wide range of anatomical and functional parameters, but both also have several limitations. Echocardiography can provide information on anatomy and physiology, while color Doppler can perform qualitative flow assessment. The quality of the image is greatly influenced by the acoustic window as well as the proficiency of the operator. CMR has become common in managing many congenital conditions, owing to its ability to visualize structures not well seen by echocardiography without ionizing radiation. It can provide a non-invasive three-dimensional cardiovascular anatomy, volumes, and function evaluation. In patients with rTOF, CMR has emerged as the imaging method of choice, playing an essential role in postoperative follow-up and evaluation [6]. Flow can be analyzed using two-dimensional phase-contrast (2D PC) MRI, providing flow volumes and velocity measurements perpendicular to a single plane placed in the vessel of interest. Each plane of interest must be individually planned to obtain flow measurements, and separate breath-holding scans must be performed. Furthermore, 2D PC MRI encounters challenges in accurately quantifying flow due to the heart's motion in relation to the imaging plane. It may provide an incomplete evaluation of blood flow due to technical limitations, particularly in cases of complex CHDs.

Recently, four-dimensional flow MRI (4D-flow MRI) has emerged as a promising and non-invasive imaging technique that can provide a comprehensive quantitative evaluation of flow in an entire volume within the chest in a single short imaging session [7]. That is, 4D-flow MRI illustrates 3D blood flow patterns and hemodynamics by utilizing velocity encoding (VENC) in three spatial directions. This imaging technique enhances our comprehension of blood flow properties in both normal and pathological conditions, offering comprehensive 3D visualization of anatomy and velocity. Consequently, it facilitates precise measurements of vessel lengths and provides valuable hemodynamic data [3]. This technique evaluates intricate flow patterns, such as helical or vortical flow, and measures advanced fluid dynamic parameters, including pressure difference maps, turbulent kinetic energy, and viscous energy loss [8]. Furthermore, 4D-flow MRI has been demonstrated to be effective for the qualitative and quantitative evaluation of pulmonary hemodynamics in TOF patients [9].

Blood pressure measurements play a crucial role in diagnosing cardiovascular disease and detecting irregular blood flow in large vessels [10]. Pressure gradients serve as significant clinical indicators for various cardiovascular conditions. In clinical practice, catheter measurements are considered the gold standard for assessing in vivo pressure gradients. Although this method is reliable and safe, it is an invasive procedure that carries potential complications and involves exposure to radiation for catheter guidance. An alternative approach for estimating pressure gradients is to utilize the simplified Bernoulli equation, which relies on the measurement of maximum velocity (V_{max}) obtained from standard clinical Doppler ultrasound (US). However, V_{max} measurements using ultrasound are prone to error due to limited acoustic windows and technical difficulties in aligning Doppler interrogation with peak velocities. The exponentiation of velocities in the Bernoulli equation makes pressure difference estimation highly sensitive to errors in V_{max} measurements. Another non-invasive and user-independent method for deriving relative pressure gradients is using time-resolved (CINE), three-directionally encoded phase contrast (PC) MRI to measure the time-resolved velocity field. By solving the Navier–Stokes (NS) equation, assuming blood is an incompressible, laminar Newtonian fluid, pressure gradients can be calculated accurately [11].

Furthermore, the NS equation can be transformed by taking the divergence into a pressure Poisson equation (PPE) [12]. Based on derived pressure gradients, pressure differences or pressure drops and relative pressure fields can be obtained by solving the PPE. For example, Bock et al. [11] applied the iterative approach to solving the PPE to obtain pressure differences or drops. On the other hand, Ebbers and Farnebäck [12] suggested using a multigrid-based PPE solver to compute relative pressures. However, no study to date has compared the iterative approach by Bock et al. [11] with the multigrid-based solver approach introduced by Ebbers and Farnebäck [12].

In addition to different pressure mapping methods, several ways exist to report these pressures. Numerous studies use 2D analysis planes positioned in the vessel of interest [13,14]. Some studies report the summation of pressures over the total volume [1,15]. In comparison, others report pressure values along a centerline that mimics a virtual catheter [16]. Many authors have used volumes and centerlines in literature, but no study has reported pressures using both. Therefore, this study aimed to compare pressure mapping methods and investigate whether one is more stable. We hypothesized that pressure drop is a more stable pressure mapping method than relative pressures to detect altered hemodynamics.

2. Materials and Methods

In vitro validation using a stenosis phantom was performed, followed by an in vivo validation between (i) healthy subjects and patients with rTOF who reported using volumes and centerlines and (ii) volume and centerline data of the controls and patients. The study diagram is presented in Figure 1.

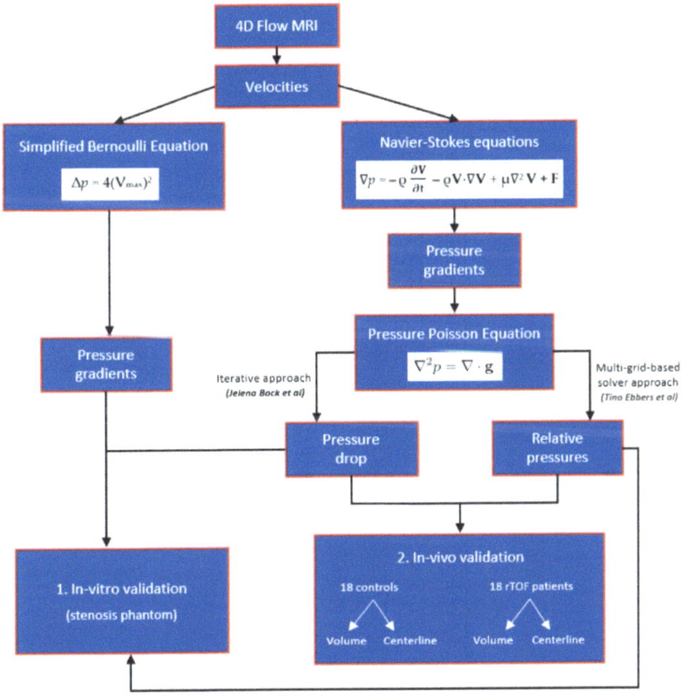

Figure 1. A schematic representation of the study. This figure illustrates the methodology employed in the study to estimate pressure gradients and derive pressure drops and relative pressure fields. Velocity measurements were used to estimate pressure gradients through the simplified Bernoulli and Navier–Stokes equations. The Navier–Stokes equation was further transformed into a pressure

Poisson equation (PPE) by taking the divergence. Pressure drops were obtained using an iterative approach to solving the PPE, as suggested by Bock et al. [11]. Ebbers and Farnebäck [12] proposed a multigrid-based PPE solver for computing relative pressures. The study involved in vitro validation followed by in vivo validation to compare these different pressure mapping methods. In vitro validation was initially performed using a stenosis phantom. Subsequently, in vivo validation was conducted, which involved comparing (i) healthy subjects and patients with repaired tetralogy of Fallot (rTOF) using volumes and centerlines and (ii) volume and centerline data of the controls and patients.

2.1. Pressure Gradients, Pressure Drop, and Relative Pressures Estimation

Pressure gradients $\nabla p = (\frac{\partial p}{\partial x}, \frac{\partial p}{\partial y}, \frac{\partial p}{\partial z})$ assuming a viscous, incompressible fluid was calculated from the three-directional velocity vector fields $\mathbf{V} = (u, v, w)$ using the NS equations:

$$\nabla p = -\rho\frac{\partial \mathbf{V}}{\partial t} - \rho\mathbf{V}\cdot\nabla\mathbf{V} + \mu\nabla^2\mathbf{V} + \mathbf{F} \quad (1)$$

where p is the pressure, ρ is the fluid density, $\mathbf{V}$ is the measured three-directional velocity, μ is the dynamic fluid viscosity, and $\mathbf{F}$ represents the body forces. The terms on the right-hand side from left to right are transient inertia, convective inertia, viscous resistance, and body forces, respectively. As phantom and human subjects were placed in a horizontal position in the scanner, the body forces were neglected for all calculations.

By taking the divergence, the NS equation can be transformed into the PPE [12]:

$$\nabla^2 p = \nabla\cdot\mathbf{g} \quad (2)$$

where $\mathbf{g}$ denotes the pressure gradient field obtained from the right-hand side of Equation (1) while dropping the body force term. Pressure drop was calculated by applying the approach used by Bock et al. [11], where the pressure is initialized by region-growing integration of the pressure gradient with a user-defined reference point ($\Delta p = 0$), and all calculations were performed independently for each time frame. On the contrary, relative pressures were computed using the approach introduced by Ebbers and Farnebäck [12]. This method did not require a user-defined reference point and utilized a multigrid-based PPE solver that works directly on the structure-defined computational domain. For pressure gradient estimation, blood properties were set to 3.2×10^{-3} Pa s and 1060 kg/m^3 for viscosity and density, respectively. The pressure gradient Δp across the stenosis in the in vitro model was determined by applying the simplified Bernoulli equation, as used in clinical routine [17]:

$$\Delta p = 4(V_{max})^2 \quad (3)$$

V_{max} represents the peak velocity at the maximum narrowing of the stenosis of the phantom.

2.2. In Vitro Model

A simple stenosis phantom ($\varnothing_1$ 20.4 $\pm$ 0.5 mm, stenosis $\varnothing_2$ 10 $\pm$ 0.5 mm, EOA = 0.78 cm^2) with a contraction coefficient = 1 (i.e., EOA/AVA, where the anatomic valve area (AVA) is the cross-sectional area of the stenosis phantom) by the potential flow theory was filled with glycerol and water, and connected to a pump under constant flow conditions (Chemflo Unit, MP Pumps, Inc., Fraser, MI, USA). To increase the signal-to-noise ratio (SNR), the fluid was doped with a contrast agent (Magnevist®, Bayer Schering Pharma AG, Leverkusen, Germany) at a concentration of 1.08 mmol/L. Then, 4D-flow measurements were performed on a 1.5 T System (Aera, Siemens AG, Erlangen, Germany). The measurement parameters were as follows: VENC = 1–4 m/s along all three velocity encoding directions, spatial resolution $1.0 \times 1.0 \times 1.0$ mm^3, field of view (FOV) = 350×350 mm^2, flip angle = 15°, TE/TR = 2.7–3.1/5.6–5.9 ms, and scan time = 10 min.

2.3. *In Vivo Model*

2.3.1. Study Population

A total of 36 subjects, including 18 rTOF patients (age: 32 ± 10, 6 females) and 18 controls (age: 37 ± 14, 7 females), were recruited retrospectively. The reported values represent the group mean ± standard deviation. All subjects were registered with the Calgary Cardiovascular Imaging Registry (CIROC). The study was authorized by the board of the University of Calgary Research Ethics, and all subjects gave informed consent. The research activities were conducted in accordance with the Helsinki Declaration. Commercial software (cardioDITM, Cohesic Inc., Calgary, AB, Canada) was used to manage the study's routine capture of patient informed consent, health questionnaires, and standardized collection of MRI-related variables. Patients were required to meet the criteria of being ≥18 years old and having a documented history of rTOF at the time of the examination to qualify for participation. However, patients with implantable devices, severe renal impairment (eGFR 30 mL/min/1.73 m^2), or other known contraindications for MRI were excluded [18]. In the control group, subjects were 18 years or older without any history of cardiovascular disease, diabetes, or uncontrolled hypertension. The controls who were unable to complete the MRI scan were also excluded. Before scanning, demographic data such as age, gender, height, weight, and heart rate were collected. The Mosteller formula converted volume and mass measurements to body surface area measurements.

2.3.2. Cardiac Magnetic Resonance Imaging Protocol

Cardiac imaging was examined utilizing 3T MRI scanners (Skyra and Prisma, Siemens, Erlangen, Germany) [18]. A standardized clinical imaging protocol was performed on all subjects, which involved retrospective electrocardiographic gating and time-resolved balanced steady-state free precession (SSFP) cine imaging of the LV in four-chamber, three-chamber, two-chamber, and short-axis views at end-expiration. In addition, a contrast-enhanced 3D magnetic resonance angiogram (MRA) of the cardiovascular structures was obtained using a gadolinium contrast volume of 0.2 mmol/kg (Gadovist®, Bayer Inc., Mississauga, ON, Canada). For a period of 5–10 min following contrast administration, time-resolved three-dimensional phase-contrast MRI with three-directional velocity encoding and retrospective ECG-gating (known as 4D-flow, Siemens WIP 785A) was conducted to determine the in vivo blood flow velocities throughout the entire heart [1,15]. This whole-heart protocol has been previously described in our reports [1,19,20]. In brief, 4D-flow data was collected during free breathing using the navigator gating of diaphragmatic motion and the following sequence parameters were used: bandwidth = 455–495 Hz/Pixel; pulse repetition time = 4.53–5.07 ms; echo time = 2.01–2.35 ms; flip angle = 15 degrees, spatial resolution = 2.0–3.5 × 2.0–3.5 × 2.5–3.5 mm; temporal resolution = 25–35 ms; and VENC = 150–250 cm/s. The total acquisition time ranged from 5 to 10 min, depending on the heart rate and respiratory navigator efficiency. The number of phases varied from 25 to 30 depending on the clinical scan operator.

2.3.3. Cardiac Imaging and 4D-Flow Analysis

A blinded reader assessed standard cardiac images on the same day of acquisition using dedicated software, cvi42 version 5.11.5 (Circle Cardiovascular Imaging Inc., Calgary, AB, Canada). This was carried out to determine left and right end-diastolic volume (LVEDV; RVEDV), LV and RV end-systolic volume (LVSEV; RVESV), as well as LV and RV ejection fraction (LVEF; RVEF) [1,21].

Following the acquisition, all 4D-flow MRI data were pre-processed with the "Velomap tool", a MATLAB tool developed by Bock et al. in 2007 that is widely used in the flow MRI community [22]. The pre-processing tool was used to perform the following tasks: corrections for Maxwell terms, eddy currents, and aliasing (see Figure 3A). In addition, a 3D phase-contrast (PC) angiogram (PC MRA) was created after pre-processing (see Figure 3B). Using an in-house MATLAB-based tool called "4D-Flow Analysis Tool" [20], the angiogram was used to segment (see Figure 3C) the aorta (Ao), pulmonary artery (PA), left ventricle (LV), and right ventricle (RV). A pressure mapping tool, "4D Flow Pressure Mapping Tool", created in MATLAB 2019a (MathWorks, Natick, MA, USA), was used to compute the pressure drop and relative pressures for each segmented vessel. Using the approach proposed by Elbaz et al. [16], 4D virtual catheter mathematical models for probing pressures were constructed as tubes with an automatically derived radius along the centerlines of the segmented vessels.

2.4. In Vitro Data Analysis

The in vitro data analysis was conducted using MATLAB 2019a (Mathworks, Natick, MA, USA) [23]. As shown in Figure 2, first, MRI velocities were used to calculate pressure gradients using the simplified Bernoulli equation. The pressure drop was then computed using the velocities by placing the reference point at the outflow tract. Then, the relative pressures were calculated from the velocities [11,12,24,25]. This procedure was followed for all VENCs (1–4 m/s). Finally, the influence of noise and filtering on both pressure mapping methods was investigated. The original pressure data were subjected to five levels of white Gaussian noise (SNR 5, 15, 30, 45, and 135) for each VENC, followed by a Gaussian filter. Additionally, the Reynolds number, $Re = \rho uD/\mu$, at the inlet and stenosis regions for each VENC (1–4 m/s), was calculated, where ρ is the density, u is the average velocity, D is the diameter, and μ is the dynamic viscosity [26].

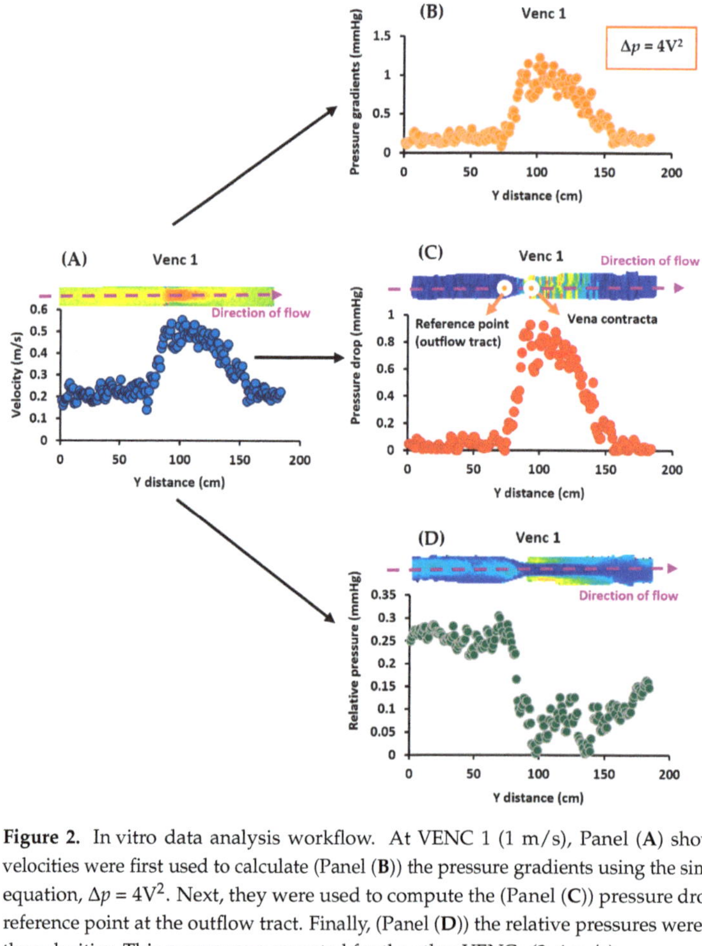

Figure 2. In vitro data analysis workflow. At VENC 1 (1 m/s), Panel (**A**) shows that the MRI velocities were first used to calculate (Panel (**B**)) the pressure gradients using the simplified Bernoulli equation, $\Delta p = 4V^2$. Next, they were used to compute the (Panel (**C**)) pressure drop by placing the reference point at the outflow tract. Finally, (Panel (**D**)) the relative pressures were calculated using the velocities. This process was repeated for the other VENCs (2–4 m/s).

2.5. In Vivo Data Analysis

The analysis was performed in MATLAB 2019a (Mathworks, Natick, MA, USA). As illustrated in Figure 3D, the reference points for the Ao, PA, LV, and RV were set at the left ventricular outflow tract (LVOT), RVOT, mitral valve, and tricuspid valve, respectively, for the pressure drop calculation. The pressure mapping tool divided the Ao into three sections: ascending aorta (AAo), aortic arch, and descending aorta (DAo). MPA, RPA, and LPA were the three divided sections of the PA. The entirety of the RV and LV was used for the evaluation. A variety of hemodynamic parameters were computed for each section at peak systole, including mean pressure drop (PDmean), mean relative pressure (RPmean), maximum pressure drop (PDmax), maximum relative pressure (RPmax), and standard deviation pressure drop (PDstd) and standard deviation relative pressure (RPstd). PDmean and RPmean were defined as the mean pressure drop and mean relative pressures in each region, respectively. Likewise, PDmax and RPmax were defined as the maximum pressure drop and maximum relative pressures in any voxel in each region, respectively. Finally, PDstd and RPstd were defined as the standard deviation pressure drop and standard deviation relative pressures in each region, respectively.

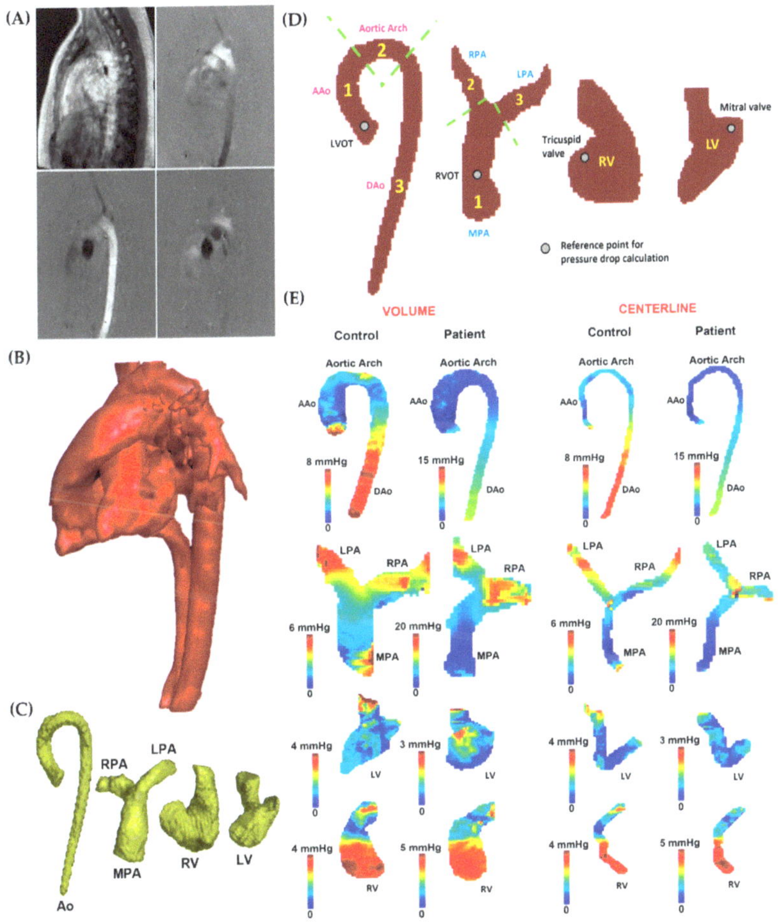

Figure 3. In vivo data analysis workflow. Panel (**A**): Initially, the 4D-flow velocity data in each direction (Vx, Vy, and Vz) were subjected to correction for eddy currents, noise, and aliasing. Subsequently, a phase contrast-magnetic resonance angiogram (PC-MRA) was generated (Panel (**B**)) to isolate the

specific vessels (Panel (**C**)), including the aorta (Ao), pulmonary artery (PA), left ventricle (LV), and right ventricle (RV). Panel (**D**): After segmentation, the aorta was divided into ascending aorta (AAo), aortic arch, and descending aorta (DAo), while the left ventricular outflow tract (LVOT) was used as the reference point for the pressure drop calculation. The PA was divided into the main pulmonary artery (MPA), right pulmonary artery (RPA), and left pulmonary artery (LPA); the right ventricular outflow tract (RVOT) was used as the reference point for the pressure drop calculation. The entirety of the LV and RV was used for the analysis; the mitral and tricuspid valves were used as reference points for the pressure drop calculation. Panel (**E**) illustrates the pressure drop measurements in the AAo, aortic arch, DAo, MPA, LPA, RPA, LV, and RV of the control and a patient, reported within vessel volumes and centerlines.

2.6. Statistical Analysis

The statistical analysis was conducted using SPSS 25 (SPSS, Chicago, IL, USA). To assess normality, the Shapiro–Wilk and Kolmogorov–Smirnov tests were utilized. However, since the data did not follow a normal distribution, a Mann–Whitney U test was performed for both study demographics, as well as for the pressure drop and relative pressure comparison between the two groups. The results are presented as the group mean ± standard deviation with a p-value < 0.05 to determine statistical significance. Furthermore, Spearman's correlation was used to analyze the relationship of pressure drop and relative pressures with the left ventricular ejection fraction (LVEF), right ventricular ejection fraction (RVEF), indexed left ventricular end-diastolic volume (LVEDVi), indexed left ventricular systolic volume (LVESVi), indexed right ventricular end-diastolic volume (RVEDVi), and indexed right ventricular end-systolic volume (RVESVi). The body surface area was utilized for indexation. A p-value < 0.01 was considered statistically significant. Furthermore, scatter plots were employed for each VENC to investigate the impact of the five levels of Gaussian noise and filters on both pressure mapping methods. Finally, Bland–Altman plots were used for each VENC to assess the differences between original pressure data and data with all five Gaussian noise levels, as well as between original pressure data and data with the Gaussian filter.

3. Results

3.1. In Vitro Data

Pressure gradients calculated using the simplified Bernoulli equation followed the same profiles as MRI velocities, as shown in Figure 4. Furthermore, the pressure drop profiles followed the trends of the velocities and pressure gradients, with the maximum pressure drop occurring at the location of the vena contracta. Pressure drop and pressure gradients produced similar results for each VENC. Relative pressures, on the other hand, did not follow the profiles of velocities, pressure drop, or pressure gradients; instead, different results were observed for each VENC. Supplementary Materials presents the scatter and Bland–Altman plots, illustrating the impact of Gaussian noise and filters on pressure drop and relative pressures. Finally, Re for each VENC (1–4 m/s) at the inlet and stenosis regions were 1148 and 1225, 2500 and 2683, 3243 and 3776, and 3784 and 4770, respectively.

3.2. Patient Characteristics

Table 1 shows the clinical parameters as well as the demographic data collected for the 18 patients and 18 controls who participated in the study. The controls had a higher age at the scan than patients (37 ± 14 years vs. 32 ± 10 years, $p = 0.39$). As displayed in Table 1, RVEF (56 ± 4% vs. 46 ± 10%, $p = 0.00$), RVEDV (172 ± 60 mL vs. 253 ± 95 mL, $p = 0.02$), RVEDVi (88 ± 20 mL/m^2 vs. 135 ± 47 mL/m^2, $p = 0.00$), RVESV (77 ± 31 mL vs. 142 ± 74 mL, $p = 0.01$), and RVESVi (39 ± 12 mL/m^2 vs. 76 ± 39 mL/m^2, $p \leq 0.001$) were statistically significantly different between the patients and controls. As expected, rTOF patients had significantly larger RV volumes and lower RVEF compared with healthy controls.

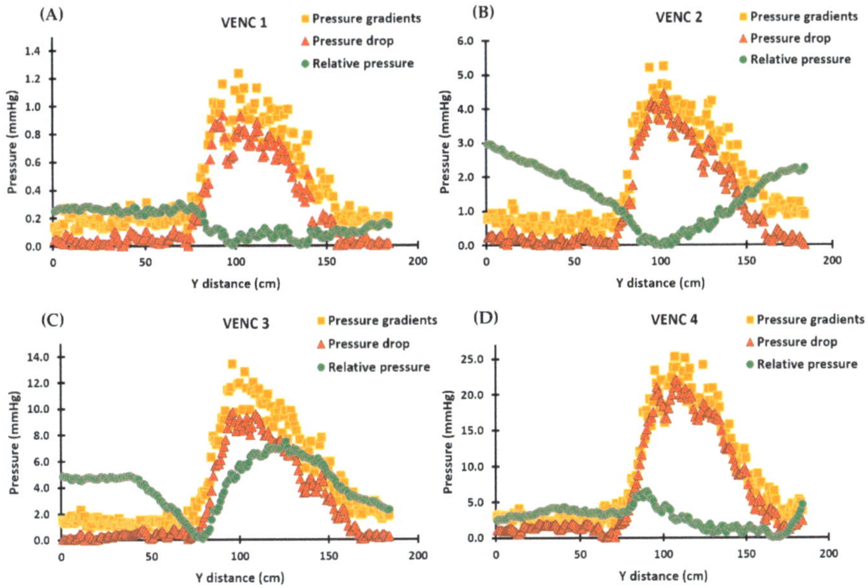

Figure 4. In vitro validation results demonstrating the calculated pressure gradients, pressure drop, and relative pressures for each VENC. Panels (**A–D**): The profiles of the pressure gradients are calculated using the simplified Bernoulli equation and are similar to those of MRI velocities. The pressure drop profiles follow the velocities and pressure gradients, with the maximum pressure drop occurring at the vena contracta location. The relative pressures, however, do not follow the velocity, pressure drop, or pressure gradient profiles and instead align with the pressure recovery phenomenon in Panels (**A**,**B**). In Panels (**C**,**D**), the relative pressures no longer align with the pressure recovery phenomenon and exhibit increased instability. For each VENC, the pressure drop and pressure gradient results show comparable results, while the profiles of relative pressures undergo a significant change and become increasingly unstable.

Table 1. Subject data baseline characteristics.

Characteristic	Patients (n = 18)	Controls (n = 18)	p-Value
Age at scan (year)	32 ± 10	37 ± 14	0.39
Sex (f/m)	6/12	7/11	0.73
BSA (m^2)	1.87 ± 0.19	1.88 ± 0.32	0.83
HR (bpm)	73 ± 10	67 ± 12	0.08
BP systolic (mmHg)	109 ± 8	107 ± 12	0.63
BP diastolic (mmHg)	60 ± 10	58 ± 12	0.40
LVEF (%)	58 ± 9	61 ± 3	0.11
LVEDV (mL)	151 ± 41	132 ± 49	0.14
LVEDVi (mL/m^2)	80 ± 17	110 ± 48	0.09
LVESV (mL)	65 ± 26	52 ± 20	0.10
LVESVi (mL/m^2)	35 ± 13	43 ± 19	0.26
RVEF (%)	46 ± 10	56 ± 4	0.00
RVEDV (mL)	253 ± 95	172 ± 60	0.02
RVEDVi (mL/m^2)	135 ± 47	88 ± 20	0.00
RVESV (mL)	142 ± 74	77 ± 31	0.01
RVESVi (mL/m^2)	76 ± 39	39 ± 12	<0.001

BSA: body surface area; HR: heart rate; BP: blood pressure; LVEDVi: indexed left ventricular end diastolic volume; LVESVi: indexed left ventricular end systolic volume; LVEF: left ventricular ejection fraction; LVEDV: left ventricular end diastolic volume; LVESV: left ventricular end diastolic volume; RVEDVi: indexed right ventricular end diastolic volume; RVESVi: indexed right ventricular end systolic volume; RVEF: right ventricular ejection fraction; RVEDV: right ventricular end diastolic volume; RVESV: right ventricular end systolic volume.

3.3. *In Vivo Data*

Mean, max, and STD pressure measurements of the aorta, LV, PA, and the RV reported as volumes are shown in Figure 5. For the left side of the heart, the controls demonstrated a higher PDmean in the AAo, aortic arch, DAo, and LV. Statistically significant differences between the two cohorts for PDmean were only found in the AAo (3.017 ± 1.461 mmHg vs. 1.928 ± 0.990 mmHg, $p < 0.05$). On the other hand, a higher RPmean was seen in the AAo and the LV of the controls. However, a higher RPmean was seen in the patients' aortic arch and the DAo (Figure 5A(i)). Statistically significant differences between the two cohorts for RPmean were only observed in the LV (1.080 ± 0.427 mmHg vs. 0.717 ± 0.422 mmHg, $p < 0.05$). Moreover, the controls demonstrated a higher PDmax in the AAo, aortic arch, DAo, and LV. There was no significant statistical difference observed between the two cohorts for PDmax. On the contrary, a higher RPmax for the controls was observed in the AAo and the LV, and it was also higher for the patients in the aortic arch and DAo (Figure 5A(ii)). Statistically significant differences between the two cohorts for RPmax were only found in the LV (4.129 ± 1.788 mmHg vs. 2.432 ± 1.491 mmHg, $p < 0.05$). Lastly, a higher PDstd was observed in the AAo, DAo, and the LV of the controls and it was also slightly higher in the aortic arch of the patients. There was no significant statistical difference identified between the two cohorts for PDstd. In contrast, a higher RPstd was seen in the AAo and the LV of the controls, and it was also higher in the aortic arch and the DAo of the patients (Figure 5A(iii)). Again, statistically significant differences between the two cohorts for RPstd were only observed in the LV (0.802 ± 0.384 mmHg vs. 0.435 ± 0.335 mmHg, $p < 0.05$).

For the right side of the heart, patients demonstrated a higher PDmean in all the vessels, including the LPA, RPA, MPA, and RV. Statistically significant differences between the two cohorts were observed for PDmean at RPA (0.511 ± 0.564 mmHg vs. 2.690 ± 2.788 mmHg, $p < 0.05$), MPA (1.176 ± 0.572 mmHg vs. 2.574 ± 1.894 mmHg, $p < 0.05$), and RV (0.271 ± 0.169 mmHg vs. 0.641 ± 0.675 mmHg, $p < 0.05$). Similarly, patients demonstrated a higher RPmean in all four vessels (Figure 5B(i)). Statistically significant differences between the two cohorts for RPmean were only found in the MPA (0.830 ± 0.359 mmHg vs. 2.289 ± 1.399 mmHg, $p < 0.05$). Additionally, PDmax was seen to be higher in patients in each vessel, including LPA, RPA, MPA, and the RV. Statistically significant differences between the two cohorts were witnessed for PDmax in the RPA (1.506 ± 1.851 mmHg vs. 5.209 ± 4.582 mmHg, $p < 0.05$), MPA (4.893 ± 4.815 mmHg vs. 9.778 ± 10.098 mmHg, $p < 0.05$), and RV (2.483 ± 2.176 mmHg vs. 6.915 ± 7.614 mmHg, $p < 0.05$). Similarly, RPmax was higher in the patients' LPA, RPA, MPA, and RV (Figure 5B(ii)). Statistically significant differences between the two cohorts for RPmax were observed in the LPA (2.222 ± 1.294 mmHg vs. 5.180 ± 4.190 mmHg, $p < 0.05$) and MPA (2.073 ± 0.742 mmHg vs. 8.380 ± 7.330 mmHg, $p < 0.05$). Finally, patients demonstrated a higher PDstd in every vessel. Statistically significant differences between the two cohorts were found for PDstd at the RPA (0.294 ± 0.330 mmHg vs. 1.100 ± 0.991 mmHg, $p < 0.05$), MPA (0.949 ± 0.57 mmHg vs. 1.950 ± 2.156 mmHg, $p < 0.05$), and RV (0.336 ± 0.270 mmHg vs. 0.896 ± 0.817 mmHg, $p < 0.05$). Likewise, a higher RPstd was observed in the patients' RPA, LPA, MPA, and RV (Figure 5B(iii)). Statistically significant differences between the two cohorts for RPstd were observed in the LPA (0.429 ± 0.258 mmHg vs. 0.889 ± 0.612 mmHg, $p < 0.05$) and MPA (0.447 ± 0.196 mmHg vs. 1.632 ± 1.279 mmHg, $p < 0.05$).

Figure 6 illustrates mean, max, and STD pressure measurements of the aorta, LV, PA, and RV reported as centerlines. For the left side of the heart, the controls demonstrated a higher PDmean in the AAo, aortic arch, DAo, and LV. Statistically significant differences between the two cohorts for PDmean was only identified in the AAo (2.559 ± 1.148 mmHg vs. 1.682 ± 1.164 mmHg, $p < 0.05$). On the other hand, patients demonstrated a higher RPmean in the AAo, aortic arch, and DAo, whereas a slightly higher RPmean was seen in the LV of the controls (Figure 6A(i)). There was no significant statistical difference observed between the two cohorts for RPmax. Moreover, the controls demonstrated a higher PDmax in the AAo, aortic arch, DAo, and LV. Then again, patients demonstrated a higher RPmax in the AAo, aortic arch, and DAo, whereas a slightly higher RPmax was seen in the LV of the controls (Figure 6A(ii)). No significant statistical difference was witnessed between the two cohorts for PDmax and RPmax. Lastly, a higher PDstd was observed in the AAo, aortic arch, DAo, and LV of the controls. Statistically significant differences between the two cohorts for PDstd were only found in the LV (1.105 ± 0.442 mmHg vs. 1.037 ± 0.603 mmHg, $p < 0.05$). In contrast, a higher RPstd was seen in the patients' AAo, aortic arch, and DAo, and a slightly higher RPstd in the LV of the controls (Figure 6A(iii)). Statistically significant differences between the two cohorts for RPstd were only observed in the LV (0.582 ± 0.247 mmHg vs. 0.518 ± 0.379 mmHg, $p < 0.05$).

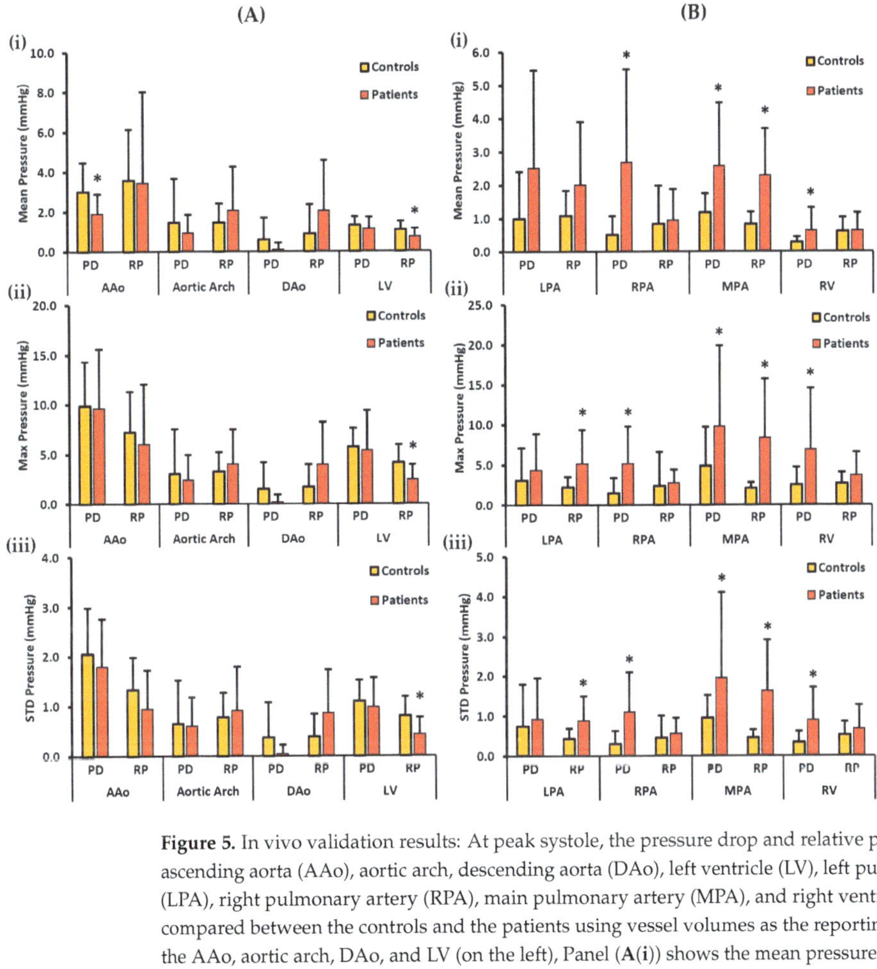

Figure 5. In vivo validation results: At peak systole, the pressure drop and relative pressures in the ascending aorta (AAo), aortic arch, descending aorta (DAo), left ventricle (LV), left pulmonary artery (LPA), right pulmonary artery (RPA), main pulmonary artery (MPA), and right ventricle (RV) were compared between the controls and the patients using vessel volumes as the reporting method. For the AAo, aortic arch, DAo, and LV (on the left), Panel (**A**(**i**)) shows the mean pressure drop (PD) and mean relative pressures (RP). Panel (**A**(**ii**)) shows the maximum PD and RP. Panel (**A**(**iii**)) shows the standard deviation of the PD and RP measurements. For the LPA, RPA, MPA, and RV (on the right), Panel (**B**(**i**)) shows the mean PD and RP. Panel (**B**(**ii**)) shows the max PD and RP, and Panel (**B**(**iii**)) shows the standard deviation of the PD and RP measurements. *: $p < 0.05$ between the controls and the patients. The controls showed a significantly higher PDmean in the AAo, and a significantly higher RPmean, RPmax, and RPstd in the LV compared to the patients. On the other hand, the patients showed a significantly higher PDmean, PDmax, and PDstd in the RPA, MPA, and the RV; a significantly higher RPmean, RPmax, and RPstd in the MPA; and a significantly higher RPmax and RPstd in the LPA compared to the controls.

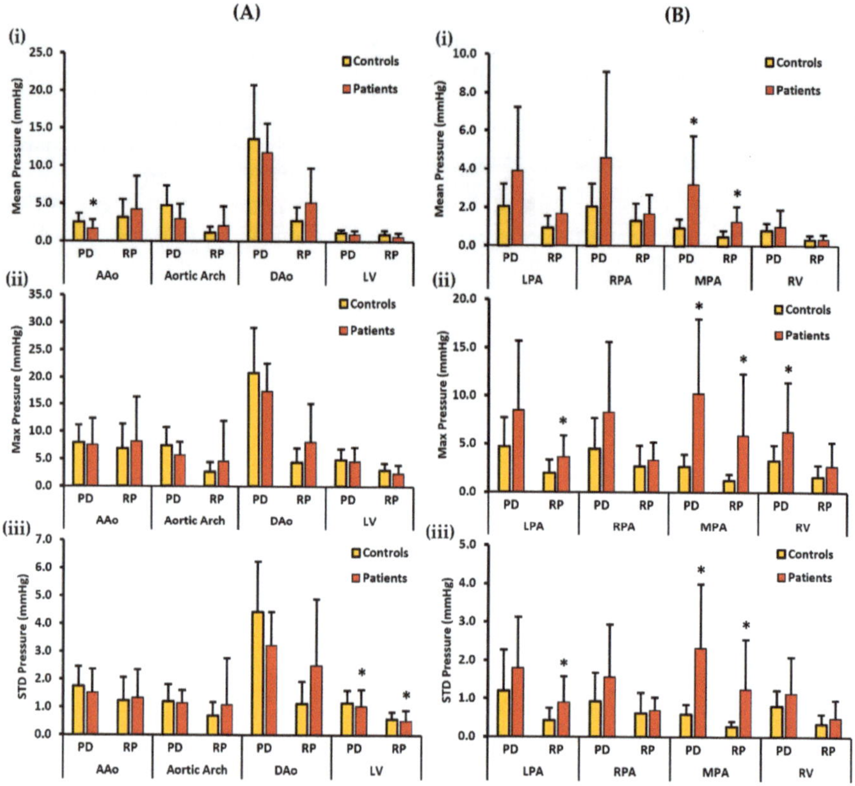

Figure 6. In vivo validation results: At peak systole, the pressure drop and relative pressures in the ascending aorta (AAo), aortic arch, descending aorta (DAo), left ventricle (LV), left pulmonary artery (LPA), right pulmonary artery (RPA), main pulmonary artery (MPA), and right ventricle (RV) were compared between the controls and the patients using the centerlines as the reporting method. For the AAo, aortic arch, DAo, and LV (on the left), Panel (**A(i)**) shows the mean pressure drop (PD) and mean relative pressures (RP). Panel (**A(ii)**) shows the maximum PD and RP. Panel (**A(iii)**) shows the standard deviation of the PD and RP measurements. For the LPA, RPA, MPA, and RV (on the right), Panel (**B(i)**) shows the mean PD and RP. Panel (**B(ii)**) shows the max PD and RP, and Panel (**B(iii)**) shows the standard deviation of the PD and RP measurements. *: $p < 0.05$ between the controls and the patients. The controls showed a significantly higher PDmean in the AAo, and a significantly higher PDstd and RPstd in the LV compared to the patients. Conversely, the patients show a significantly higher PDmean, PDmax, and PDstd in the MPA and a significantly higher PDmax in the RV than the controls. Furthermore, the patients showed a significantly higher RPmean, RPmax, and RPstd in the MPA and a significantly higher RPmax and RPstd in the LPA than the controls.

For the right side of the heart, the patients demonstrated a higher PDmean in all the vessels, including the LPA, RPA, MPA, and RV. Statistically significant differences between the two cohorts were observed for PDmean only in the MPA (0.925 ± 0.469 mmHg vs. 3.211 ± 2.534 mmHg, $p < 0.05$). Similarly, patients demonstrated a higher RPmean in all vessels (Figure 6B(i)). Statistically significant differences between the two cohorts for RPmean were identified in the MPA (0.477 ± 0.318 mmHg vs. 1.259 ± 0.795 mmHg, $p < 0.05$). PDmax was also seen to be higher in patients in each vessel, including LPA, RPA, MPA, and the RV. Statistically significant differences between the two cohorts were observed for PDmax in MPA (2.656 ± 1.281 mmHg vs. 10.233 ± 7.740 mmHg, $p < 0.05$) and RV (3.303 ± 1.566 mmHg vs. 6.274 ± 5.140 mmHg, $p < 0.05$). Similarly, RPmax was found to be higher in each vessel of the patients (Figure 6B(ii)). Statistically significant differences between the two cohorts were witnessed for RPmax in the LPA (1.986 ± 1.340 mmHg vs. 3.682 ± 2.195 mmHg, $p < 0.05$) and MPA (1.256 ± 0.634 mmHg vs. 5.893 ± 6.384 mmHg, $p < 0.05$). Finally, patients demonstrated a higher PDstd in each vessel. Statistically significant differences between the two cohorts for PDstd were seen in the MPA (0.595 ± 0.251 mmHg vs. 2.330 ± 1.666 mmHg, $p < 0.05$). Likewise, patients demonstrated a higher RPstd in all vessels (Figure 6B(iii)). Statistically significant differences between the two cohorts for RPstd were observed in the LPA (0.435 ± 0.312 mmHg vs. 0.903 ± 0.676 mmHg, $p < 0.05$) and MPA (0.273 ± 0.136 mmHg vs. 1.239 ± 1.312 mmHg, $p < 0.05$).

Figure 7 depicts the mean, maximum, and standard deviation pressure measurements of the controls' aorta, LV, PA, and RV. For the left side of the heart, volumes demonstrated a higher PDmean in the AAo and the LV. On the other hand, centerlines demonstrated a higher PDmean in the aortic arch and the DAo. Statistically significant differences between the two cohorts for PDmean were observed in the aortic arch (1.465 ± 2.208 mmHg vs. 4.780 ± 2.635 mmHg, $p < 0.05$), DAo (0.597 ± 1.114 mmHg vs. 13.577 ± 7.211 mmHg, $p < 0.05$) and the LV (1.321 ± 0.418 mmHg vs. 0.992 ± 0.351 mmHg, $p < 0.05$). RPmean, on the contrary, was slightly higher in the AAo, aortic arch, and LV volumes. However, the centerline demonstrated a higher RPmean in the DAo of the controls. Therefore, statistically significant differences between the two cohorts for RPmean were found only in the DAo (0.885 ± 1.483 mmHg vs. 2.768 ± 1.793 mmHg, $p < 0.05$). Furthermore, volumes demonstrated a higher PDmax in the AAo and the LV. Centerlines, on the other hand, demonstrated a higher PDmax in the aortic arch and the DAo. Statistically significant differences between the two cohorts for PDmax were seen in the aortic arch (3.028 ± 4.496 mmHg vs. 7.456 ± 3.343 mmHg, $p < 0.05$), and the DAo (1.498 ± 2.701 mmHg vs. 20.812 ± 8.262 mmHg, $p < 0.05$).

On the contrary, a higher RPmax was observed in the volumes of the AAo, aortic arch, and LV, and in the centerline of the DAo. Statistically significant differences between the two cohorts for RPmax were identified in the DAo (1.715 ± 2.262 mmHg vs. 4.367 ± 2.609 mmHg, $p < 0.05$) and the LV (4.129 ± 1.788 mmHg vs. 2.992 ± 1.256 mmHg, $p < 0.05$). Finally, a higher PDstd was observed in the volume of the AAo. The centerlines demonstrated a higher PDstd in the aortic arch and DAo. PDstd was almost equal in the volume and centerline of the LV. Statistically significant differences between the two cohorts for PDstd was observed in the aortic arch (0.576 ± 0.874 mmHg vs. 1.163 ± 0.611 mmHg, $p < 0.05$), and the DAo (0.374 ± 0.703 mmHg vs. 4.429 ± 1.807 mmHg, $p < 0.05$). In contrast, volumes showed a higher RPstd in the AAo, aortic arch, and LV. RPstd was higher in the centerline of the DAo. Statistically significant differences between the two cohorts for RPstd were seen in the DAo (0.388 ± 0.455 mmHg vs. 1.128 ± 0.790 mmHg, $p < 0.05$) and the LV (0.802 ± 0.384 mmHg vs. 0.582 ± 0.247 mmHg, $p < 0.05$).

For the right side of the heart, centerlines demonstrated a higher PDmean in the LPA, RPA, and RV. A higher PDmean was found in the volume of the MPA. Statistically significant differences between the two cohorts were witnessed for PDmean in the LPA (0.994 ± 1.414 mmHg vs. 2.053 ± 1.166 mmHg, $p < 0.05$), RPA (0.511 ± 0.564 mmHg vs. 2.049 ± 1.191 mmHg, $p < 0.05$), and RV (0.271 ± 0.169 mmHg vs. 0.808 ± 0.374 mmHg, $p < 0.05$). Alternatively, volumes demonstrated a higher RPmean in the LPA, MPA, and RV. The centerline showed a higher RPmean in the RPA. Statistically significant differences between the two cohorts were perceived for RPmean in the RPA (0.847 ± 1.132 mmHg vs. 1.311 ± 0.888 mmHg, $p < 0.05$), MPA (0.830 ± 0.359 mmHg vs. 0.477 ± 0.318 mmHg, $p < 0.05$), and RV (0.608 ± 0.418 mmHg vs. 0.329 ± 0.224 mmHg, $p < 0.05$). Moreover, a higher PDmax was noted in the centerline of the LPA, RPA, and RV. A higher PDmax was observed in the volume of the MPA. Statistically significant differences between the two cohorts were observed for PDmax in the LPA (3.082 ± 4.042 mmHg vs. 4.702 ± 2.997 mmHg, $p < 0.05$), MPA (4.893 ± 4.815 mmHg vs. 2.656 ± 1.281 mmHg, $p < 0.05$), and RPA (1.506 ± 1.851 mmHg vs. 4.504 ± 3.160 mmHg, $p < 0.05$). Conversely, volumes demonstrated a higher RPmax in the LPA, MPA, and RV. The centerline demonstrated a higher

RPmax in the RPA. Statistically significant differences between the two cohorts were found for RPmax in the RPA (2.376 ± 4.235 mmHg vs. 2.710 ± 2.144 mmHg, $p < 0.05$), MPA (2.073 ± 0.742 mmHg vs. 1.256 ± 0.634 mmHg, $p < 0.05$), and RV (2.674 ± 1.391 mmHg vs. 1.597 ± 1.222 mmHg, $p < 0.05$). Lastly, centerlines demonstrated a higher PDstd in the LPA, RPA, and RV. Volumes demonstrated a higher PDstd in the MPA. Statistically significant differences between the two cohorts were seen for PDstd in the LPA (0.744 ± 1.060 mmHg vs. 1.196 ± 1.072 mmHg, $p < 0.05$), RPA (0.294 ± 0.330 mmHg vs. 0.926 ± 0.737 mmHg, $p < 0.05$), and RV (0.336 ± 0.270 mmHg vs. 0.808 ± 0.410 mmHg, $p < 0.05$). On the other hand, a higher RPstd was noted in the volumes of the MPA and RV, and in the centerline of the RPA. RPstd was almost equal in the volume and centerline of the LPA. Statistically significant differences between the two cohorts were identified for RPstd in the RPA (0.450 ± 0.556 mmHg vs. 0.621 ± 0.527 mmHg, $p < 0.05$), MPA (0.447 ± 0.196 mmHg vs. 0.273 ± 0.136 mmHg, $p < 0.05$), and RV (0.524 ± 0.333 mmHg vs. 0.338 ± 0.259 mmHg, $p < 0.05$).

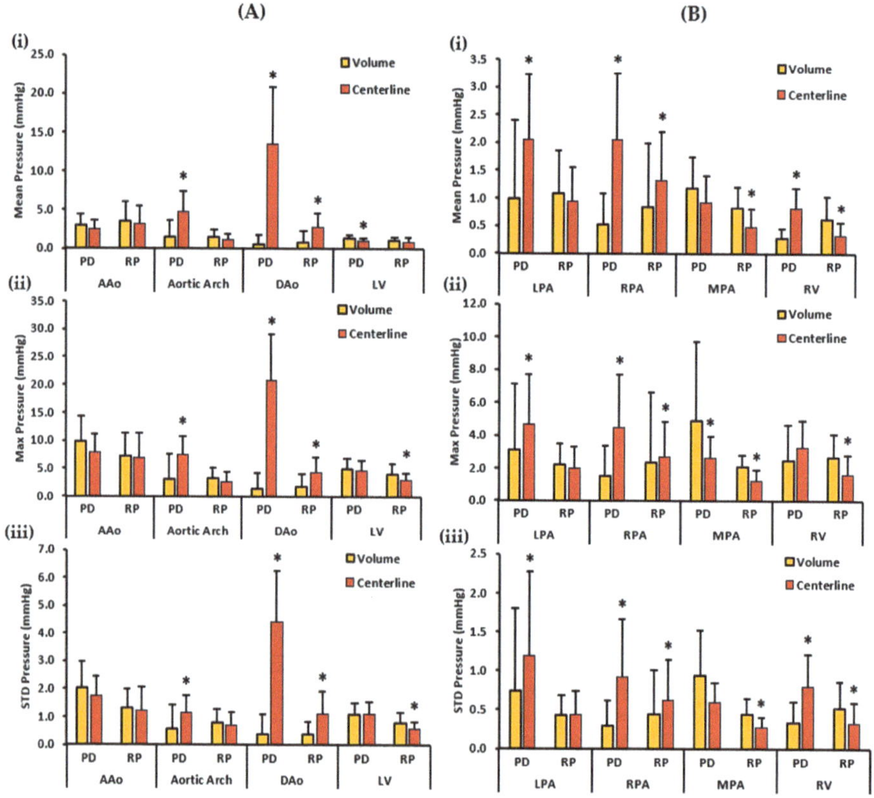

Figure 7. In vivo validation results: At peak systole, the pressure drop and relative pressures in the ascending aorta (AAo), aortic arch, descending aorta (DAo), left ventricle (LV), left pulmonary artery (LPA), right pulmonary artery (RPA), main pulmonary artery (MPA), and right ventricle (RV) of the controls were compared between the reporting methods (volumes and centerlines). For the AAo, aortic arch, DAo, and LV (on the left), Panel (**A**(**i**)) shows the mean pressure drop (PD) and mean relative pressures (RP). Panel (**A**(**ii**)) shows the maximum PD and RP. Panel (**A**(**iii**)) shows the standard deviation of the PD and RP measurements. For the LPA, RPA, MPA, and RV (on the right), Panel (**B**(**i**)) shows the mean PD and RP. Panel (**B**(**ii**)) shows the max PD and RP, and Panel (**B**(**iii**)) the shows standard deviation of the PD and RP measurements. *: $p < 0.05$ between the controls and

the patients. The volumes exhibited a significantly higher PDmean, RPmax, and RPstd in the LV compared to the centerlines. Furthermore, the volumes demonstrate a significantly higher PDmax, RPmean, RPmax, and RPstd in the MPA and a significantly higher RPmean, RPmax, and RPstd in the RV compared to the centerlines. On the contrary, the centerlines exhibited a significantly higher PDmean, PDmax, and PDstd in the aortic arch and DAo, and a significantly higher RPmean, RPmax, and RPstd in the DAo compared to the volumes. Additionally, the centerlines showed a higher PDmean, PDmax, and PDstd in the LPA and RPA, and a significantly higher PDmean and PDstd in the RV compared to the volumes. The centerlines also showed a higher RPmean, RPmax, and RPstd in the RPA compared to the volumes.

Figure 8 depicts the patients' mean, maximum, and standard deviation pressure measurements of the aorta, LV, PA, and RV. For the left side of the heart, the volumes demonstrated a higher PDmean in the AAo and the LV. On the other hand, the centerlines demonstrated a higher PDmean in the aortic arch and the DAo. Statistically significant differences between the two cohorts for PDmean were obtained in the aortic arch (0.934 ± 0.931 mmHg vs. 2.984 ± 1.992 mmHg, $p < 0.05$) and DAo (0.107 ± 0.331 mmHg vs. 11.800 ± 3.832 mmHg, $p < 0.05$). RPmean, on the contrary, was higher in the AAo, and DAo centerlines. However, RPmean was almost equal in the volume and centerline of the aortic arch and LV. Statistically significant differences between the two cohorts for RPmean were found only in the DAo (2.057 ± 2.523 mmHg vs. 5.149 ± 4.591 mmHg, $p < 0.05$). Furthermore, volumes demonstrated a higher PDmax in the AAo and the LV. Centerlines, on the other hand, demonstrated a higher PDmax in the aortic arch and the DAo. Statistically significant differences between the two cohorts for PDmax were observed in the aortic arch (2.417 ± 2.531 mmHg vs. 5.695 ± 2.394 mmHg, $p < 0.05$), and the DAo (0.205 ± 0.709 mmHg vs. 17.370 ± 5.167 mmHg, $p < 0.05$). On the contrary, a higher RPmax was identified in the centerlines of the AAo, aortic arch, and DAo. RPmax was almost equal in the volume and centerline of the LV. Statistically significant differences between the two cohorts for RPmax were observed only in the DAo (3.995 ± 4.206 mmHg vs. 8.100 ± 7.057 mmHg, $p < 0.05$). Finally, a higher PDstd was observed in the volume of the AAo. Centerlines demonstrated a higher PDstd in the aortic arch, DAo, and LV. Statistically significant differences between the two cohorts for PDstd were witnessed in the aortic arch (0.602 ± 0.577 mmHg vs. 1.149 ± 0.457 mmHg, $p < 0.05$), and the DAo (0.049 ± 0.178 mmHg vs. 3.230 ± 1.194 mmHg, $p < 0.05$). In contrast, centerlines showed a higher RPstd in all the vessels, including AAo, aortic arch, DAo, and LV. Statistically significant differences between the two cohorts for RPstd were observed only in the DAo (0.869 ± 0.867 mmHg vs. 2.507 ± 2.370 mmHg, $p < 0.05$).

For the right side of the heart, the centerlines demonstrated a higher PDmean in all vessels, including LPA, RPA, MPA, and RV. Nevertheless, no statistically significant differences were found between the two cohorts for PDmean. Alternatively, volumes demonstrated a higher RPmean in the LPA, MPA, and RV. The centerline showed a higher RPmean in the RPA. Statistically significant differences between the two cohorts were observed for RPmean in the RPA (0.887 ± 0.931 mmHg vs. 1.669 ± 0.987 mmHg, $p < 0.05$) and MPA (2.289 ± 1.399 mmHg vs. 1.259 ± 0.795 mmHg, $p < 0.05$). Moreover, a higher PDmax was noted in the centerline of the LPA, RPA, MPA, and RV. Statistically significant differences between the two cohorts were found for PDmax only in the LPA (4.357 ± 4.509 mmHg vs. 8.489 ± 7.183 mmHg, $p < 0.05$). Conversely, volumes demonstrated a higher RPmax in the LPA, MPA, and RV. The centerline demonstrated a higher RPmax in the RPA. Statistically significant differences between the two cohorts were observed for RPmax in the RPA (1.676 ± 1.621 mmHg vs. 3.338 ± 1.867 mmHg, $p < 0.05$), and MPA (8.380 ± 7.330 mmHg vs. 5.893 ± 6.384 mmHg, $p < 0.05$). Lastly, the centerlines demonstrated a higher PDstd in all vessels, including the LPA, RPA, MPA, and RV. Statistically significant differences between the two cohorts were identified for PDstd only in the LPA (0.927 ± 1.034 mmHg vs. 1.794 ± 1.335 mmHg, $p < 0.05$). On the other hand, a higher RPstd was noted in the volumes of the MPA and RV; and in the centerline of the RPA. RPstd was almost equal in the volume and centerline of the LPA. Statistically significant differences between the two cohorts were perceived for RPstd only in the RPA (0.411 ± 0.388 mmHg vs. 0.695 ± 0.338 mmHg, $p < 0.05$). Spearman's correlation was also calculated between standard clinical measurements and the two pressure mapping methods. However, no significant or strong correlation between the two was discovered.

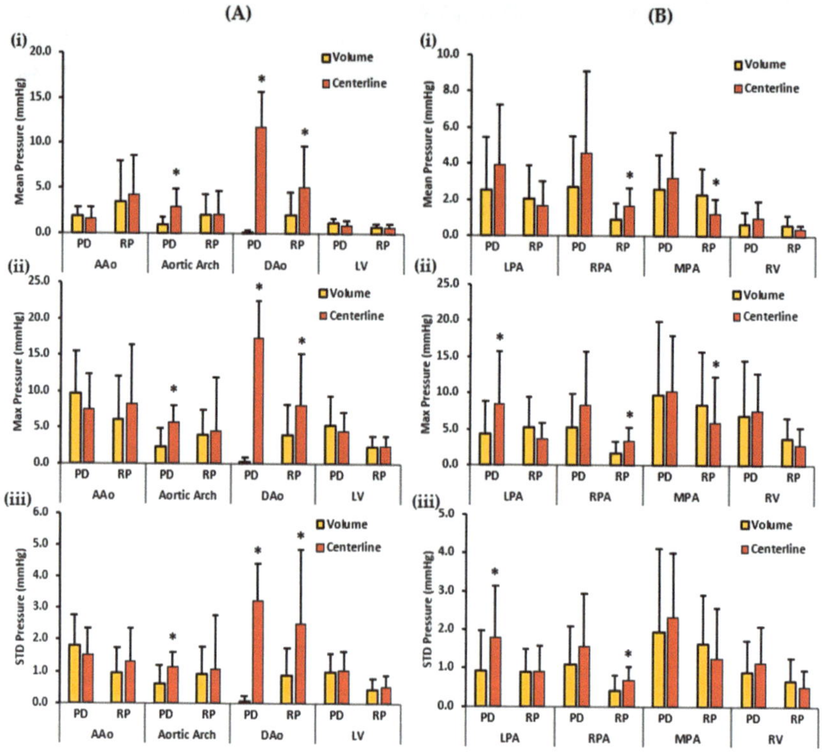

Figure 8. In vivo validation results: At peak systole, the pressure drop and relative pressures in the ascending aorta (AAo), aortic arch, descending aorta (DAo), left ventricle (LV), left pulmonary artery (LPA), right pulmonary artery (RPA), main pulmonary artery (MPA), and right ventricle (RV) of the patients were compared between the reporting methods (volumes and centerlines). For the AAo, aortic arch, DAo, and LV (on the left), Panel (**A**(**i**)) shows the mean pressure drop (PD) and mean relative pressures (RP). Panel (**A**(**ii**)) shows the maximum PD and RP. Panel (**A**(**iii**)) shows the standard deviation of the PD and RP measurements. For the LPA, RPA, MPA, and RV (on the right), Panel (**B**(**i**)) shows the mean PD and RP. Panel (**B**(**ii**)) shows the max PD and RP, and Panel (**B**(**iii**)) shows the standard deviation of the PD and RP measurements. *: $p < 0.05$ between controls and patients. Volumes exhibited significantly higher RPmean and RPmax in MPA compared to the centerlines. On the contrary, the centerlines indicate significantly higher PDmean, PDmax, and PDstd in the aortic arch and DAo; and significantly higher RPmean, RPmax, and RPstd in the DAo compared to volumes. Additionally, the centerlines showed a higher PDmax and PDstd in the LPA. The centerlines also showed a higher RPmean, RPmax, and RPstd in the RPA compared to the volumes.

4. Discussion

Our main findings showed: (1) the pressure drop and pressure gradient profiles followed velocity profiles whereas the relative pressure profiles aligned with the pressure recovery phenomenon; (2) a significantly higher pressure drop and relative pressure measurements in the MPA of the rTOF patients compared to the controls, reported as volumes and centerlines; and (3) the volumes and centerlines are separate means of reporting pressures and do not reflect pressures in the same way.

Previous studies have demonstrated that 2D- and 4D-flow measurements in TOFs have a reasonable agreement [27–29]. In the context of this study, we evaluated flow only using 4D-flow MRI, and we derived the advanced hemodynamic parameter—pressure mapping. The term "pressure gradient" is often misused in the clinical literature, including clinical guidelines [30–32], where it is inaccurately used to describe the "pressure drop." Similarly, the term "relative pressure" is misused to refer to the "pressure gradient". Therefore, this study aimed to compare and solidify the understanding that pressure gradient, pressure drop, and relative pressures are all different pressure mapping methods that represent different information.

The simplified Bernoulli equation, which calculates the maximum pressure drop across a valve based on the maximum flow velocity at the vena contracta, has been found to overestimate the actual pressure drop in cases involving valve stenosis or downstream obstructions [33–35]. This discrepancy is primarily attributed to the phenomenon of pressure recovery, where a portion of the kinetic energy is converted back into pressure downstream of the stenosis. However, some kinetic energy is lost as thermal and acoustic energy. While this loss is minimal in laminar flow, it becomes significant under turbulent conditions. The simplified Bernoulli equation assumes that pressure recovery is negligible [36–40]. Our in vitro findings support this evidence. Pressure gradients estimated using the simplified Bernoulli equation overestimated the pressures compared to the pressure drop measurements; moreover, the profiles of the relative pressure measurements aligned with the pressure recovery phenomenon. However, as the number of VENCs increased, the RP profiles became unstable and failed to accurately depict the pressure recovery phenomenon. On the other hand, pressure gradient and pressure drop profiles remained stable and represented the maximum pressure drop at the obstruction.

The Reynolds numbers at the inlet and stenosis regions of the pipe indicated that the flow altered from being laminar to transitional to turbulent. The flow of a steady laminar stream in a circular pipe becomes unstable when the Reynolds number reaches 2000, and the transition into fully turbulent flow occurs at a Reynolds number of approximately 4000 [41]. Furthermore, as the SNR increased, the scatter plots revealed a progressive resemblance between the pressure measurements and the original pressure data. Additionally, the applied Gaussian filter demonstrated an efficient reduction of noise. Consistent results were observed across all five VENCs and for both pressure mapping methods. Moreover, with the increasing number of VENCs and turbulence, Bland–Altman analysis of the pressure drop method exhibited better agreement between the original and filtered pressure data compared to the relative pressure measurements. The pressure drops proved to be more robust under noise, as seen in Supplementary Materials.

Our in vivo study demonstrated a significantly higher PDmean, PDmax, PDstd, RPmean, RPmax, and RPstd in the MPA in the rTOF patients compared to the controls. The outcomes of this investigation were consistent with a study carried out by Sotelo et al. that showed pressure maps in rTOF patients had increased and that statistically significant pressure differences in the pulmonary system compared to healthy volunteers were present [42]. The elevated pressure measurements in the MPA could be attributed to the presence of pulmonary regurgitation (PR) in this patient group. PR and potentially residual or recurrent pulmonary stenosis are the most commonly seen complications in patients with rTOF [6]. Furthermore, our study also observed a higher PDmean, PDmax, PDstd, RPmean, RPmax, and RPstd in the LPA, RPA, and RV in the rTOF patients compared to the controls. The observed results could be attributed to anomalies that are predominantly located on the right side of the heart among individuals with rTOF. Following TOF surgery, several significant concerns may appear, including right ventricle enlargement and dysfunction, pulmonary regurgitation, as well as stenosis in both the right and left pulmonary arteries [43]. Patients with repaired TOFs have irregular flow patterns in their right chambers; the majority of the inflow through the tricuspid valve is directed towards the RV's apex, mainly due to the PR jet [44].

Furthermore, this study observed higher values of PDmean, PDmax, and PDstd in the AAo, aortic arch, DAo, and LV in the controls compared to the patients. On the contrary, the controls showed higher values of RPmean, RPmax, and RPstd in the LV and in the volumes of AAo compared to the patients. The observed findings can be primarily attributed to the malformations predominantly occurring on the right side of the heart. Furthermore, the discrepancy may be attributed to lower velocities observed in these vessels among patients with rTOF compared to those in the control group. This indicates that individuals with rTOF may experience a disparity in velocities between the left and right cardiac chambers, resulting in heightened velocities on the right side and elevated pressure values in the RPA, LPA, MPA, and RV. Due to ventricular–ventricular interaction, the LV often develops a functional deterioration over time when there are impaired RV mechanics, as they share myofibers. Control left hearts are expected to function better than rTOF left hearts [45].

Unfortunately, studies investigating pressure mapping in rTOF patients are scarce. However, a substantial body of literature exists in the context of aortic valve disease and aortopathy [25,36,46]. Additionally, there is a limited literature on the assessment of 4D-flow-based pressure drop and relative pressures in rTOF patients. On the other hand, numerous studies demonstrated an elevated ventricular kinetic energy (KE), vorticity, wall shear stress (WSS), and turbulent KE in the patient cohort [47–49]. A study by Hirtler et al. reported a significantly higher intracardiac vorticity in patients with rTOF, as well as the association of a higher right vorticity with the regurgitant flow in the MPA [50]. WSS and energy loss are also known to be linked with pulmonary hemodynamic changes in the MPA and the RPA [51]. RV KE was also higher in the rTOF patients than in the healthy subjects [52], indicating that the RV must work harder in the rTOF patients to produce the same cardiac output as in the healthy subjects [53]. Vorticity, WSS, and KE were not evaluated or investigated in relation to the pressure drop or relative pressures in the current study. The association between pressure mapping and the aforementioned hemodynamic parameters is unclear and should be investigated. This study also compared the pressure drop and relative pressures to standard clinical measures, including LVEF, RVEF, LVEDVi, LVESVi, RVEDVi, and RVESVi. As no significant or strong correlation was observed between the pressure mapping methods and the standard clinical measures, this suggests that pressure mapping is an independent local measurement providing additional insights into the atypical flow patterns present in this group of patients.

Another novel finding of this study is that the volumes and centerlines are different methods and do not report pressures in the same way. Centerlines demonstrated a higher PDmean, PDmax, PDstd, RPmean, RPmax, and RPstd in the DAo and the RPA in both the controls and the patients. Moreover, the centerlines also demonstrated a higher PDmean, PDmax, and PDstd in the aortic arch, LPA, and RV of the controls and the patients; and were higher in the MPA of the patients. The study that proposed the 4D virtual catheter technique that mathematically mimics clinical invasive catheterization found that patients with bicuspid aortic valves had a higher viscous energy loss than healthy participants along the center volume of the aorta vessel [16]. A reason for the higher centerline measurements rather than volume measurements is the number of voxels. Centerline measurements are more local as they consider voxels only along the center of the vessel's volume. On the other hand, volume measurements are global; the reason for lower volume measurements could be owed to the effect of averaging over a greater number of voxels.

The repair of a TOF can produce multiple changes in the patient's hemodynamics, even in the case of a successful procedure [45,54]. The right ventricular function can be drastically impaired by pulmonary regurgitation by producing a chronic right-ventricular load increment, dilation of the right chambers, and a reduction in right heart performance. Pressure mapping based on 4D-flow MRI can overcome the limitations of a standard Doppler and characterize, more specifically, the anatomic locations with abnormal flow, pressure load, and vessel/chamber remodeling. These contributions may have implications for the indication of re-intervention during follow-up. Altered flow hemodynamics in rTOF are associated with a >3-fold increase in adverse cardiovascular events. For example, a peak right-ventricular outflow tract pressure gradient $\geq$ 25 mmHg has an HR = 3.69 [54]. Francois et al. [47] reported that rTOF patients showed an unbalanced flow distribution in the inferior and superior vena cava during the cardiac cycle, being greater during diastole. An increment of vortical flow in the right atria and right ventricle during diastole was also reported, along with the presence of altered flow in the pulmonary artery. All these observations were performed in consideration of their clinical relevance to the patients to evaluate the post-surgical alterations of geometry and hemodynamics. Furthermore, Hirtler et al. [50] demonstrated that the right ventricle and atrial vorticity were associated with chamber volumes and were also directly impacted by pulmonary regurgitation development.

There exist several possible limitations to our study. A significant limitation of this study is that only a small number of the rTOF subjects and controls were examined. More patients and healthy controls should be recruited to better understand how pressure mapping may have an impact on several vessels among patients in this cohort. An additional limitation of the study is the absence of data on inter- and intra-observer variability in calculating pressure maps. Because the NS equations calculate pressure gradients from both spatial and temporal velocity gradients, an insufficient spatial or temporal resolution will result in underestimating the pressure gradients and will lead to the loss of important fundamental or diagnostic pressure gradient features. Furthermore, a limited spatial resolution restricts small vessel analyses as high jet velocities and strong gradients at the jet boundary dominate stenotic blood flow. When the spatial resolution is sufficient, these strong gradients can be computed accurately. Recently introduced advanced imaging acceleration techniques, such as k-t under-sampling, compressed sensing, or radial under-sampling, are promising and have helped significantly reduce total scan times, allowing for a greater flexibility in spatial and temporal resolution selection [55]. Radial under-sampling overcomes the limitations of Cartesian 4D velocity mapping CMR by providing ample volume coverage with a high spatial resolution in reasonable scan times. Furthermore, radial acquisitions are also preferable to Cartesian acquisitions because they are less susceptible to motion artifacts [47]. Another limitation is that no validation against gold-standard invasive catheterization was performed in this study.

To extract the peak velocity, velocity maximum intensity projections (MIPs) were generated by masking the pre-processed 4D-flow MRI velocity field with the 3D segment. However, in our study, no noise filter was used to account for any false values caused by residual velocity aliasing or noise voxels. Rose et al. used a noise filter to exclude noise from the peak velocity assessment in their study, which used a vectorial vector containing voxel-wise velocity data [56].

Our in vitro validation results showed that pressure gradients, the pressure drop, and the relative pressure profiles are stable at lower VENCs. Conversely, with increasing turbulence and VENCs, the relative pressure profiles changed entirely and did not represent the pressure recovery phenomenon. Therefore, we recommend using the iterative approach over the multigrid-based PPE solver at higher VENCs. We did not evaluate the pressure maps according to VENC ranges in the patient population. However, this can be addressed by the multi-VENC acquisition strategies reported by Ha et al. [57]. Moreover, the reference points for the pressure drop calculation were manually selected in this study. A future alternative could be using a less operator-dependent automated reference detection approach using machine learning methods to reduce the variability in reference selection.

In our study, we included an experimental evaluation of pressure measurements. We acknowledged that particle image velocimetry is the experimental gold standard [58–60]. The conservation of mass principle is widely used in 4D-flow MRI as a quality control, and we used it in our data [13,61]. Theoretical values from cellar rotational flow [62], the Poiseuille and Lamb–Oseen equation [63], and the Hagen–Poiseuille equation [64] have been used for validation purposes of 4D-flow-derived metrics. In our study, we considered only the experimental component as a validation and comparison of pressure methods. Furthermore, several studies proposed the integration of computational fluid dynamics (CFD) with 4D-flow MRI to overcome the spatial and temporal limitations of 4D-flow MRI [40,65,66]. In particular, finite element methods can be used to discretize the in vivo 4D-flow velocity field and estimate the pressure maps more accurately [10,67]. More recently, deep learning super-resolution approaches have been proposed to better capture the spatiotemporal characteristics of the 4D-flow MRI velocity field [68]. Our study did not include any of these novel approaches, but it should be considered in future assessments.

5. Conclusions

In conclusion, this study demonstrated that flow hemodynamics in rTOF can exhibit altered pressure maps. This study's results suggest that pressure mapping could be an independent biomarker for monitoring rTOF. Following the in vitro validation method, the pressure drops proved to be a more stable pressure mapping method than the method using relative pressures, as the flow loses its laminarity and becomes more turbulent. Further in vivo validation and longitudinal studies are needed to standardize a pressure mapping method that may provide further insight into rTOF patients' hemodynamics to improve patient care and clinical decisions.

Supplementary Materials: The following supporting information can be downloaded at: https://www.mdpi.com/article/10.3390/fluids8070196/s1, This file displays the Bland–Altman plot findings regarding the impact of Gaussian noise and filter on the pressure drop and relative pressure measurements. Figure S1: Measured pressure drop against distance, Y, for VENC 1.; Figure S2: Measured pressure drop against distance, Y, for VENC 2; Figure S3: Measured pressure drop against distance, Y, for VENC 3; Figure S4: Measured pressure drop against distance, Y, for VENC 4; Figure S5: Measured relative pressures against distance, Y, for VENC 1; Figure S6: Measured relative pressures against distance, Y, for VENC 2; Figure S7: Measured relative pressures against distance, Y, for VENC 3; Figure S8: Measured relative pressures against distance, Y, for VENC 4; Figure S9: Bland-Altman plots for pressure drop against distance, Y, for VENC 1; Figure S10: Bland-Altman plots for pressure drop against distance, Y, for VENC 2; Figure S11: Bland-Altman plots for pressure drop against distance, Y, for VENC 3; Figure S12: Bland-Altman plots for pressure drop against distance, Y, for VENC 4; Figure S13: Bland-Altman plots for relative pressures against distance, Y, for VENC 1; Figure S14: Bland-Altman plots for relative pressures against distance, Y, for VENC 2; Figure S15: Bland-Altman plots for relative pressures against distance, Y, for VENC 3; Figure S16: Bland-Altman plots for relative pressures against distance, Y, for VENC 4.

Author Contributions: Conceptualization, S.I.A. and J.G.; methodology, S.I.A. and J.G.; software, S.I.A. and J.G.; validation, S.I.A., D.P., K.A.M. and J.G.; formal analysis, S.I.A.; investigation, S.I.A. and J.G.; resources, J.G.; data curation, S.I.A.; writing—original draft preparation, S.I.A. and J.G.; writing—review and editing, D.P., K.A.M. and J.G.; visualization, S.I.A. and J.G.; supervision, J.G.; project administration, J.G.; funding acquisition, J.G. All authors have read and agreed to the published version of the manuscript.

Funding: This research was funded by The University of Calgary, URGC SEM #1054341; J.G. start-up funding. Unrestricted research funding was provided by Siemens Healthineers. We acknowledge the support of the Natural Science and Engineering Research Council of Canada/Conseil de recherche en science naturelles et en génie du Canada, RGPIN-2020-04549 and DGECR-2020-00204.

Institutional Review Board Statement: The study was conducted according to the guidelines of the Declaration of Helsinki and approved by the Conjoint Health Research Ethics Board of the University of Calgary (REB13-0902 was approved on 18 June 2014 and is currently active).

Informed Consent Statement: Written informed consent was obtained from all subjects involved in the study.

Data Availability Statement: The anonymized data presented in this study are available upon request from the corresponding author. The data are not publicly available due to privacy and ethical restrictions.

Conflicts of Interest: The authors declare no conflict of interest.

References

1. Hudani, A.; White, J.A.; Greenway, S.C.; Garcia, J. Whole-Heart Assessment of Turbulent Kinetic Energy in the Repaired Tetralogy of Fallot. *Appl. Sci.* **2022**, *12*, 10946. [CrossRef]
2. Loke, Y.-H.; Capuano, F.; Kollar, S.; Cibis, M.; Kitslaar, P.; Balaras, E.; Reiber, J.H.C.; Pedrizzetti, G.; Olivieri, L. Abnormal Diastolic Hemodynamic Forces: A Link Between Right Ventricular Wall Motion, Intracardiac Flow, and Pulmonary Regurgitation in Repaired Tetralogy of Fallot. *Front. Cardiovasc. Med.* **2022**, *9*, 929470. [CrossRef] [PubMed]
3. Zhuang, B.; Sirajuddin, A.; Zhao, S.; Lu, M. The role of 4D flow MRI for clinical applications in cardiovascular disease: Current status and future perspectives. *Quant. Imaging Med. Surg.* **2021**, *11*, 4193–4210. [CrossRef] [PubMed]
4. van der Ven, J.P.; Bosch, E.V.D.; Bogers, A.J.; Helbing, W.A. Current outcomes and treatment of tetralogy of Fallot. *F1000Research* **2019**, *8*, 1530. [CrossRef] [PubMed]
5. Elsayed, A.; Gilbert, K.; Scadeng, M.; Cowan, B.R.; Pushparajah, K.; Young, A.A. Four-dimensional flow cardiovascular magnetic resonance in tetralogy of Fallot: A systematic review. *J. Cardiovasc. Magn. Reson.* **2021**, *23*, 59. [CrossRef]
6. Vasanawala, S.S.; Hanneman, K.; Alley, M.T.; Hsiao, A. Congenital heart disease assessment with 4D flow MRI. *J. Magn. Reson. Imaging* **2015**, *42*, 870–886. [CrossRef]
7. Warmerdam, E.G.; Krings, G.J.; Leiner, T.; Grotenhuis, H.B. Three-dimensional and four-dimensional flow assessment in congenital heart disease. *Heart* **2020**, *106*, 421–426. [CrossRef]
8. Robinson, J.D.; Rose, M.J.; Joh, M.; Jarvis, K.; Schnell, S.; Barker, A.J.; Rigsby, C.K.; Markl, M. 4-D flow magnetic-resonance-imaging-derived energetic biomarkers are abnormal in children with repaired tetralogy of Fallot and associated with disease severity. *Pediatr. Radiol.* **2019**, *49*, 308–317. [CrossRef]
9. Tsuchiya, N.; Nagao, M.; Shiina, Y.; Miyazaki, S.; Inai, K.; Murayama, S.; Sakai, S. Circulation derived from 4D flow MRI correlates with right ventricular dysfunction in patients with tetralogy of Fallot. *Sci. Rep.* **2021**, *11*, 11623. [CrossRef]
10. Nolte, D.; Urbina, J.; Sotelo, J.; Sok, L.; Montalba, C.; Valverde, I.; Osses, A.; Uribe, S.; Bertoglio, C. Validation of 4D Flow based relative pressure maps in aortic flows. *Med. Image Anal.* **2021**, *74*, 102195. [CrossRef]
11. Bock, J.; Frydrychowicz, A.; Lorenz, R.; Hirtler, D.; Barker, A.J.; Johnson, K.M.; Arnold, R.; Burkhardt, H.; Hennig, J.; Markl, M. In vivo noninvasive 4D pressure difference mapping in the human aorta: Phantom comparison and application in healthy volunteers and patients. *Magn. Reson. Med.* **2011**, *66*, 1079–1088. [CrossRef]
12. Ebbers, T.; Farnebäck, G. Improving computation of cardiovascular relative pressure fields from velocity MRI. *J. Magn. Reson. Imaging* **2009**, *30*, 54–61. [CrossRef]
13. Dyverfeldt, P.; Bissell, M.; Barker, A.J.; Bolger, A.F.; Carlhäll, C.-J.; Ebbers, T.; Francios, C.J.; Frydrychowicz, A.; Geiger, J.; Giese, D.; et al. 4D flow cardiovascular magnetic resonance consensus statement. *J. Cardiovasc. Magn. Reson.* **2015**, *17*, 72. [CrossRef]
14. Stankovic, Z.; Allen, B.D.; Garcia, J.; Jarvis, K.B.; Markl, M. 4D flow imaging with MRI. *Cardiovas Diagn Ther* **2014**, *4*, 173–192.
15. Hudani, A.; Ali, S.I.; Patton, D.; Myers, K.A.; Fine, N.M.; White, J.A.; Greenway, S.; Garcia, J. 4D-Flow MRI Characterization of Pulmonary Flow in Repaired Tetralogy of Fallot. *Appl. Sci.* **2023**, *13*, 2810. [CrossRef]
16. Elbaz, M.S.M.; Scott, M.B.; Barker, A.J.; McCarthy, P.; Malaisrie, C.; Collins, J.D.; Bonow, R.O.; Carr, J.; Markl, M. Four-dimensional Virtual Catheter: Noninvasive Assessment of Intra-aortic Hemodynamics in Bicuspid Aortic Valve Disease. *Radiology* **2019**, *293*, 541–550. [CrossRef]
17. Hatle, L.; Brubakk, A.; Tromsdal, A.; Angelsen, B. Noninvasive assessment of pressure drop in mitral stenosis by Doppler ultrasound. *Heart* **1978**, *40*, 131–140. [CrossRef]
18. Kramer, C.M.; Barkhausen, J.; Bucciarelli-Ducci, C.; Flamm, S.D.; Kim, R.J.; Nagel, E. Standardized cardiovascular magnetic resonance imaging (CMR) protocols: 2020 update. *J. Cardiovasc. Magn. Reson.* **2020**, *22*, 17. [CrossRef]
19. Garcia, J.; Sheitt, H.; Bristow, M.S.; Lydell, C.; Howarth, A.G.; Heydari, B.; Prato, F.S.; Drangova, M.; Thornhill, R.E.; Nery, P.; et al. Left atrial vortex size and velocity distributions by 4D flow MRI in patients with paroxysmal atrial fibrillation: Associations with age and CHA_2DS_2-VASc risk score. *J. Magn. Reson. Imaging* **2019**, *51*, 871–884. [CrossRef]
20. Geeraert, P.; Jamalidinan, F.; Burns, F.; Jarvis, K.; Bristow, M.S.; Lydell, C.; Tobon, S.S.H.; Alonso, B.D.C.; Fedak, P.W.M.; White, J.A.; et al. Hemodynamic Assessment in Bicuspid Aortic Valve Disease and Aortic Dilation: New Insights From Voxel-By-Voxel Analysis of Reverse Flow, Stasis, and Energetics. *Front. Bioeng. Biotechnol.* **2022**, *9*, 725113. [CrossRef]
21. Hong, Z.M.; Garcia, J. Pulmonary Artery Remodeling and Advanced Hemodynamics: Magnetic Resonance Imaging Biomarkers of Pulmonary Hypertension. *Appl. Sci.* **2022**, *12*, 3518. [CrossRef]
22. Bock, M.; Kreher, J.; Hennig, B.; Markl, J. Optimized Pre-Processing of Time-Resolved 2 D and 3 D Phase Contrast MRI Data. In Proceedings of the 15th Annual Meeting of ISMRM, Berlin, Germany, 19–25 May 2007.
23. Garcia, J.; Markl, M.; Schnell, S.; Allen, B.; Entezari, P.; Mahadevia, R.; Malaisrie, S.C.; Pibarot, P.; Carr, J.; Barker, A.J. Evaluation of aortic stenosis severity using 4D flow jet shear layer detection for the measurement of valve effective orifice area. *Magn. Reson. Imaging* **2014**, *32*, 891–898. [CrossRef] [PubMed]
24. Donati, F.; Figueroa, C.A.; Smith, N.P.; Lamata, P.; Nordsletten, D.A. Non-invasive pressure difference estimation from PC-MRI using the work-energy equation. *Med. Image Anal.* **2015**, *26*, 159–172. [CrossRef] [PubMed]

25. Hassanabad, A.F.; Burns, F.; Bristow, M.S.; Lydell, C.; Howarth, A.G.; Heydari, B.; Gao, X.; Fedak, P.W.; White, J.A.; Garcia, J. Pressure drop mapping using 4D flow MRI in patients with bicuspid aortic valve disease: A novel marker of valvular obstruction. *Magn. Reson. Imaging* **2020**, *65*, 175–182. [CrossRef] [PubMed]
26. Munson, B.; Yound, D.; Okiishi, T.; Huebsch, W. *Fundamentals of Fluid Mechanics*, 6th ed.; Wiley: Hoboken, NJ, USA, 2009.
27. Hsiao, A.; Alley, M.T.; Massaband, P.; Herfkens, R.J.; Chan, F.P.; Vasanawala, S.S. Improved cardiovascular flow quantification with time-resolved volumetric phase-contrast MRI. *Pediatr. Radiol.* **2011**, *41*, 711–720. [CrossRef]
28. Gabbour, M.; Schnell, S.; Jarvis, K.; Robinson, J.; Markl, M.; Rigsby, C.K. 4-D flow magnetic resonance imaging: Blood flow quantification compared to 2-D phase-contrast magnetic resonance imaging and Doppler echocardiography. *Pediatr. Radiol.* **2015**, *45*, 804–813. [CrossRef]
29. Rizk, J. 4D flow MRI applications in congenital heart disease. *Eur. Radiol.* **2021**, *31*, 1160–1174. [CrossRef]
30. Baumgartner, H.; Hung, J.; Bermejo, J.; Chambers, J.B.; Evangelista, A.; Griffin, B.P.; Iung, B.; Otto, C.M.; Pellikka, P.A.; Quiñones, M. Echocardiographic Assessment of Valve Stenosis: EAE/ASE Recommendations for Clinical Practice. *J. Am. Soc. Echocardiogr.* **2008**, *22*, 1–23. [CrossRef]
31. Nishimura, R.A.; Otto, C.M.; Bonow, R.O.; Carabello, B.A.; Erwin, J.P.; Guyton, R.A.; O'gara, P.T.; Ruiz, C.E.; Skubas, N.J.; Sorajja, P.; et al. 2014 AHA/ACC Guideline for the Management of Patients With Valvular Heart Disease: Executive Summary. *Circulation* **2014**, *129*, 2440–2492. [CrossRef]
32. Kazemi, A.; Padgett, D.A.; Callahan, S.; Stoddard, M.; Amini, A.A. Relative pressure estimation from 4D flow MRI using generalized Bernoulli equation in a phantom model of arterial stenosis. *Magn. Reson. Mater. Phys. Biol. Med.* **2022**, *35*, 733–748. [CrossRef]
33. Yoganathan, A.P.; Valdes-Cruz, L.M.; Schmidt-Dohna, J.; Jimoh, A.; Berry, C.; Tamura, T.; Sahn, D.J. Doppler Echocardiography Continuous-wave Doppler velocities and gradients across fixed tunnel obstructions: Studies in vitro and in vivo. *Circulation* **1987**, *76*, 657–666. [CrossRef]
34. Donati, F.; Myerson, S.; Bissell, M.M.; Smith, N.P.; Neubauer, S.; Monaghan, M.J.; Nordsletten, D.A.; Lamata, P. Beyond Bernoulli Improving the Accuracy and Precision of Non-invasive Estimation of Peak Pressure Drops. *Circ. Cardiovasc. Imaging* **2017**, *10*, e005207. [CrossRef]
35. Pibarot, P.; Garcia, D.; Dumesnil, J.G. Energy loss index in aortic stenosis. *Circulation* **2013**, *127*, 1101–1104. [CrossRef]
36. Geeraert, P.; Jamalidinan, F.; Hassanabad, A.F.; Sojoudi, A.; Bristow, M.; Lydell, C.; Fedak, P.W.; White, J.A.; Garcia, J. Bicuspid aortic valve disease is associated with abnormal wall shear stress, viscous energy loss, and pressure drop within the ascending thoracic aorta. *Medicine* **2021**, *100*, e26518. [CrossRef]
37. Bach, D.S. Echo/Doppler Evaluation of Hemodynamics after Aortic Valve Replacement: Principles of Interrogation and Evaluation of High Gradients. *JACC Cardiovasc. Imaging* **2010**, *3*, 296–304. [CrossRef]
38. Reddy, Y.N.V.; Miranda, W.R.; Nishimura, R.A. Measuring Pressure Gradients after Transcatheter Aortic Valve Implantation: Rethinking the Bernoulli Principle. *J. Am. Heart Assoc.* **2021**, *10*, e022515. [CrossRef]
39. Levine, R.A.; Jimoh, A.; Cape, E.G.; McMillan, S.; Yoganathan, A.P.; Weyman, A.E. "Seminar on in vitro studies of cardiac flow and their applications for clinical doppler echocardiography-III" Pressure recovery distal to a stenosis: Potential cause of gradient "verestimation" by Doppler echocardiography. *J. Am. Coll. Cardiol.* **1989**, *13*, 706–715. [CrossRef]
40. Casas, B.; Lantz, J.; Dyverfeldt, P.; Ebbers, T. 4D Flow MRI-based pressure loss estimation in stenotic flows: Evaluation using numerical simulations. *Magn. Reson. Med.* **2016**, *75*, 1808–1821. [CrossRef]
41. White, F.M. *Viscous Fluid Flow*, 2nd ed.; McGraw-Hill: New York, NY, USA, 1991.
42. Urbina, J.; Sotelo, J.; Andía, M.; Tejos, C.; Hurtado, D.; Irarrázabal, P.; Uribe, S. Relative pressure measurement in thoracic aorta and pulmonary artery of healthy volunteers and repaired Tetralogy of Fallot patients using the 4D Flow sequence of cardiac magnetic resonance. *Rev. Chil. De Radiol.* **2012**, *18*, 157–162.
43. Kilner, P.J. Imaging congenital heart disease in adults. *Br. J. Radiol.* **2011**, *84*, S258–S268. [CrossRef]
44. Valbuena-López, S.; Refoyo, E.; Rosillo, S.; Guzmán, G. Advanced Cardiovascular Magnetic Resonance Techniques in Grown-Up Congenital Heart Disease. *Curr. Cardiovasc. Imaging Rep.* **2018**, *11*, 9. [CrossRef]
45. Geva, T. Repaired tetralogy of Fallot: The roles of cardiovascular magnetic resonance in evaluating pathophysiology and for pulmonary valve replacement decision support. *J. Cardiovasc. Magn. Reson.* **2011**, *13*, 9. [CrossRef]
46. Garcia, J.; Barker, A.J.; Markl, M. The Role of Imaging of Flow Patterns by 4D Flow MRI in Aortic Stenosis. *JACC: Cardiovasc. Imaging* **2019**, *12*, 252–266. [CrossRef] [PubMed]
47. François, C.J.; Srinivasan, S.; Schiebler, M.L.; Reeder, S.B.; Niespodzany, E.; Landgraf, B.R.; Wieben, O.; Frydrychowicz, A. 4D cardiovascular magnetic resonance velocity mapping of alterations of right heart flow patterns and main pulmonary artery hemodynamics in tetralogy of Fallot. *J. Cardiovasc. Magn. Reson.* **2012**, *14*, 16. [CrossRef] [PubMed]
48. Ota, H.; Higuchi, S.; Sun, W.; Ueda, T.; Takase, K.; Tamura, H. Four-Dimensional Flow Magnetic Resonance Imaging for Cardiovascular Imaging: From Basic Concept to Clinical Application. *Cardiovasc. Imaging Asia* **2018**, *2*, 85–96. [CrossRef]
49. Schäfer, M.; Browne, L.P.; Morgan, G.J.; Barker, A.J.; Fonseca, B.; Ivy, D.; Mitchell, M.B. Reduced proximal aortic compliance and elevated wall shear stress after early repair of tetralogy of Fallot. *J. Thorac. Cardiovasc. Surg.* **2018**, *156*, 2239–2249. [CrossRef]
50. Hirtler, D.; Garcia, J.; Barker, A.; Geiger, J. Assessment of intracardiac flow and vorticity in the right heart of patients after repair of tetralogy of Fallot by flow-sensitive 4D MRI. *Eur. Radiol.* **2016**, *26*, 3598–3607. [CrossRef]

51. Hu, L.; Ouyang, R.; Sun, A.; Wang, Q.; Guo, C.; Peng, Y.; Qin, Y.; Zhang, Y.; Xiang, Y.; Zhong, Y. Pulmonary artery hemodynamic assessment of blood flow characteristics in repaired tetralogy of Fallot patients versus healthy child volunteers. *Quant. Imaging Med. Surg.* **2020**, *10*, 921–933. [CrossRef]
52. Jeong, D.; Anagnostopoulos, P.V.; Roldan-Alzate, A.; Srinivasan, S.; Schiebler, M.; Wieben, O.; François, C.J. Ventricular kinetic energy may provide a novel noninvasive way to assess ventricular performance in patients with repaired tetralogy of Fallot. *J. Thorac. Cardiovasc. Surg.* **2015**, *149*, 1339–1347. [CrossRef]
53. Jeong, D.; Roldan-Alzate, A.; François, C.J. Right ventricular kinetic energy: 4D flow MRI analysis of healthy volunteers and repaired Tetralogy of Fallot. *J. Cardiovasc. Magn. Reson.* **2014**, *16*, O46. [CrossRef]
54. Latus, H.; Stammermann, J.; Voges, I.; Waschulzik, B.; Gutberlet, M.; Diller, G.; Schranz, D.; Ewert, P.; Beerbaum, P.; Kühne, T.; et al. Impact of Right Ventricular Pressure Load After Repair of Tetralogy of Fallot. *J. Am. Heart Assoc.* **2022**, *11*, e022694. [CrossRef]
55. Markl, M.; Schnell, S.; Barker, A. 4D Flow Imaging: Current Status to Future Clinical Applications. *Curr. Cardiol. Rep.* **2014**, *16*, 481. [CrossRef]
56. Rose, M.J.; Jarvis, K.; Chowdhary, V.; Barker, A.J.; Allen, B.D.; Robinson, J.D.; Markl, M.; Rigsby, C.K.; Schnell, S. Efficient method for volumetric assessment of peak blood flow velocity using 4D flow MRI. *J. Magn. Reson. Imaging* **2016**, *44*, 1673–1682. [CrossRef]
57. Ha, H.; Kim, G.B.; Kweon, J.; Kim, Y.-H.; Kim, N.; Yang, D.H.; Lee, S.J. Multi-VENC acquisition of four-dimensional phase-contrast MRI to improve precision of velocity field measurement. *Magn. Reson. Med.* **2016**, *75*, 1909–1919. [CrossRef]
58. Brindise, M.C.; Rothenberger, S.; Dickerhoff, B.; Schnell, S.; Markl, M.; Saloner, D.; Rayz, V.L.; Vlachos, P.P. Multi-modality cerebral aneurysm haemodynamic analysis: In vivo 4D flow MRI, in vitro volumetric particle velocimetry and in silico computational fluid dynamics. *J. R. Soc. Interface* **2019**, *16*, 20190465. [CrossRef]
59. Medero, R.; Falk, K.; Rutkowski, D.; Johnson, K.; Roldán-Alzate, A. In Vitro Assessment of Flow Variability in an Intracranial Aneurysm Model Using 4D Flow MRI and Tomographic PIV. *Ann. Biomed. Eng.* **2020**, *48*, 2484–2493. [CrossRef]
60. Kitajima, H.D.; Sundareswaran, K.S.; Teisseyre, T.Z.; Astary, G.W.; Parks, W.J.; Skrinjar, O.; Oshinski, J.N.; Yoganathan, A.P. Comparison of Particle Image Velocimetry and Phase Contrast MRI in a Patient-Specific Extracardiac Total Cavopulmonary Connection. *J. Biomech. Eng.* **2008**, *130*, 041004. [CrossRef]
61. Zhong, L.; Schrauben, E.M.; Garcia, J.; Uribe, S.; Grieve, S.M.; Elbaz, M.S.; Barker, A.J.; Geiger, J.; Nordmeyer, S.; Marsden, A.; et al. Intracardiac 4D Flow MRI in Congenital Heart Disease: Recommendations on Behalf of the ISMRM Flow & Motion Study Group. *J. Magn. Reson. Imaging* **2019**, *50*, 677–681. [CrossRef]
62. García, J.; LaRose, E.; Pibarot, P.; Kadem, L. On the Evaluation of Vorticity Using Cardiovascular Magnetic Resonance Velocity Measurements. *J. Biomech. Eng.* **2013**, *135*, 124501. [CrossRef]
63. Sotelo, J.; Urbina, J.; Valverde, I.; Mura, J.; Tejos, C.; Irarrazaval, P.; Andia, M.E.; Hurtado, D.E.; Uribe, S. Three-dimensional quantification of vorticity and helicity from 3D cine PC-MRI using finite-element interpolations. *Magn. Reson. Med.* **2018**, *79*, 541–553. [CrossRef]
64. Fukuyama, A.; Isoda, H.; Morita, K.; Mori, M.; Watanabe, T.; Ishiguro, K.; Komori, Y.; Kosugi, T. Influence of Spatial Resolution in Three-dimensional Cine Phase Contrast Magnetic Resonance Imaging on the Accuracy of Hemodynamic Analysis. *Magn. Reson. Med. Sci.* **2017**, *16*, 311–316. [CrossRef] [PubMed]
65. Sadeghi, R.; Tomka, B.; Khodaei, S.; Daeian, M.; Gandhi, K.; Garcia, J.; Keshavarz-Motamed, Z. Impact of extra-anatomical bypass on coarctation fluid dynamics using patient-specific lumped parameter and Lattice Boltzmann modeling. *Sci. Rep.* **2022**, *12*, 9718. [CrossRef] [PubMed]
66. Canstein, C.; Cachot, P.; Faust, A.; Stalder, A.; Bock, J.; Frydrychowicz, A.; Küffer, J.; Hennig, J.; Markl, M. 3D MR flow analysis in realistic rapid-prototyping model systems of the thoracic aorta: Comparison with in vivo data and computational fluid dynamics in identical vessel geometries. *Magn. Reson. Med.* **2008**, *59*, 535–546. [CrossRef] [PubMed]
67. Lamata, P.; Pitcher, A.; Krittian, S.; Nordsletten, D.; Bissell, M.; Cassar, T.; Barker, A.; Markl, M.; Neubauer, S.; Smith, N.P. Aortic relative pressure components derived from four-dimensional flow cardiovascular magnetic resonance. *Magn. Reson. Med.* **2014**, *72*, 1162–1169. [CrossRef]
68. Fathi, M.F.; Perez-Raya, I.; Baghaie, A.; Berg, P.; Janiga, G.; Arzani, A.; D'souza, R.M. Super-resolution and denoising of 4D-Flow MRI using physics-Informed deep neural nets. *Comput. Methods Programs Biomed.* **2020**, *197*, 105729. [CrossRef]

fluids

Article

Inlet and Outlet Boundary Conditions and Uncertainty Quantification in Volumetric Lattice Boltzmann Method for Image-Based Computational Hemodynamics

Huidan Yu [1,2,*], Monsurul Khan [1,†], Hao Wu [1], Chunze Zhang [3], Xiaoping Du [1], Rou Chen [1,‡], Xin Fang [4], Jianyun Long [4] and Alan P. Sawchuk [2]

1 Department of Mechanical and Energy Engineering, Indiana University-Purdue University Indianapolis, Indianapolis, IN 46202, USA; khan212@purdue.edu (M.K.); hw51@iu.edu (H.W.); duxi@iu.edu (X.D.); rouchen@cjlu.edu.cn (R.C.)
2 Department of Surgery, Division of Vascular Surgery, Indiana University School of Medicine, Indianapolis, IN 46202, USA; asawchuk@iupui.edu
3 Southwest Institute of Water Transport Engineering, Chongqing Jiaotong University, Chongqing 400074, China; zhangchunze@whu.edu.cn
4 Department of Vascular Surgery, The Affiliated Hangzhou First People's Hospital, Zhejiang University School of Medicine, Hangzhou 310006, China; fangxin324@hotmail.com (X.F.); longjianyun1208@126.com (J.L.)
* Correspondence: whyu@iupui.edu
† Current address: School of Mechanical Engineering, Purdue University, West Lafayette, IN 47907, USA.
‡ Current address: College of Metrology and Measurement Engineering, China Jiliang University, Hangzhou 310004, China.

Citation: Yu, H.; Khan, M.; Wu, H.; Zhang, C.; Du, X.; Chen, R.; Fang, X.; Long, J.; Sawchuk, A.P. Inlet and Outlet Boundary Conditions and Uncertainty Quantification in Volumetric Lattice Boltzmann Method for Image-Based Computational Hemodynamics. *Fluids* **2022**, *7*, 30. https://doi.org/10.3390/fluids7010030

Academic Editors: Mehrdad Massoudi and Goodarz Ahmadi

Received: 3 November 2021
Accepted: 30 December 2021
Published: 10 January 2022

Abstract: Inlet and outlet boundary conditions (BCs) play an important role in newly emerged image-based computational hemodynamics for blood flows in human arteries anatomically extracted from medical images. We developed physiological inlet and outlet BCs based on patients' medical data and integrated them into the volumetric lattice Boltzmann method. The inlet BC is a pulsatile paraboloidal velocity profile, which fits the real arterial shape, constructed from the Doppler velocity waveform. The BC of each outlet is a pulsatile pressure calculated from the three-element Windkessel model, in which three physiological parameters are tuned by the corresponding Doppler velocity waveform. Both velocity and pressure BCs are introduced into the lattice Boltzmann equations through Guo's non-equilibrium extrapolation scheme. Meanwhile, we performed uncertainty quantification for the impact of uncertainties on the computation results. An application study was conducted for six human aortorenal arterial systems. The computed pressure waveforms have good agreement with the medical measurement data. A systematic uncertainty quantification analysis demonstrates the reliability of the computed pressure with associated uncertainties in the Windkessel model. With the developed physiological BCs, the image-based computation hemodynamics is expected to provide a computation potential for the noninvasive evaluation of hemodynamic abnormalities in diseased human vessels.

Keywords: volumetric lattice Boltzmann method; image-based computational hemodynamics; three-element Windkessel model; boundary conditions; uncertainty quantification

1. Introduction

With the recent advances in medical imaging, computational power, and mathematical algorithms, image-based computational hemodynamics (ICHD) has emerged [1–7] as a new capability giving rise to the potential for computation-aided diagnostics and therapeutics in a patient-specific environment for cardiovascular diseases. Based on radiological imaging data, such as computed tomography angiography (CTA) images and Doppler ultrasound (DUS) velocity waveforms, ICHD enables noninvasive and patient-specific quantification of pulsatile hemodynamics in human vessels. Such data, including velocity vector, pressure, vorticity vector, and wall-shear stress (WSS) in the entire artery segment with fine spatial and temporal resolutions, are not readily available from the current standard clinical measurements. Through further postprocessing of the pulsatile hemodynamic data, either the assessment of the true hemodynamic abnormality or the prediction of potential therapeutic/surgical outcomes from an interventional treatment may aid in clinical decision-making for various cardiovascular diseases.

A typical ICHD from medical data to medical insights mainly consists of three steps. They are (1) image extractions of a three-dimensional anatomical geometry of the diseased artery from CTA data and one-dimensional velocity waveforms from DUS images at inlet and outlets, (2) computation of pulsatile hemodynamics employing physical parameters together with the flow environment, and (3) post-processing of the computed pulsatile hemodynamics with analysis, visualization, and parametrization to the key insights of the disease assessment and potential therapeutic outcomes. Since only a segment of the blood circulation system is being computed, boundary conditions (BCs) are required to be applied at the inlet(s) and outlet(s) of the vessel segment to represent the remaining vascular network. In general, the introduction of the inlet BC is relatively straightforward, imposing parabolic-like flow profiles at the cross-section of the inlet. Usually, an inlet cross-section of a human vessel is not a perfect circle. Therefore, neither a steady Poiseuille-Hagen nor an oscillating Womersley velocity profile can be directly constructed based on a DUS-measured velocity waveform. The choices of the outflow BC in ICHD vary among zero pressure or zero traction conditions, resistance or impedance conditions, reduced-order models which can be an open or closed loop, and reduced-order one-dimensional wave propagation equations [8–10]. To capture the interaction between the local three-dimensional artery segment and the one-dimensional global circulation, the three-dimensional flow solver must be coupled to a reduced-order lumped parameter network model. Among them, the three-element Windkessel model [11–15] (WK3) has been commonly used to construct such a network, in which a Windkessel circuit is adopted to model the distal vasculature with one capacitor, modeling vessel compliance, and two resistors, modeling proximal and distal flow resistances, respectively. Evidence has shown that the WK3 can well reproduce physiological pressure waveforms [16,17] in large vessels.

In this work, we present the physiological inlet and outlet BCs in volumetric lattice Boltzmann method (VLBM) [18] for ICHD together with uncertainty quantification (UQ). The lattice Boltzmann method (LBM) [19,20] is a class of computational fluid dynamics (CFD) methods for solving complex flows. Instead of directly solving a set of nonlinear partial differential equations, i.e., Navier-stokes (NS) equations, the LBM is a discretized kinetic model on a regular lattice to solve the dynamics of incompressible fluid flow. Due to its particulate nature and local dynamics, the LBM has its advantages over the NS-based CFD methods, especially in dealing with complex boundaries [18,21], incorporating microscopic interactions [22,23] in multiphase flows, and implementing GPU (Graphics Processing Unit) parallel computing [21,24]. Nevertheless, the LBM has not been extensively used for ICHD so far and the majority of attempts have imposed non-physiological BCs. An example is the zero pressure BC [25–28] at the outlets. The zero pressure BC, although easy to be implemented, is well known to lead to unrealistic hemodynamics, in part because of its inability to capture physiologic levels of pressure [6]. Few other studies have used the fully developed BC [18,29] at the outlets, which is also inappropriate for a pulsatile flow in arbitrary flow domains. In this paper, we develop the physiological velocity BC at the inlet based on the DUS waveform and pressure BC at each outlet via WK3 model tuned by the corresponding DUS waveform and then integrated them into the VLBM. We study six aortorenal arterial systems, with given CTA image data and DUS velocity waveforms of each, for noninvasive quantification of pulsatile hemodynamics. To demonstrate the accuracy of the computation, we compare the computed pressure waves with the corresponding invasive pressure measurements during digital subtraction angiography (DSA) in the clinic. Meanwhile, we perform uncertainty quantification to demonstrate the reliability of the computation.

2. Methods and Materials

We have previously developed and validated a VLBM solver [30] for solving image-based pore-scale porous media flows. The solver synergistically employs the traditional node-based LBM for image segmentation and the cell-based VLBM [18] for CFD, enabling a seamless connection between these two parts and unified GPU parallelization for fast computation [21,31,32].

The VLBM is formulated on a cell-based mesh. Fluid particles are uniformly distributed in lattice cells, as opposed to sitting at lattice nodes in conventional LBM. As schematized in Figure 1, an arbitrary boundary (black line) separates a fluid domain (without dots) from a solid boundary structure (with dots). Three distinct cells are characterized through a volumetric parameter, i.e., the occupation of solid volume $\Delta V_s(x)$ in the cell with total volume $\Delta V(x)$, defined as $\mathcal{P}(x) \equiv \Delta V_s(x)/\Delta V(x)$. Thus three different cells, fluid cell ($\mathcal{P} = 0$), solid cell ($\mathcal{P} = 1$), and boundary cell ($0 < \mathcal{P} < 1$), can be distinguished.

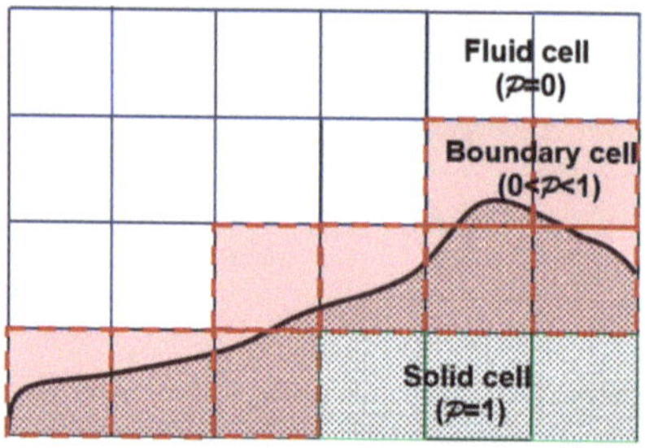

Figure 1. Three types of lattice cells in VLBM: fluid cell ($\mathcal{P} = 0$), solid cell ($\mathcal{P} = 1$), and boundary cell ($0 < \mathcal{P} < 1$). The solid line represents an arbitrary boundary of the flow domain.

On a lattice space with b directions of discrete molecular velocity, VLBM deals with the time evolution of the particle population, $n_i(x,t)$, corresponding to the ith velocity e_i

$$n_i(x + e_i\delta t, t + \delta t) = n_i(x,t) - \frac{\left[n_i(x,t) - n_i^{eq}(x,t)\right]}{\tau}; i = 0, \ldots, b \tag{1}$$

where $n_i^{eq}(x,t)$ and τ are the corresponding equilibrium particle population and relaxation time, respectively. The resulting density $\rho(x,t)$ and velocity $u(x,t)$ in the fluid domain are

$$\rho(x,t) = \sum n_i(x,t)/[1 - \mathcal{P}(x,t)] \tag{2}$$

and

$$u(x,t) = \sum e_i n_i(x,t)/\sum n_i(x,t) \tag{3}$$

The pressure field $p(x,t)$ is then calculated from:

$$p(x,t) - p_0 = c_s^2[\rho(x,t) - \rho_0] \tag{4}$$

where p_0 and ρ_0 are reference pressure and density, respectively. In this work, we adapted the VLBM solver to ICHD, named *InVascular*, based on medical imaging data.

The implementation flow chart of *InVascular* is shown in Figure 2. It starts with the image segmentation from CTA image data to extract the anatomical geometry using the conventional LBM with D3Q7 lattice model [33]. A distance field [34] governed by a level set equation [35] is solved in which the zero-level distance represents the morphological boundary of the vessel segment. From the distance field, $\mathcal{P}(x)$ of each cell is calculated and then, together with the inlet and outlet BCs, fed to VLBM [18] with D3Q19 lattice model. Both image segmentation and computational hemodynamics (dashed part in Figure 2) are carried out on a unified mesh and connected seamlessly. Thus the state-of-the-art GPU parallelism can be efficiently utilized. The detailed formulation of LBM for image segmentation and VLBM computational hemodynamics, as well as the connection between them, and the GPU parallelization are referred to in our published papers [21,30]. In this paper, we focus on the integration of the physiological inlet and outlet BCs with the VLBM, as highlighted in Figure 2.

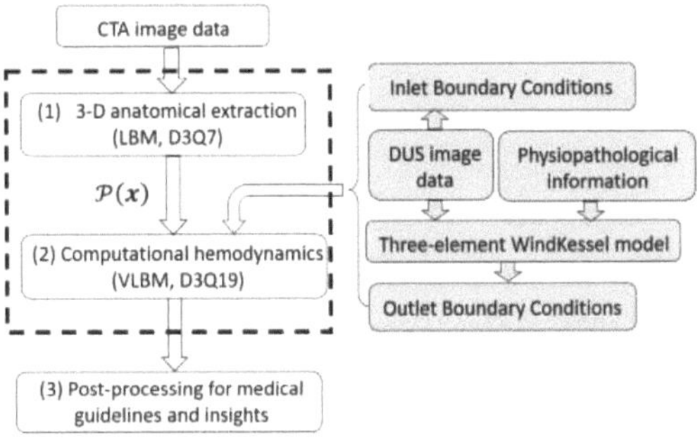

Figure 2. Flow chart of *InVascular*: (1) 3-D anatomical extraction of vessel segment from CTA image data; (2) ICHD with the inputs of $\mathcal{P}(x)$ and inlet and outlet BCs based on DUS image data as well as three-element WK model; and (3) post-processing for intepretation and medical insights. The unified LBM (dashed part) is accelerated by GPU parallelism.

2.1. Physiological Inlet and Outlet Boundary Conditions

In this part, we present the algorithms to construct the physiological inlet and outlet BCs based on the DUS velocity waveforms. The inlet BC is a velocity profile, and the output BC is a pressure calculated from WK3. Both velocity and pressure BCs are introduced into VLBM.

2.1.1. Implementation of Velocity and Pressure BCs in VLBM

In the VLBM, we employed the non-equilibrium extrapolation boundary condition developed by Guo et al. [36] as follows.

$$n_i(x_b,t) - n_i^{eq}(x_b,t) = n_i\left(x_f,t\right) - n_i^{eq}\left(x_f,t\right) \tag{5}$$

for i-th direction where x_b and x_f are the boundary cell and its next fluid cell along that direction, respectively. If the velocity, $u(x_b,t)$, is known at the boundary cell, the velocity BC is:

$$n_i(x_b,t) = n_i^{eq}\left(\rho\left(x_f,\ t\right), u(x_b,t)\right) + n_i\left(x_f,t\right) - n_i^{eq}\left(x_f,t\right) \tag{6}$$

Whereas if the pressure $p(x_b,t)$ is given at the boundary cell, the pressure BC reads:

$$n_i(x_b,t) = n_i^{eq}\left(\rho(x_b,t), u\left(x_f,\ t\right)\right) + n_i\left(x_f,t\right) - n_i^{eq}\left(x_f,t\right) \tag{7}$$

where $\rho(x_b,t)$ is calculated from Equation (4). We use the velocity BC and pressure BC at the inlet and each outlet, respectively.

In *InVascular*, the inlet and outlet BCs are based on the patient's DUS data, as shown by the shaded part in Figure 2. We present the introduction of the inlet and outlet BCs in the following subsections.

2.1.2. Lumen-Fitted Velocity BC Profile at an Inlet

The DUS measured velocity waveform, $u_{in}(t)$, has been commonly used as the inflow BC [6] in ICHD. For a pipe with its radius of R, the typical way to introduce the pulsatile velocity to drive the flow into the pipe is to construct a parabolic profile of Poiseuille flow, $u(r,t) = u_{in}(r,t)(1 - r^2/R^2)$, in which r is the distance to the pipe center. However, real arterial lumens are often not perfectly circular. To use this parabolic velocity profile, one needs to extend the inlet from noncircular to circular, which may introduce an unrealistic inflow.

We present an algorithm, as illustrated in Figure 3, for an irregular paraboloid-like velocity profile that fits the real inlet cross-section. The velocity waveform of $u_{in}(t)$ is digitized from the patient's DUS shown in Figure 3a. It should be noted that, for a blood flow, the inflow velocity is pulsatile thus the irregular velocity profile needs to be constructed at every time point and the time resolution should be fine enough, determined through a temporal convergence check. To refine the temporal resolution, we use linear interpolation.

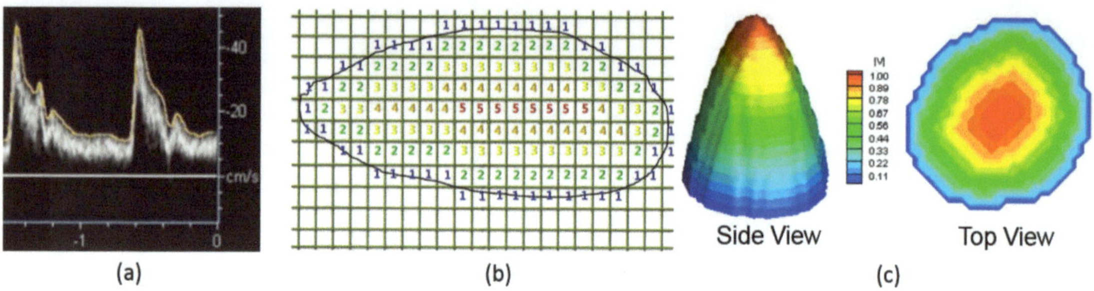

Figure 3. Illustration of inlet boundary condition from DUS image data for an irregular artery plane. (**a**) A generic DUS image recording velocity magnitude waveform $u_{in}(t)$. (**b**) An example of indexing to construct an irregular paraboloidal velocity profile on the inlet plane. (**c**) Normalized velocity distribution on inlet plane varying from u_{in} at the center to zero at the edge.

Assume the inlet plane is perpendicular to the z-direction, i.e., the direction of the bloodstream, and it is located at $z = z_0$. On the plane, each cell has known $\mathcal{P}(i, j, z_0)$ with $i = 1, \ldots, N_x$ and $j = 1, \ldots, N_y$. The algorithm to generate an irregular paraboloidal velocity profile at time t includes the following steps, schematized in Figure 3b.

(1) Declare a matrix $N_x \times N_y$, i.e., L_{ij} $(i = 1, \ldots, N_x,\ j = 1, \ldots, N_y)$ and initialize it as $L_{ij} = 0$.
(2) Loop i from 1 to N_x and j from 1 to N_y, if
   a. a cell's $\mathcal{P}$ is neither 0 nor 1 (i.e., a boundary cell), assign $L_{ij} = 0$ for this cell and define its velocity magnitude 0,
   b. a cell's $\mathcal{P}$ is 0 (i.e., a fluid cell) and the L_{ij} value of any neighboring cell is 0, assign $L_{ij} = 1$ for this cell,
   c. a cell's $\mathcal{P}$ is 0 and the L_{ij} value of any neighboring cell is 1, assign $L_{ij} = 2$ for this cell,
   d. continue until all the fluid cells are assigned. The last index of the cell labeling is $L_{ij} = M$.
(3) Loop i from 1 to N_x and j from 1 to N_y and define velocity magnitude as $u_{ij}(t) = L_{ij} \times u_{in}(t)/M$.

The largest velocity $u_{in}(t)$ is recognized at the cell labeled as $L_{ij} = M$. The velocity reduces radially from $u_{in}(t)$ at label M (center) to zero at label $L_{ij} = 0$ (wall). Figure 3c shows two views of paraboloid-like velocity distribution on an irregular inlet plane at a time point. The inlet velocity profile is introduced in VLBM through Equation (6).

2.1.3. WK3-Based Pressure BC at an Outlet

For the outlet BC, we use the popular WK3 model in an open vessel loop to calculate the pressure, $p(t)$, on the outlet plane. It has been well-known that WK3 is the best outlet BC model among other physiologically relevant zero-dimenensional outflow models to simulate the peripheral vasculature [37] and has been popularly used when significant compliance is located in the modeled distal vasculature [12]. As shown in Figure 4, WK3 is an analogy to an electrical circuit, which models the distal vasculature with one capacitor, modeling vessel compliance and two resistors, modeling proximal and distal resistance. The flow rate (Q) and the mean pressure (p) over the outlet plane are related by the following ordinary differential equation [12]

$$\frac{dp}{dt}+\frac{1}{RC}p=r\frac{dQ}{dt}+\frac{1}{RC}(r+R)Q \quad (8)$$

where r and R represent the proximal and distal resistances, and C is the compliance of the distal vasculature. Specifically, r is used to absorb the incoming waves and reduce artificial wave reflections. Equation (8) has an analytical solution.

$$p(t)=e^{-t/(RC)}\int_0^t e^{s/(RC)}\left[rdQ(s)/ds+\left(r+\frac{RQ(s)}{RC}\right)\right]ds+p_{t=0} \quad (9)$$

where $p_{t=0}$ is the initial pressure at the outlet.

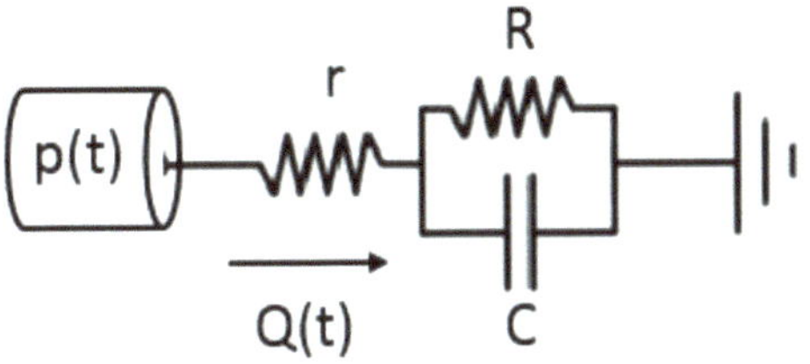

Figure 4. WK3 model consists of one capacitor (C), modeling vessel compliance, and two resistors (r and R), modeling proximal and distal resistance, respectively.

In Equation (9), the three elements, r, C, and R, specified at each outlet, must be tuned to obtain the physiological values for the total outflow rate Q_t and target systolic (p_{sys}) and diastolic (p_{dia}) pressure, with the mean arterial pressure, $p_m = (p_{sys}+2p_{dia})/3$, based on patient's clinical data. For an aortorenal system, see Figure 5 below, we use brachial pressure for a pressure target and DUS velocity waveform for the target flow rate (Q) with the understanding that the capacitor and resistors have independent functionalities in the WK3 circuit: a capacitor reflects the pulsatility of blood flow whereas a resistor determines the flow rate [15]

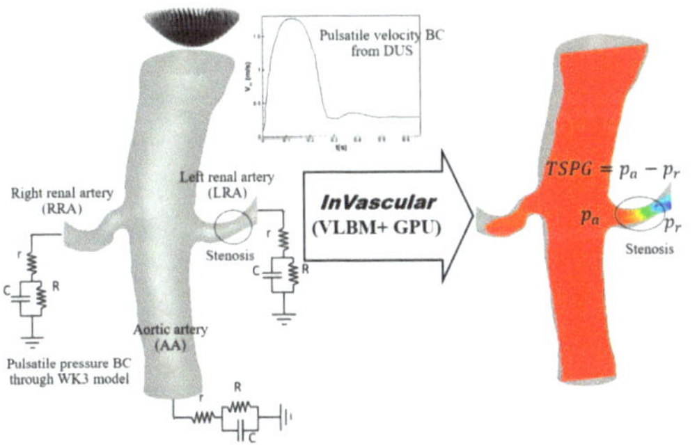

Figure 5. Integration of *InVascular* with velocity BC from DUS at inlet and pressure BCs through the WK3 model at outlets for quantification of TSPG ($\equiv p_a - p_r$) in an aortorenal system extracted from patient's CTA.

The integration of the WK3 model [15] and VLBM is described as follows

(1) Determine the total resistance in the arterial segment

a. Assume the total system compliance $C_t = 0.1\ \text{cm}^5/\text{dynes}$.

b. Calculate the total resistance $R_t(= r+R) = p_m/Q_t$.

(2) Determine r and R at each outlet based on the published works: the proximal resistance r weights 28% [38,39] and 5.6% [40] out of the total resistance in the renal artery and abdominal aorta, respectively.

(3) Tune r and R based on DUS flow rate at each outlet.

   a. Integrate the pressure BC from the WK3, Equation (9), into VLBM, Equation (7), and run *InVascular*. In one pulsation, r, R, and C remain the same but $Q\,(t)$ at each outlet is obtained from the simulation.

   b. Once a simulation is done, check if the flow rate at each outlet matches that calculated from DUS imaging data. If yes, r and R are determined; If not, adjust R_t and repeat (1) b, (2), and (3).

(4) Determine the compliance C at each outlet.

   a. Distribute C_t to each outlet proportional to the corresponding mean flow rate.

   b. Check if the mean arterial pressure $\overline{p}_{in}^{CHD}$ matches $\overline{p}_{in}$ at the inlet. If not, adjust C_t in (1) a. and repeat (1) and (2).

The outlet BC at each outlet is introduced in VLBM through Equations (4) and (7) after the pressure is obtained from Equation (9) at each time step.

2.2. Uncertainty Quantification

There is no doubt that uncertainty always exists in any modeling and simulation process [41]. In the process shown in the flowchart of *InVascular* in Figure 2, uncertainties come from noises introduced during image scanning and the extraction of the arterial segmentation, the use of empirical blood properties, parameters involved in the boundary conditions, and so on. The uncertainties in the input variables will affect the output (hemodynamics) of *InVascular*. Following the common practice in uncertainty quantification (UQ), we treat the parameters with uncertainty as random variables and quantify their effects on the output variables. In this study, we use the common UQ method: the First Order Second Moment (FOSM) [42] method.

Denote output variables by $\mathbf{Y} = (Y_1, Y_2, \ldots, Y_m)^{\mathrm{T}}$ and input variables by $\mathbf{X} = (X_1, X_2, \ldots, X_n)^{\mathrm{T}}$, where m and n are the numbers of output and input variables, respectively. If the elements of $\mathbf{X}$ are non-normally distributed and dependent, Rosenblatt transformation [43] can be used to transform $\mathbf{X}$ into independent and normal variables.

Suppose the black-box models of *InVascular* are given by:

$$Y_j = g_j(\mathbf{X}),\quad j = 1, 2, \ldots, m) \tag{10}$$

Linearizing a model at the mean values, $\boldsymbol{\mu} = (\mu_1, \mu_2, \ldots, \mu_n)$, of $\mathbf{X}$, yields

$$Y_j \approx g_j(\boldsymbol{\mu}) + \nabla^{\mathrm{T}}(\mathbf{X} - \boldsymbol{\mu}),\ j = 1, 2, \ldots, m \tag{11}$$

where $\nabla = \left(\frac{\partial g}{\partial X_1}, \frac{\partial g}{\partial X_2}, \ldots, \frac{\partial g}{\partial X_n}\right)^T$ is the gradient of $g_j(\mathbf{X})$ at $\boldsymbol{\mu}$. Since Y_j is approximated as a linear combination of $\boldsymbol{x}$, it is also normally distributed, denoted by $N\left(\mu_{Y_j}, \sigma_{Y_j}^2\right)$ with μ_{Y_j} and σ_{Y_j}, the mean and standard deviation of Y_j, respectively. The two parameters are given as follows.

$$\mu_{Y_j} = g_j(\mathbf{X}),\ j = 1, 2, \ldots, m \tag{12}$$

$$\sigma_{Y_j}^2 = \sum_{i=1}^{n}\left(\frac{\partial g_j}{\partial X_i}\right)^2 \sigma_i^2,\ j = 1, 2, \ldots, m \tag{13}$$

where σ_i is the standard deviation of X_i. The covariance between output variables Y_j and Y_k is calculated by

$$C_{jk} = \sum_{i=1}^{n}\left(\frac{\partial g_j}{\partial X_i}\right)\left(\frac{\partial g_k}{\partial X_i}\right)\sigma_i^2 \tag{14}$$

Then the joint probability density function (PDF) of output $\mathbf{Y} = (Y_1, Y_2, \ldots, Y_m)^T$ is determined by the mean vector $\boldsymbol{\mu}_\mathbf{Y} = \left(\mu_{Y_1}, \mu_{Y_2}, \ldots, \mu_{Y_m}\right)^T$ and covariance matrix $\Sigma_Y = \left(C_{jk}\right)_{j,k=1,2,\ldots,m}$. Since $g_j(\mathbf{X})$ is a black box, its gradient is evaluated numerically by the finite difference method. The total computational cost, measured by the number of function (model) calls, is $n+1$. The efficiency is high since the number of function calls is linearly proportional to the dimensionality of input variables.

2.3. Materials

We studied six human aortorenal arterial systems. The medical data of each case included CTA images and DUS waveforms at the inlet and outlets, obtained from the electronic medical libraries in Indiana University Methodist Hospital in Indianapolis, Indiana, USA, and Hangzhou First People's Hospital, Hangzhou, China. The CTA resolutions are approximately 0.752×2.5 mm^3 (US cases) and 0.652×0.6mm^3 (China cases).

We show one representative case in Figure 5 to demonstrate the integration of VLBM and physiological BCs at inlet and outlets, for *InVascular*. The aortorenal arterial system, anatomically extracted from the patient's CTA data, consists of the aortic artery (AA), left renal artery (LRA), and right renal artery (RRA). The inlet BC based on the DUS velocity waveform and outlet BCs of WK3 are imposed at the inlet and three outlets, respectively. A minor lumen reduction (circled, about 20% lumen reduction) is seen in the LRA. The DUS images are available at the AA inlet to construct a paraboloidal velocity profile and outlets of AA, LRA, and RRA to tune the r, R, and C parameters. The physical flow domain is $63 \times 116 \times 84$ mm^3. The cardiac cycle is 0.68 s with a time resolution of 6.79 ms. The density and kinematic viscosity are 1.06×10^3 kg/m^3 and 3.3×10^{-6} m/s^2, respectively. The dimensionless relaxation time τ in VLBM is 0.5079.

The WK3 parameters r, R, and C at the three outlets are listed in Table 1. The pulsatile pressure waveforms in AA, LRA, and RRA were invasively measured during a clinical intervention, which are used to validate the computed pressure below. As seen in Figure 6 below, for a given pressure waveform, the pressure values at the peak and the end of the waveform are called systolic blood pressure, p_{sys}, and diastolic blood pressure, p_{dia}, respectively. The mean arterial pressure (MAP) is defined as $MAP = \left(p_{sys} - p_{dia}\right)/3 + p_{dia}$.

Table 1. Values of resistance and compliance parameters, *r*, *R*, and *C*, in WK3 model at corresponding outlets tuned from the DUS data.

Outlet	r (dynes×s/cm^5)	R (dynes×s/cm^5)	10^5C (cm^5/dynes)
AA	88.0	2773.1	1.8
LRA	2982.4	7666.03	0.36
RRA	5972.8	15358.7	0.32

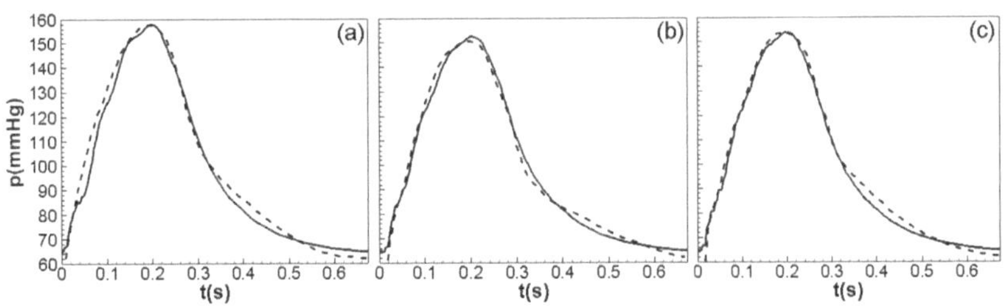

Figure 6. Comparisons of pressure waveforms in (**a**) AA, (**b**) RRA, and (**c**) LRA between noninvasive computation (solid line) and invasive measurement (dashed line).

The spatial and temporal convergence checks are exhibited in Table 2. The relative errors are the normalized differences of the mean arterial pressure (MAP) and systolic pressure (P_{sys}) between two successive grids and cycles, respectively. To balance the accuracy and the computation cost, we chose 200 × 368 × 265 as the resolution for the simulation and run 10 cycles to produce the computational results.

Table 2. Spatial (left) and temporal (right) convergence check. The spatial reoslution is represented by the grid number along the flow direction. MAP and p_{sys} stand for mean arterial presure and systolic pressure, respectively. The relative error is the normalized difference of the correponding pressure values between two sucsessive grids and cycles.

Spatial			Temporal		
Grid	**MAP(mmHg)**	**Relative Error (%)**	**Cycle**	**P_{sys} (mmHg)**	**Relative Error (%)**
170	100		1	150	
180	87.5	12.5	3	154	2.7
190	89	1.71	5	152	−1.3
200	90	0.34	10	155	2.0
210	90.15	0.19	15	155	0
220	90.20	0.05	20	155	0

3. Results

In this section, we demonstrate the applicability and reliability of *InVascular* in two aspects. First, we use the representative study case to show the computed pulsatile pressure, velocity, and vorticity fields in the arterial system. The noninvasively computed pressure waveforms are compared with the invasively measured ones at three locations. Second, we perform a systematic UQ analysis for the representative case and for all six cases to study how the r, R, and C parameters impact the computed pressure.

3.1. Pulsatile Hemodynamics in an Aortorenal Arterial System

We first demonstrate the accuracy of *InVascular* for the quantification of the pulsatile pressure field. Figure 6 shows the comparisons of the cyclic pressure waveforms in (a) AA, (b) RRA, and (c) LRA between noninvasive computation (solid lines) and invasive measurements (dashed lines).

The computed pressure waveforms agree very well with the medical measured waveforms. The pressure contours on (a) the AA-LRA plane, (b)AA-RRA plane, and (c) representative cross-sections are plotted in Figure 7. The trans-stenotic pressure gradient ($TSPG$) in the LRA can be calculated through ether MAP or p_{sys}. The comparison of the $TSPGs$ between noninvasive computation and invasive measurement is shown in Table 3. Again, both are in good agreement.

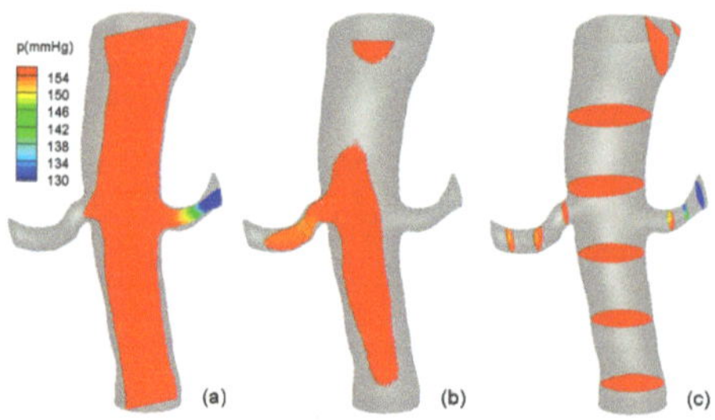

Figure 7. Systolic pressure contours (**a**) the AA-LRA plane, (**b**) AA-RRA plane, and (**c**) representative cross-sections.

Table 3. Comparison of *TSPG* in LRA and RRA based on *MAP* andp_{sys}.

TSPG	*MAP*		p_{sys}	
	Computed	**Measured**	**Computed**	**Measured**
$p_a - p_r$, left	2.5	2.6	4.1	4.0
$p_a - p_r$, right	2.0	2.0	4.0	4.0

We found that the fully-developed BC and DUS-based velocity BC, which are commonly used in the LBM, cannot capture the physiological pressure waveform. Figure 8 shows the cyclic evolution of the systolic pressure at a representative location in the arterial system under three different BCs with identical computation environments and conditions. Neither the fully-developed BC (long dash) nor DUS-based velocity BC (short dash) is convergent. The pressure (left scale) asymptotically increases with time and exceeds the human blood pressure after a few cardiac cycles, whereas the WK3-based pressure BC (solid line) leads to a convergent systolic pressure ($p_{sys} \approx 154.8$ mmHg) (right vertical scale) after 10 cardiac cycles.

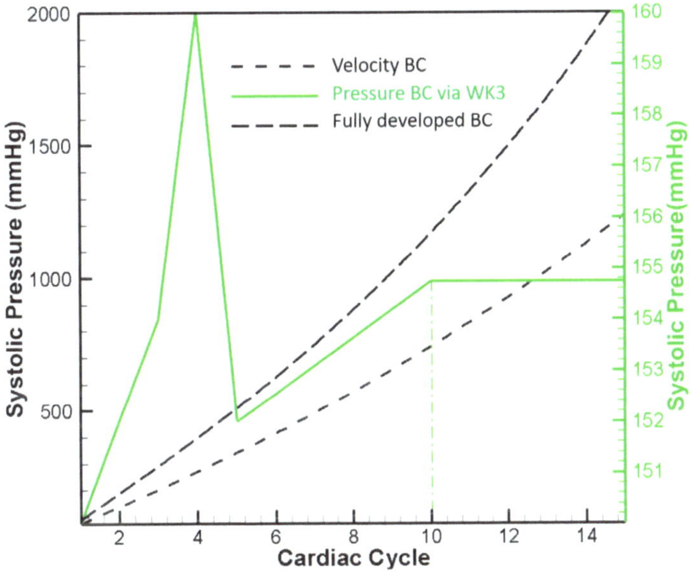

Figure 8. Cyclic evolution of systolic pressure in a representative location using three different BCs: velocity BC, pressure BC via WK3, and fully-developed BC.

Besides the pressure field, *InVascular* simultaneously computes the pulsatile velocity field, from which the vorticity and shear stress fields can be calculated. Figures 9 and 10 show the velocity field with magnitude contours and streamlines and vorticity contours, respectively, at t = (a) 0.1, (b) 0.23, and (c) 0.63 in seconds in one cardiac cycle corresponding to systole (heart contraction, flow acceleration), diastole (heart relaxation, flow deceleration), and the end of diastole respectively. In Figure 9, flow in AA is stronger at systole (a) than at diastole (b) but remains intensive in LRA and RRA at both time points and is better organized at systole than at diastole. Whereas at the end of diastole, the flow is weak but chaotic. The vorticity contours shown in Figure 10 are similarly intensive in (a) and (b) with a large degree of skewness in AA, demonstrating the complexity of the flow in the real arteries. At the end of diastole, vorticity contours are much smaller and chaotic.

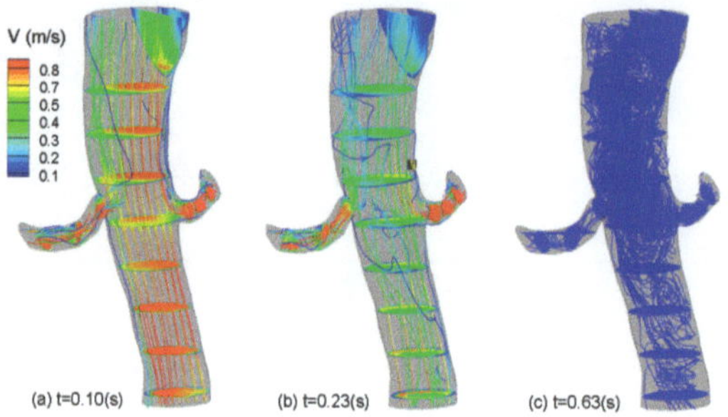

Figure 9. Velocity contours and streamlines at t = (**a**) 0.10 (systole), (**b**) 0.23, and (**c**) 0.63 (end of diastole) in seconds.

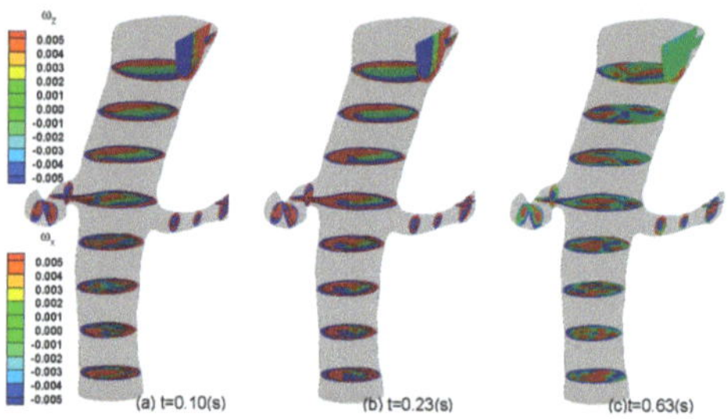

Figure 10. Z-component (vertically up) and x-component (horizontally right) vorticity contours at t = (**a**) 0.10, (**b**) 0.23, and (**c**) 0.63 in seconds.

3.2. Impact of r, C, and R Parameters on Pressure Quantification

Although the WK3 has been popularly used to model the physiological BC at each outlet of the artery segment (see Figure 5), its parameters reflecting resistances, r and R, and compliance, C, are determined empirically [38–40], which is subjected to uncertainty. To demonstrate the impact of the uncertainty in r-C-R parameters on the quantification of proximal and distal pressure, we performed a UQ analysis using FOSM. The input variables are the r, R, and C parameters in WK3, defined in Table 4. The elements of $\mathbf{X}$ are independently and normally distributed. The output variables are the pressure values in AA, LRA, and RRA, defined in Table 5. In this study, we assumed that the standard deviation of a random input variable is 3% of its mean. We performed UQ for five cases. The input distributions for the representative case in Section 3.1 are shown in Table 4.

Table 4. Input distributions for the representative case in Section 3.1.

Artery	Parameter	Variables	Mean	Standard Deviation	Distribution Type
AA	r(dynes×s/cm^5)	X_1	108.12	3.24	Normal
AA	R(dynes×s/cm^5)	X_2	3386.38	101.59	Normal
LRA	r(dynes×s/cm^5)	X_3	2879.76	86.39	Normal
LRA	R(dynes×s/cm^5)	X_4	7386.06	221.58	Normal
RRA	r(dynes×s/cm^5)	X_5	3306.39	99.19	Normal
RRA	R(dynes×s/cm^5)	X_6	8505.96	255.18	Normal
AA	C(cm^5/dynes)	X_7	1.0×10^{-5}	3.0×10^{-7}	Normal
LRA	C(cm^5/dynes)	X_8	5.4×10^{-6}	1.62×10^{-7}	Normal
RRA	C(cm^5/dynes)	X_9	4.8×10^{-6}	1.43×10^{-7}	Normal

Table 5. UQ results.

Artery	Output Variable	Mean μ_{y_1}	Standard Deviation σ_{y_1}	95% Confidence Interval
AA	Y_1 (mmHg)	155.80	1.37	[153.05, 158.55]
LRA	Y_2 (mmHg)	141.72	1.12	[139.49, 143.95]
RRA	Y_3 (mmHg)	144.61	1.12	[142.37, 147.86]

UQ results are given in Table 5. All the model output variables are normally distributed. For example, $Y_1 \sim N(\mu_{Y_1}, \sigma_{Y_1}) = N(155.80 \text{ mmHg}, 1.37)$. With these results, we know complete information about the simulation predictions, including the 95% confidence intervals of the model predictions. The formula for 95% confidence interval is $\mu_{Y_i} \pm 2\sigma_{Y_i}$. For example, the 95% confidence interval of Y_1 is [153.051, 158.55] mmHg unit. This means that the chance the actual value of Y_1 falling into the interval is 95%, or we have 95% confidence that the actual value of Y_1 is between 153.05 mmHg and 158.56 mmHg. The results of mean and standard deviation from five patients are also given in Table 6. The 95% confidence intervals of the model predictions of five patients are in Table 7.

Table 6. Mean and standard deviation of five patient cases.

Case	Y_1 (mmHg)	Y_2 (mmHg)	Y_3 (mmHg)
1	$N(156.80, 1.37^2)$	$N(141.72, 1.11^2)$	$N(144.61, 1.12^2)$
2	$N(163.61, 2.06^2)$	$N(154.25, 1.93^2)$	$N(56.42, 0.09^2)$
3	$N(157.22, 1.54^2)$	$N(152.35, 1.45^2)$	$N(153.85, 1.47^2)$
4	$N(109.71, 0.93^2)$	$N(106.49, 0.89^2)$	$N(74.50, 0.56^2)$
5	$N(123.21, 0.97^2)$	$N(117.34, 0.89^2)$	$N(102.17, 1.24^2)$

Table 7. 95% confidence intervals of model predictions of five patient cases.

Case	Y_1 (mmHg)	Y_2 (mmHg)	Y_3 (mmHg)
1	[153.05, 158.55]	[139.49, 143.95]	[142.37, 146.86]
2	[159.48, 167.74]	[150.38, 158.12]	[56.25, 56.59]
3	[154.14, 160.30]	[149.45,155.25]	[150.92,156.79]
4	[107.84, 111.58]	[104.72,108.27]	[73.37,75.62]
5	[121.28, 125.15]	[115.57, 119.11]	[99.69,104.64]

4. Discussion

We have presented the physiological inlet and outlet BCs for ICHD and integrated them into our in-house computational platform, *InVascular*. Using the unified LBM modeling for image segmentation and computational hemodynamics, *InVascular* seamlessly integrates the anatomical extraction of the interested arterial segment and quantification of pulsatile hemodynamics and achieves fast computation via GPU parallel computing. The inlet BC is a pulsatile velocity. A paraboloidal velocity profile is constructed based on the DUS velocity waveform, which fits the real shape of the arterial lumen (usually noncircular). Each outlet BC is a pulsatile pressure determined by WK3 during the simulation. The inlet velocity and outlet pressure BCs are introduced in the VLBM via a non-equilibrium extrapolation BC scheme. Using *InVascular*, we performed UQ analysis to quantify the impact of input variations caused by uncertainties. We applied *InVascular* into a human aortorenal arterial system extracted from medical CTA imaging data and demonstrated the applicability and reliability of *InVascular* for a real-world flow system. Six cases were studied. The pressure waveforms in AA, LRA, and RRA computed from *InVascular* have excellent agreements with the invasive measurements. The pulsatile velocity and then vorticity fields are shown as well. Due to the lack of available data, the validation of the velocity quantification has not been conducted. A systematic UQ analysis focuses on the impact of the variation of r, R, and C parameters on the quantification of the pressure field. Results include joint probability density of the computed pressure, which also provides the uncertainty or the confidence of the prediction. Due to the suitability of LBM for GPU parallel computing, *InVascular* features exceptionally fast computation speed. With a great potential to further speed up through parallel optimization and/or multiple GPU cards, the computation time is expected to be around 10 min per patient case. Such a computation capability is critically important for the clinical use of *InVascular*, enabling massive numerical analysis through parametrization to assess the true degree of existing arterial stenosis, either severe for immediate therapeutics or mild to avoid unnecessary intervention, within clinic permitted time.

Author Contributions: Conceptualization, H.Y. and X.D.; methodology, H.Y., X.D., M.K. and C.Z.; formal analysis, M.K., H.W. and R.C.; investigation, H.Y., X.D., M.K. and H.W.; data curation, M.K. and H.Y.; resources: A.P.S., X.F. and J.L.; writing—original draft preparation, M.K., H.Y. and X.D.; writing—review and editing, H.Y. and X.D.; visualization, M.K. and H.W.; supervision, H.Y., X.D. and A.P.S.; project administration, H.Y.; funding acquisition, H.Y., X.D. and A.P.S. All authors have read and agreed to the published version of the manuscript.

Funding: This research was funded by NSF through grant CBET 1803845. This work used the Extreme Science and Engineering Discovery Environment (XSEDE), which is supported by National Science Foundation Grant No. ACI-1548562. The 1st and corresponding author would like to also acknowledge the IUPUI University Fellowship and IUPUI MEE Graduate Fellowship.

Institutional Review Board Statement: The IRB approval (#1309233521R003 | N) was obtained for the patients enrolled at Indiana University. The study (#116-01) was approved by the Ethics Committee of Hangzhou First People's Hospital. The investigation conformed to the principles outlined in the Declaration of Helsinki.

Informed Consent Statement: It only involved a retrospective analysis of clinically indicated procedures; therefore, informed consent was not required.

Data Availability Statement: The data presented in this study are available on request from the corresponding author. The data are not publicly available.

Conflicts of Interest: The authors declare no conflict of interest.

Nomenclatures

AA	Aortic Artery
BC	Boundary Condition
CFD	Computational Fluid Dynamics
CTA	Computed Tomography Angiography
DUS	Doppler Ultrasound
FOSM	First-Order Second Moment
GPU	Graphic Processing Unit
ICHD	Image-Based Computational Hemodynamics
LBM	Lattice Boltzmann Method
LRA	Left Renal Artery
MAP	Mean Arterial Pressure
N-S	Navier-Stokes
RRA	Right Renal Artery
TSPG	Trans-Stenotic Pressure Gradient
UQ	Uncertainty Quantification
VLBM	Volumetric Lattice Boltzmann Method
WK3	Three-Element Windkessel Model
WSS	Wall-Shear Stress

References

1. Taylor, C.A.; Draney, M.T. Experimental and Computational Methods in Cardiovascular Fluid Mechanics. *Annu. Rev. Fluid Mech.* **2004**, *36*, 197–231. [CrossRef]
2. Withey, D.J.; Koles, Z.J. A Review of Medical Image Segmentation: Methods and Available Software. *Int. J. Bioelectromagn.* **2008**, *10*, 125–148.
3. Taylor, C.A.; Figueroa, C. Patient-Specific Modeling of Cardiovascular Mechanics. *Annu. Rev. Biomed. Eng.* **2009**, *11*, 109–134. [CrossRef] [PubMed]
4. Shi, Y.; Lawford, P.; Hose, R. Review of Zero-D and 1-D Models of Blood Flow in the Cardiovascular System. *Biomed. Eng. Online* **2011**, *10*, 33. [CrossRef] [PubMed]
5. Zhang, J.M.; Zhong, L.; Su, B.; Wan, M.; Yap, J.S.; Tham, J.P.; Chua, L.P.; Ghista, D.N.; Tan, R.S. Perspective on CFD Studies of Coronary Artery Disease Lesions and Hemodynamics: A Review. *Int. J. Numer. Methods Biomed. Eng.* **2014**, *30*, 659–680. [CrossRef]
6. Marsden, A.L.; Esmaily-Moghadam, M. Multiscale Modeling of Cardiovascular Flows for Clinical Decision Support. *Appl. Mech. Rev.* **2015**, *67*, 030804. [CrossRef]
7. Morris, P.D.; Narracott, A.; von Tengg-Kobligk, H.; Silva Soto, D.A.; Hsiao, S.; Lungu, A.; Evans, P.; Bressloff, N.W.; Lawford, P.V.; Hose, D.R. Computational Fluid Dynamics Modelling in Cardiovascular Medicine. *Heart* **2016**, *102*, 18–28. [CrossRef]
8. Formaggia, L.; Lamponi, D.; Quarteroni, A. One-Dimensional Models for Blood Flow In Arteries. *J. Eng. Math.* **2003**, *47*, 251–276. [CrossRef]
9. Vignon-Clementel, I.E.; Figueroa, C.; Jansen, K.; Taylor, C. Outflow Boundary Conditions For 3D Simulations of Non-Periodic Blood Flow and Pressure Fields in Deformable Arteries. *Comput. Methods Biomech. Biomed. Eng.* **2010**, *13*, 625–640. [CrossRef]
10. Vignon-Clementel, I.E.; Figueroa, C.A.; Jansen, K.E.; Taylor, C.A. Outflow Boundary Conditions for Three-Dimensional Finite Element Modeling of Blood Flow and Pressure in Arteries. *Comput. Methods Appl. Mech. Eng.* **2006**, *195*, 3776–3796. [CrossRef]
11. Vignon-Clementel, I.E.; Marsden, A.L.; Feinstein, J.A. A Primer on Computational Simulation in Congenital Heart Disease for the Clinician. *Prog. Pediatric Cardiol.* **2010**, *30*, 3–13. [CrossRef]
12. Alastruey, J.; Parker, K.; Peiró, J.; Sherwin, S. Lumped Parameter Outflow Models for 1-D Blood Flow Simulations: Effect on Pulse Waves and Parameter Estimation. *Commun. Comput. Phys.* **2008**, *4*, 317–336.
13. Stergiopulos, N.; Young, D.; Rogge, T. Computer Simulation of Arterial Flow with Applications to Arterial and Aortic Stenoses. *J. Biomech.* **1992**, *25*, 1477–1488. [CrossRef]
14. Reymond, P.; Merenda, F.; Perren, F.; Rufenacht, D.; Stergiopulos, N. Validation of A One-Dimensional Model of The Systemic Arterial Tree. *Am. J. Physiol.-Heart Circ. Physiol.* **2009**, *297*, H208–H222. [CrossRef] [PubMed]
15. Bonfanti, M.; Balabani, S.; Greenwood, J.P.; Puppala, S.; Homer-Vanniasinkam, S.; Díaz-Zuccarini, V. Computational Tools for Clinical Support: A Multi-Scale Compliant Model for Haemodynamic Simulations in an Aortic Dissection Based on Multi-Modal Imaging Data. *J. R. Soc. Interface* **2017**, *14*, 20170632. [CrossRef] [PubMed]
16. Pirola, S.; Cheng, Z.; Jarral, O.A.; O'Regan, D.P.; Pepper, J.R.; Athanasiou, T.; Xu, X.Y. On the Choice of Outlet Boundary Conditions for Patient-Specific Analysis of Aortic Flow Using Computational Fluid Dynamics. *J. Biomech.* **2017**, *60*, 15–21. [CrossRef] [PubMed]
17. Morbiducci, U.; Gallo, D.; Massai, D.; Consolo, F.; Ponzini, R.; Antiga, L.; Bignardi, C.; Deriu, M.A.; Redaelli, A. Outflow Conditions for Image-Based Hemodynamic Models of the Carotid Bifurcation: Implications for Indicators of Abnormal Flow. *J. Biomech. Eng.* **2010**, *132*, 091005. [CrossRef]
18. Yu, H.; Chen, X.; Wang, Z.; Deep, D.; Lima, E.; Zhao, Y.; Shawn, D. Mass-Conserved Volumetric Lattice Boltzmann Method for Complex Flows with Willfully Moving Boundaries. *Phys. Rev. E* **2014**, *89*, 063304. [CrossRef]
19. Succi, S.; Foti, E.; Higuera, F. Three-Dimensional Flows in Complex Geometries with the Lattice Boltzmann Method. *Europhys. Lett.* **1989**, *10*, 433–438. [CrossRef]
20. Benzi, R.S.; Succi, S.; Vergassola, M. The Lattice Boltzmann Equation—Theory and Applications. *Phys. Rep. -Rev. Sect. Phys. Lett.* **1992**, *222*, 145–197. [CrossRef]
21. An, S.; Yu, H.; Yao, J. GPU-Accelerated Volumetric Lattice Boltzmann Method for Porous Media Flow. *J. Petro. Sci. Eng.* **2017**, *156*, 546–552. [CrossRef]
22. Chen, R.; Yu, H.; Zhu, L.; Taehun, L.; Patil, R. Spatial and Temporal Scaling of Unequal Microbubble Coalescence. *AIChE J.* **2017**, *63*, 1441–1450. [CrossRef]
23. Chen, R.; Yu, H.; Zhu, L. Effects of Initial Conditions on the Coalescence of Micro-Bubbles, Proceedings of the Institution of Mechanical Engineers Part C. *J. Mech. Eng. Sci.* **2018**, *232*, 457–465. [CrossRef]
24. Wang, Z.; Zhao, Y.; Sawchuck, A.P.; Dalsing, M.C.; Yu, H.W. GPU Acceleration of Volumetric Lattice Boltzmann Method for Patient-Specific Computational Hemodynamics. *Comput. Fluids* **2015**, *115*, 192–200. [CrossRef]
25. Jain, K.; Jiang, J.; Strother, C.; Mardal, K.A. Transitional Hemodynamics in Intracranial Aneurysms—Comparative Velocity Investigations with High Resolution Lattice Boltzmann Simulations, Normal Resolution ANSYS Simulations, And MR Imaging. *Med. Phys.* **2016**, *43*, 6186–6198. [CrossRef]
26. Groen, D.; Richardson, R.A.; Coy, R.; Schiller, U.D.; Chandrashekar, H.; Robertson, F.; Coveney, P.V. Validation of Patient-Specific Cerebral Blood Flow Simulation Using Transcranial Doppler Measurements. *Front. Physiol.* **2018**, *9*, 721. [CrossRef] [PubMed]

27. Mirzaee, H.; Henn, T.; Krause, M.J.; Goubergrits, L.; Schumann, C.; Neugebauer, M.; Kuehne, T.; Preusser, T.; Hennemuth, A. MRI-Based Computational Hemodynamics in Patients with Aortic Coarctation Using the Lattice Boltzmann Methods: Clinical Validation Study. *J. Magn. Reson. Imaging* **2017**, *45*, 139–146. [CrossRef]
28. Mokhtar, N.H.; Abas, A.; Razak, N.; Hamid, M.N.A.; Teong, S.L. Effect of Different Stent Configurations Using Lattice Boltzmann Method and Particles Image Velocimetry on Artery Bifurcation Aneurysm Problem. *J. Theor. Biol.* **2017**, *433*, 73–84. [CrossRef]
29. Kang, X.; Tang, W.; Liu, S. Lattice Boltzmann Method for Simulating Disturbed Hemodynamic Characteristics of Blood Flow in Stenosed Human Carotid Bifurcation. *J. Fluids Eng.* **2016**, *138*, 121104. [CrossRef]
30. An, S.; Yu, H.; Wang, Z.; Chen, R.; Kapadia, B.; Yao, J. Unified Mesoscopic Modeling and GPU-Accelerated Computational Method for Image-Based Pore-Scale Porous Media Flows. *Int. J. Heat Mass Trans.* **2017**, *115*, 1192–1202. [CrossRef]
31. Yu, H.; Wang, Z.; Zhao, Y.; Sawchuk, A.P.; Lin, C.; Dalsing, M.C. GPU-accelerated Patient-Specific Computational Flow—From Radiological Images to in vivo Fluid Dynamics. In Proceedings of the 27th International Conference on Parallel Computational Fluid Dynamics Parallel CFD2015, Montreal, Canada, 30 June 2015.
32. Yu, H.; Chen, R.; Wang, H.; Yuan, Z.; Zhao, Y.; An, Y.; Xu, Y.; Zhu, L. GPU Accelerated Lattice Boltzmann Simulation for Rotational Turbulence. *Comput. Math. Appl.* **2014**, *67*, 445–451. [CrossRef]
33. Wolf-Gladrow, D. A Lattice Boltzmann Equation for Diffusion. *J. Stat. Phys.* **1995**, *79*, 1023–1032. [CrossRef]
34. Wang, Z.; Yan, Z.; Chen, G. Lattice Boltzmann Method of Active Contour for Image Segmentation. In Proceedings of the Image and Graphics (ICIG), 2011 Sixth International Conference, Hefei, China, 12–15 August 2011; pp. 338–343.
35. Sethian, J.A. *Level Set Methods and Fast Marching Methods: Evolving Interfaces in Computational Geometry, Fluid Mechanics, Computer Vision, and Materials Science*; Cambridge University Press: Cambridge, UK, 1999; Volume 3.
36. Guo, Z.-L.; Zheng, C.-G.; Shi, B.-C. Non-Equilibrium Extrapolation Method for Velocity And Pressure Boundary Conditions In The Lattice Boltzmann Method. *Chin. Phys.* **2002**, *11*, 366.
37. Westerhof, N.; Lankhaar, J.-W.; Westerhof, B.E. The Arterial Windkessel. *Med. Biol. Eng. Comput.* **2009**, *47*, 131–141. [CrossRef] [PubMed]
38. Holenstein, R.; Ku, D.N. Reverse Flow in The Major Infrarenal Vessels–A Capacitive Phenomenon. *Biorheology* **1988**, *25*, 835–842. [CrossRef]
39. Bax, L.; Woittiez, A.J.; Kouwenberg, H.J.; Mali, W.P.; Buskens, E.; Beek, F.J.; Braam, B.; Huysmans, F.T.; Schultze Kool, L.J.; Rutten, M.J.; et al. Stent Placement in Patients With Atherosclerotic Renal Artery Stenosis and Impaired Renal Function: A Randomized Trial. *Ann. Intern. Med.* **2009**, *150*, 840–848. [CrossRef] [PubMed]
40. Les, A.S.; Shadden, S.C.; Figueroa, C.A.; Park, J.M.; Tedesco, M.M.; Herfkens, R.J.; Dalman, R.L.; Taylor, C.A. Quantification of Hemodynamics in Abdominal Aortic Aneurysms During Rest and Exercise Using Magnetic Resonance Imaging and Computational Fluid Dynamics. *Ann. Biomed. Eng.* **2010**, *38*, 1288–1313. [CrossRef]
41. Du, X. Unified Uncertainty Analysis by the First Order Reliability Method. *J. Mech. Des.* **2008**, *130*, 091401. [CrossRef]
42. Huang, B.; Du, X. Probabilistic Uncertainty Analysis by Mean-Value First Order Saddlepoint Approximation. *J. Reliab. Eng. Syst. Saf.* **2008**, *93*, 325–336. [CrossRef]
43. Rosenblatt, M. Remarks on A Multivariate Transformation. *Ann. Math. Stat.* **1952**, *23*, 3. [CrossRef]

fluids

Article

A Mock Circulation Loop to Characterize In Vitro Hemodynamics in Human Systemic Arteries with Stenosis

Weichen Hong [1], Huidan Yu [1,2,*], Jun Chen [3,*], John Talamantes [1], Dave M. Rollins [4], Xin Fang [5], Jianyun Long [5], Chenke Xu [6] and Alan P. Sawchuk [2]

1 Department of Mechanical and Energy Engineering, Indiana University-Purdue University, Indianapolis, IN 46202, USA
2 Department of Vascular Surgery, Indiana University School of Medicine, Indianapolis, IN 46202, USA; asawchuk@iupui.edu
3 School of Mechanical Engineering, Purdue University, West Lafayette, IN 47907, USA
4 Vascular Diagnostic Center, Indiana University Health, Indianapolis, IN 46220, USA
5 Vascular Surgery, The Affiliated Hangzhou First People's Hospital, Zhejiang University School of Medicine, Hangzhou 310006, China
6 Ultrasound, The Affiliated Hangzhou First People's Hospital, Zhejiang University School of Medicine, Hangzhou 310006, China
* Correspondence: whyu@iupui.edu (H.Y.); junchen@purdue.edu (J.C.)

Abstract: Vascular disease is the leading cause of morbidity and mortality and a major cause of disability for Americans, and arterial stenosis is its most common form in systemic arteries. Hemodynamic characterization in a stenosed arterial system plays a crucial role in the diagnosis of its lesion severity and the decision-making process for revascularization, but it is not readily available in the current clinical measurements. The newly emerged image-based computational hemodynamics (ICHD) technique provides great potential to characterize the hemodynamics with fine temporospatial resolutions in realistic human vessels, but medical data is rather limited for validation requirements. We present an image-based experimental hemodynamics (IEHD) technique through a mock circulation loop (MCL) to bridge this critical gap. The MCL mimics blood circulation in human stenosed systemic arterial systems that can be either 3D-printed silicone, artificial, or cadaver arteries and thus enables in vitro measurement of hemodynamics. In this work, we focus on the development and validation of the MCL for the in vitro measurement of blood pressure in stenosed silicone arteries anatomically extracted from medical imaging data. Five renal and six iliac patient cases are studied. The pressure data from IEHD were compared with those from ICHD and medical measurement. The good agreements demonstrate the reliability of IEHD. We also conducted two parametric studies to demonstrate the medical applicability of IEHD. One was the cardiovascular response to MCL parameters. We found that blood pressure has a linear correlation with stroke volume and heart rate. Another was the effect of arterial stenosis, characterized by the volumetric reduction (VR) of the arterial lumen, on the trans-stenotic pressure gradient (TSPG). We parametrically varied the stenosis degree and measured the corresponding TSPG. The TSPG-VR curve provides a critical VR that can be used to assess the true hemodynamic severity of the stenosis. Meanwhile, the TSPG at VR = 0 can predict the potential pressure improvement after revascularization. Unlike the majority of existing MCLs that are mainly used to test medical devices involving heart function, this MCL is unique in its specific focus on pressure measurement in stenosed human systemic arteries. Meanwhile, rigorous hemodynamic characterization through concurrent IEHD and ICHD will significantly enhance our current understanding of the pathophysiology of stenosis and contribute to advancements in the medical treatment of arterial stenosis.

Keywords: image-based experimental hemodynamics; image-based computational hemodynamics; mock circulation loop; trans-stenotic pressure gradient; arterial stenosis; volumetric reduction

Citation: Hong, W.; Yu, H.; Chen, J.; Talamantes, J.; Rollins, D.M.; Fang, X.; Long, J.; Xu, C.; Sawchuk, A.P. A Mock Circulation Loop to Characterize In Vitro Hemodynamics in Human Systemic Arteries with Stenosis. *Fluids* **2023**, *8*, 198. https://doi.org/10.3390/fluids8070198

Academic Editor: D. Andrew S. Rees

Received: 18 May 2023
Revised: 27 June 2023
Accepted: 28 June 2023
Published: 29 June 2023

1. Introduction

Vascular disease, including coronary, extracranial, and peripheral arterial beds, is the leading cause of morbidity and mortality and a major cause of disability for Americans. Arterial stenosis is one of the most common vascular diseases that can lead to life- and limb-threatening consequences, including myocardial ischemia, ischemic stroke, and limb amputation. It is a condition that involves blockage of blood flow mainly due to an atherosclerotic narrowing of the arterial lumen, which commonly occurs in systemic arteries. While stenosis can be observed by imaging modalities such as computed tomography angiogram (CTA), magnetic resonance imaging (MRI), and Doppler ultrasound sonography (DUS), direct and effective noninvasive means to evaluate the true hemodynamic severity of stenosis are lacking in current clinical practice. For coronary stenosis, fractional flow reserve (FFR) [1], defined as the ratio between the distal pressure p_d and the proximal pressure p_a to the stenosis; namely, p_d/p_a, is used to determine the hemodynamic severity of myocardial ischemia [2–4]. However, trans-stenotic pressure gradient (TSPG), defined as the difference between p_a and p_d, has been popularly used for assessing the hemodynamic severity of non-coronary stenoses. Evidence, including ours, has shown that TSPG is an indicator to determine the amount of blood flow blockage caused by renal [5–7], iliac and femoral [8–10], and carotid [11] stenoses and can help guide the proper decision-making of interventional treatment. Nevertheless, the clinical application of either FFR or TSPG is rather limited [12], as they rely on the invasive pressure measurement of the local pressure values, i.e., p_d and p_a, which may expose patients to surgical complications and medical costs.

Newly emerging image-based computational hemodynamics (ICHD) [13–17] has great potential to address this unmet medical need, as seen in a recent review [16] and the references therein. Based on medical imaging data, ICHD enables noninvasive characterization of hemodynamics, including velocity, pressure, and stress, in the diseased human arterial system with a fine spatiotemporal resolution, resulting in important hemodynamic indicators, such as FFR and TSPG, for arterial stenosis. For example, image-based FFR has been a well-established clinical application of ICHD by HeartFlow Inc. (Redwood City, CA, USA). Obtained from a purely anatomical, noninvasive dataset of coronary CTA images [18] by utilizing ICHD, the FFR determines the hemodynamic severity of the coronary stenosis and then guides the decision-making of the interventional treatment for it. We have recently developed a proprietary ICHD technique [19] for a new noninvasive and patient-specific hemodynamic index that can assess the hemodynamic severity of non-coronary arterial stenosis and applied it to renal stenosis [6,7]. In addition to the application of ICHD for arterial stenosis, many studies have demonstrated the feasibility and validity of ICHD for vascular diseases caused by aneurysms [20–22]. Despite its great potential for medical applications, ICHD needs significant resources for model development and sophisticated verification and validation (V&V) before it can be translated into medical applications. First, most ICHD studies assume incompressible and Newtonian flow in rigid arterial walls [7,17,19], whereas vascular circulation is much more complicated. The anatomical flow domain is arbitrarily curved with moving walls. Blood consists of plasma, blood cells, and platelets. When severe arterial stenosis exists, the blood flow may become turbulent. Each of them needs to be modeled in ICHD. In general, more model inclusion in ICHD means higher computational costs. Models for insignificant effects will indeed introduce inaccuracies. Thus, it is crucially important to determine which and how models should be introduced in ICHD through V&V. Second, since only a segment of the blood circulation system is involved in ICHD, boundary conditions are required at the inlet(s) and outlet(s) of the vessel segment to represent the remaining vascular network. The choices of the outflow boundary condition in ICHD vary among zero pressure or zero traction conditions, resistance or impedance conditions, reduced-order models, which can be an open or closed loop, and reduced-order one-dimensional wave propagation equations [23–26]. To capture the interaction between the local 3-dimensional (3D) vessel segment and the 1-dimensional (1D) global circulation, the ICHD must be coupled to a reduced-order lumped-parameter network model. Among them, the 3-element WindKessel model [27–31] (WK3) has been commonly used to construct such a network, in which a Windkessel circuit is adopted to model the distal vasculature with one capacitor, modeling vessel compliance, and two resistors, modeling proximal and distal flow resistances, respectively. Evidence has shown that the WK3 can reproduce physiological pressure waves [32–34] in large vessels. In the state-of-the-art ICHD, the required proximal resistance, compliance of the distal vasculature, and distal resistance in the WK3 model are empirical [6], as no medical data are available.

Image-based experimental hemodynamics (IEHD) can provide a suitable test platform to address the aforementioned two needs of ICHD. A mock circulation loop (MCL) can simulate the pathophysiological environment of blood flow and measure the velocity and pressure waveforms as an alternative resource to medical measurements for the V&V of ICHD. Since its introduction by Westerhof [35] in 1971 as a simple artificial arterial system for pumping hearts in experimental labs, numerous studies have highlighted the potential of mock loop systems to create finely controlled and maintained simulated physiological conditions. Replicating these conditions in animal models or clinical settings is challenging, making MCLs advantageous for the design and testing of total artificial hearts [36–38] or ventricular assist devices [39–45]. The reliable real-time control and feedback capabilities of MCLs have demonstrated their applicability in studying human cardiovascular circulations [46,47], early stages of congestive heart failure [48,49], pediatric cardiopulmonary diseases [50–52], surgical procedures [53], and drug tests [54]. By offering safe and effective methods, MCLs allow for the investigation and analysis of the complex hemodynamics of human cardiocirculatory systems. To explore and characterize hemodynamic abnormalities in diseased human arterial circulations, researchers have worked on developing MCLs integrated with 3D-printed patient-specific models of the pulmonary artery [55,56], aorta [57,58], and coronary artery [59,60]. These patient-specific models, created through 3D printing, provide anatomically accurate features for MCLs, enabling precise simulations of the physiological hemodynamics of complex systems. In this study, we present a new MCL specifically for the characterization of in vitro hemodynamics in realistic human systemic arteries with stenosis, focusing on the reliable measurement of TSPG. We validate the accuracy of our pressure measurements using available medical and ICHD data. Subsequently, we conduct parametric studies to investigate the factors that influence blood pressure in these stenosed renal and iliac arterial systems.

The remainder of this paper is organized as follows: Section 2 includes the materials, including the medical cases and the corresponding 3D-printed silicone arterial systems, and the establishment of the MCL together with its instrumentations. Our application studies and experimental results are presented in Section 3. Finally, Section 4 concludes the paper with a summary and discussion.

2. Materials and Methods

2.1. Medical Data

We studied five renal and six iliac medical cases as listed in Table 1. The cases were from IU Health Methodist Hospital in Indianapolis, IN, USA (Renal cases I–II and Iliac cases I–V) with IRB (Institutional Review Board) approvals, #1405073181 and #1812589521, respectively, and Hangzhou First People's Hospital in Zhejiang, China (Renal cases III–V and iliac case VI) with a study approval (#116-01) by the Ethics Committee of the hospital. Each patient case contains diagnostic imaging data, including CTA DICOM (Digital Imaging and Communications in Medicine) slices and DUS M(motion)-mode velocity waveforms, together with blood pressure waveforms at specified locations invasively measured during digital subtraction angiography (DSA) for interventional treatments. The pressure waveforms are used to validate the corresponding computed ones [6,7,61] from our previous ICHD studies and the experimentally measured ones in this work.

Table 1. 11 patient cases, obtained from IU Health Methodist Hospital in Indianapolis, IN, USA (R: IV–V and I: I–V) and Hangzhou First People's Hospital in Hangzhou, Zhejiang, China (R: I–III and I:VI).

Artery	Case	Age	Gender	Stenosis	Stenting
Renal (R)	I	64	Male	Yes	No
	II	87	Male	Yes	Yes
	III	83	Male	Yes	No
	IV	74	Male	No	No
	V	75	Male	No	No
Iliac (I)	I	63	Male	Yes	Yes
	II	61	Female	Yes	Yes
	III	64	Female	Yes	Yes
	IV	53	Male	Yes	Yes
	V	76	Female	Yes	Yes
	VI	69	Male	Yes	Yes

2.2. D-Printed Silicone Arterial Systems

We used MIMICS Materialise (Materialise NV, Leuven, Belgium) to anatomically extract the aortoiliac and aortorenal arterial systems, shown in Figure 1a,b, respectively. The CTA resolution is approximately $0.75^2 \times 2.0$ mm^3 (IU Health Methodist Hospital cases) and $0.65^2 \times 0.6$ mm^3 (Hangzhou First People's Hospital cases). The CTA images are in sliced DICOM format. The 3D morphological geometry of each arterial system was output using Standard Tessellation Language (STL), a file format commonly used for 3D printing. We then use our in-house Form 3 Stereolithography 3D printer (Formlabs, Somerville, MA, USA) to fabricate each of the 3D silicone arterial systems. Three printing materials, Clear Resin, Flexible 80A Resin, and Elastic 50A Resin, as shown in Figure 1c, are available to print rigid, flexible, and elastic silicone arterial systems, respectively, for different study purposes. In this study, we use Elastic 50A Resin to print the silicone arterial systems, unless otherwise indicated.

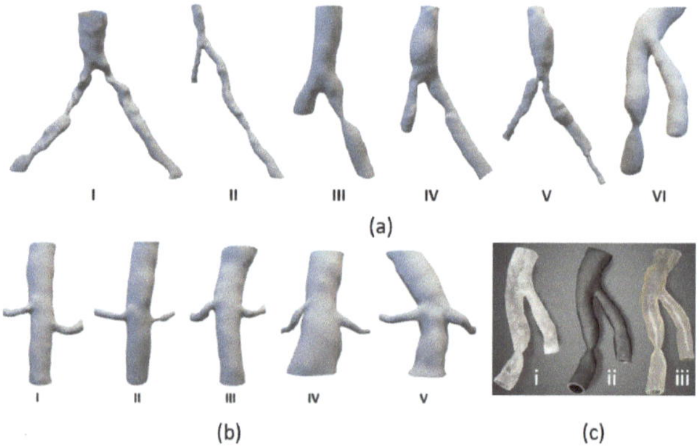

Figure 1. Antomically extracted 6 aortoiliac (**a**) and 5 aortorenal (**b**) arterial systems in STL format. (**c**) A 3D printed aortoiliac arterial system using (i) Clear, (ii) Flexible 80A, and (iii) Elastic 50A Resin. The arterial wall is more flexible from left to right.

2.3. A Mock Circulation Loop for Human Systemic Arterial Systems

The MCL is designed to mimic the blood flow in stenosed human systemic arterial systems, enabling in vitro measurement of pressure waveforms proximal or distal to the stenosis. We intend not to include coronary stenosis, so no heart model is included. An arterial system can be either 3D-printed silicone, clinically used artificial arteries, or cadaver arteries. In this work, we focus on silicone arteries. The schematic diagram and the corresponding benchtop setup of the MCL are shown in Figure 2a,b, respectively, for a stenosed arterial system. The MCL consists of a reservoir (5.5-L), a pulsatile blood pump (Figure 3a, HARVARD APPARATUS, Model 1434), a test section holding the silicone arterial system, Windkessel chambers, and resistance valves. The pulsatile blood pump mimics the ventricular ejection from the heart in large animals by pumping the fluid via an oscillation mechanism, as schematized in Figure 3b. It is equipped with controls to continuously vary heart rate (HR), stroke volume (SV), and output phase ratio (systole to diastole in one stroke) in the ranges of 0–100 beats per min (bpm), 15–100 mm (mL), and 25/75 to 50/50, respectively. The pump is placed at the same elevation level as the test section to imitate the bedrest condition, avoiding pressure drops caused by gravity. The Windkessel effect [28] in the MCL is modeled by the Windkessel chambers (fractionally filled chambers with trapped-air compliance elements). The Windkessel chamber is a cylindrical glass bottle measuring 10 cm in diameter and 20 cm in height. It is equipped with 0.5-inch inlet and outlet ports. The water column height within the chamber is adjusted between 10 cm and 20 cm, allowing for varying compliance to accommodate different scenarios. It imitates the compliance of large vessels. One Windkessle chamber is mounted upstream of the inlet of the test section, mimicking upper-body systemic compliance. At each of the arterial outlets, one Windkessel chamber and two resistance valves are adopted to model the vessel compliance and the proximal and distal flow resistances, respectively. This setting is consistent with the WK3 model used in our ICHD [6]. All the components are connected by flexible silicone tubing (with an inner diameter of 0.5 inches).

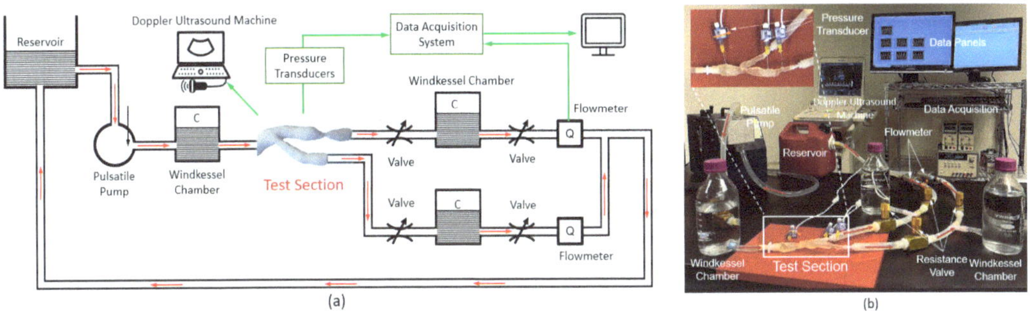

Figure 2. Illustrations of the MCL: (**a**) schematic diagram and (**b**) benchtop setup to mimic the blood flow in a 3D printed silicone arterial system for in vitro measurement of pressure and velocity waveforms. The Red arrows indicate the flow direction in the loop. The connection from a loop component to an instrument is indicated by a green arrow.

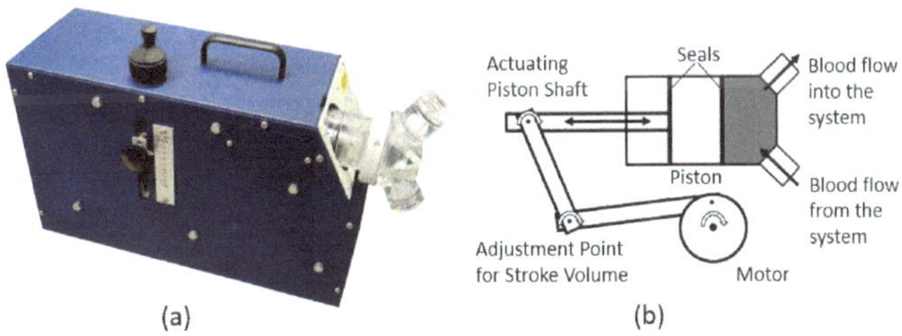

Figure 3. Harvard apparatus pulsatile blood pump: (**a**) exterior appearance and (**b**) internal linkage with reciprocating piston.

2.4. Instrumentations

The measuring instruments of the MCL include a Doppler ultrasound machine, a DAQ system, pressure transducers, and flowmeters, as shown in Figure 2. Each medical-grade pressure transducer (Deltran®, model 6069) is connected to the flow at a measuring location in a silicone aortorenal (Figure 4a) and aortoiliac (Figure 4b) arterial systems through a 21 g needle (NAD21T21WP, Qosina, Ronkonkoma, NY, USA). A voltage change detected by the deformable membrane in the pressure transducer reflects the pressure pulsation. The pressure waveform is read through a data acquisition (DAQ) system equipped with amplification by a Wheatstone bridge module (DV-10, Honeywell, Columbus, OH, USA) and digitization by an analog-to-digital converter (NI-9201, National Instruments, Austin, TX, USA). This MCL can acquire up to 8 pressure signals simultaneously. Each pressure transducer is individually calibrated using a hydrostatic water column. A magnetic inductive flowmeter (SM6004, IFM Efector Inc., Malvern, PA, USA) acquires a flowrate waveform. The range of the output current signal is from 4 to 20 $\times$ 10^3 amps(mA). The readings of the flowmeters are collected by another analog-to-digital converter (NI-9203, National Instruments) in the DAQ. The DAQ system is connected to a desktop computer (3.6 GHz CPU, RAM 8.0 GB) through an in-house program based on Labview software. The digital signals are collected at a sampling rate of 1000 Hz. The portable ultrasound machine (CX50 w/C5-1 PureWave curved transducer and L12-3 broadband linear transducer, Philips Electronics) can be used to measure the velocity waveforms of the blood flow.

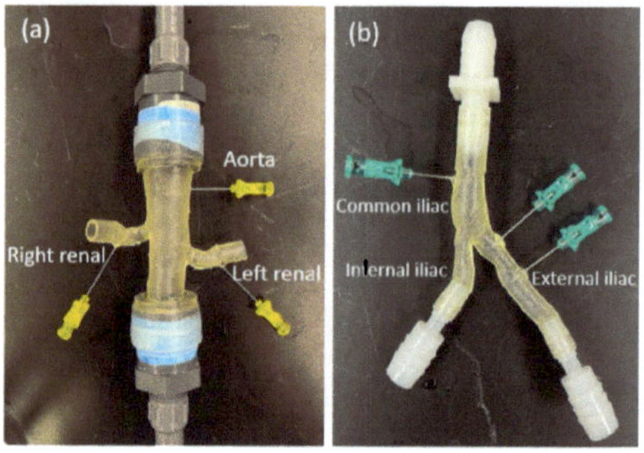

Figure 4. Silicone aortorenal (**a**) and aortoiliac (**b**) arterial systems with inserted 21 g needles for pressure measurement.

3. Application Studies and Experimental Results

We study five aortorenal and six aortoiliac arterial systems, listed in Table 1, using the MCL. The corresponding 3D aortoiliac and aortorenal silicone arterial systems anatomically extracted from patients' CTA images are shown in Figure 1a,b, respectively. Each of them has invasively measured and noninvasively computed pressure waveforms at specified locations. The numerical simulations were performed using our in-house ICHD solver. In the past 10 years or so, this solver has been continuously developed and refined [17,62–66]. Recently, it has been applied to quantify the TSPG of renal artery stenosis [6,7] and iliac artery stenosis [67]. The corresponding experimental pressure waveform was averaged over three cardiac cycles. First, we compare the pressure waveforms among the noninvasive in vitro measurements and computations and the invasive measurements to demonstrate the reliability of IEHD by reproducing the ICHD-computed pressure waveforms. The methodology of ICHD and related physical variables and boundary conditions are referred to in the references [6,17,67]. Then, we conduct two application studies to demonstrate the applicability of IEHD for medical applications. In what follows, we use the representative iliac case (Figure 1a(VI)) and renal case (Figure 1b(IV)) with HR = 78 bpm and SV = 15 mL to present the results unless otherwise indicated. We first test the effect of the fluid on the pressure measurement in the iliac case using water (viscosity of 1 cP) and glycerol aqueous solution (45 vol. % glycerol with a viscosity of 3.5 cP). As shown in Figure 5, the measured pressure waveforms do not show a noticeable difference when choosing glycerol aqueous solution or water as the working fluid (blood surrogate). Thereafter, we use water as the running fluid in this study.

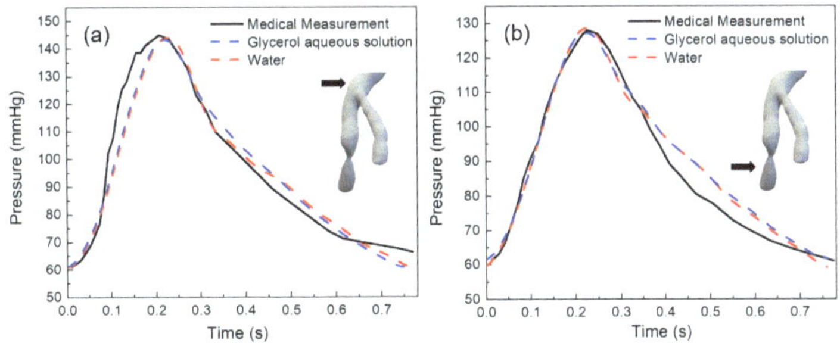

Figure 5. Pressure waveforms measured in (**a**) RCIA and (**b**) REIA of the iliac case using glycerol solution (blue) and water (red).

3.1. Reliability of In Vitro Pressure Measurement Using MCL

We first evaluate the pressure measurement using a representative iliac case, Figure 1a(VI). The arterial system includes the right common iliac artery (RCIA), the right external iliac artery (REIA), and the right internal iliac artery (RIIA), as seen in Figure 6a. The medical pressure measurement data are available at the three locations. We compare the systolic blood pressure (SBP), diastolic blood pressure (DBP), and mean arterial pressure (MAP) that are calculated from $(SBP + 2DBP)/3)$, in millimeters of mercury (mmHg), among the medical measurement (M), computation (C), and experimental measurement (E) in Figure 6b.

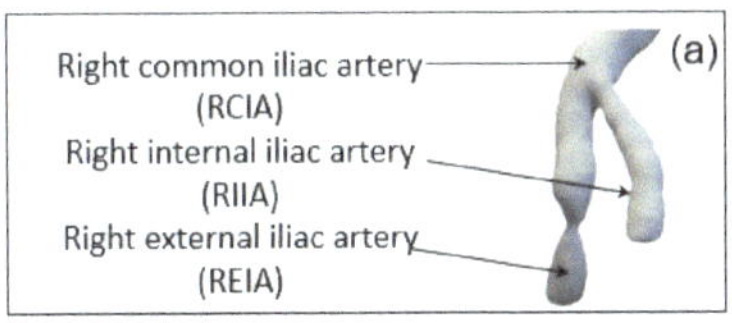

(b)

Artery	Method	Pressure (mmHg)		
		SBP	DBP	MAP
RCIA	M	145	61	89
	C	144	62	89
	E	144	61	89
REIA	M	128	61	83
	C	127	57	80
	E	128	60	83
RIIA	C	134	60	85
	E	136	64	88

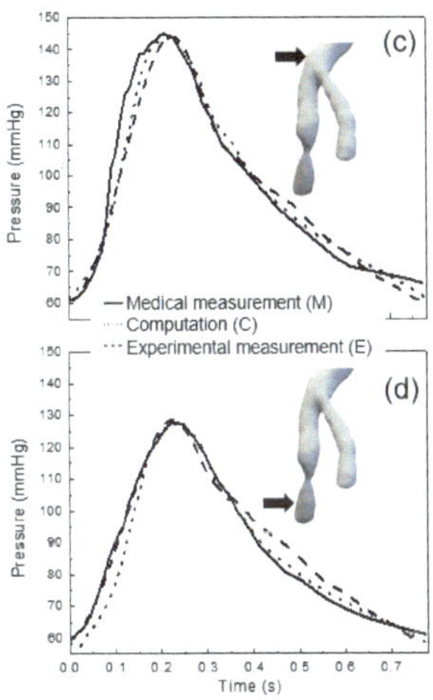

Figure 6. A comparison among medical measurement (M), computation (C), and experimental measurement (E) using iliac case VI at three locations in panel (**a**), including SBP, DBP, and MAP in panel (**b**), and pressure waveforms in a full cardiac cycle at RCIA and REIA in panels (**c**,**d**), respectively.

The relative errors of experimental measurement and computation on the medical data vary from 0 to 6.6% with a mean of 2.21% for ICHD and from 0 to 1.64% with a mean of 0.16% for IEHD. The pressure waveform comparisons among medical measurement (solid line), numerical computation (dotted line), and experimental measurement (dashed line) in RCIA and REIA are shown in Figure 6c,d, respectively. The experimental measurements agree well with the medical and computational data. Correlation scatter plots using the 6 iliac cases with 12 measurement samples and the 5 renal cases with 13 measurement samples are shown in Figure 7 for (a) iliac SBP, (b) iliac DBP, (c) renal SBP, and (d) renal DBP. In each plot, $r_{E,M}$ is the Pearson correlation coefficient between experimental (E) and medical (M) measurements is shown, and the line is diagonal, representing the equal pressure measurement between E and M. $r_{E,M}$ is calculated from the ratio of the covariance of E and M vs. the product of the standard deviation of E and the standard deviation of M. The $r_{E,M}$s for iliac SBP, iliac DBP, renal SBP, and renal DBP are 0.99 with $p < 0.05$, 0.96 with $p < 0.05$, 0.96 with $p < 0.05$, and 0.93 with $p < 0.05$. These results indicate good correlations between E and M measurements of blood pressure. Overall, the experimental pressure measurement through the MCL is reliable.

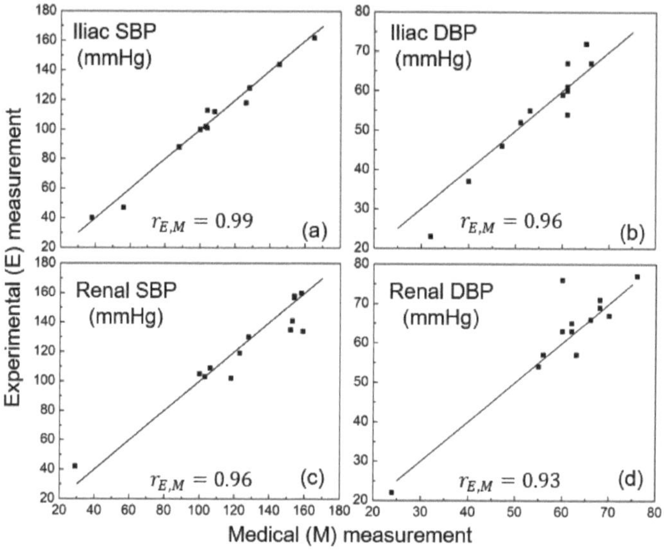

Figure 7. Correlation scatter plots of experimentally (in vitro) vs. medically (in vivo) measured SBP ((**a**,**c**) and DBP (**b**,**d**)) pressure for iliac ((**a**,**b**) and renal (**c**,**d**)) cases. r_{EM} is the Pearson correlation coefficient between experimentally (in vitro) and medically (in vivo) measured pressure.

3.2. *Parametric Studies and Experimental Results*

In this section, we perform two parametric studies and present the corresponding results.

3.2.1. Cardiovascular Responses to MCL Parameters

Cardiac output, defined as the product of HR and SV, can be adjusted by the pulsatile heart pump in the MCL. The SV, in milliliters (mL), is the volume of blood pumped out of the pulsatile pump per cardiac cycle, and the HR, in beats per minute (bpm), is the number of heartbeats per minute. The MAP, in mmHg, is an important pathophysiological indicator of cardiovascular diseases. We use a renal case, i.e., Figure 1b(IV), to study how SV and HR affect the MAP measurement when keeping the compliance and resistances the same. Figure 8 shows the effects of HR (triangles) and SV (squares) on the MAP in the aortic artery of the aortorenal arterial system. The error bars for MAP in Figure 8 exhibit a typical range of 0.5 to 0.8 mmHg, which is too small to be visually distinguished when compared to the size of the data symbols. Consequently, they were not included in the plot. We keep constant SV (55 mL) when varying HR and constant HR (77 bpm) when varying SV. The MAP shows strong linear correlations to HR and SV, with R^2 of 0.97 and 0.99, respectively. The general understanding of the result is as follows: When either HR or SV increases, the cardiac output into the MCL increases, leading to an increment in blood pressure [68,69]. Systolic pressure and diastolic pressure would both increase when either SV [69] or HR [70–72] increased. Therefore, MAP increases.

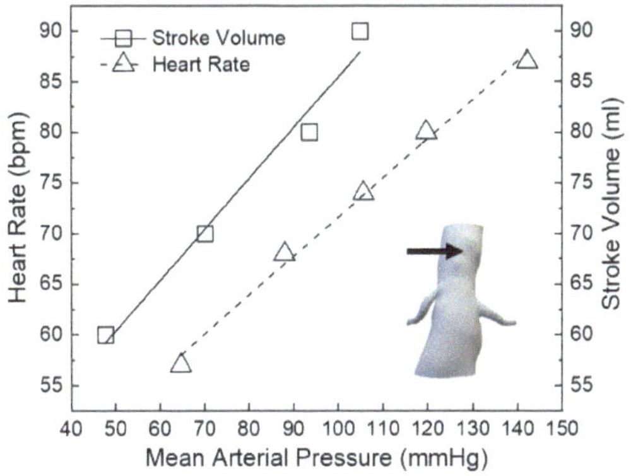

Figure 8. Effects of heart rate (**left** axis) and stroke volume (**right** axis) on mean arterial pressure in the aorta of the renal case. The measurement location is indicated by the black arrow.

3.2.2. Effects of Stenosis Degree on TSPG

As indicated in Section 1, TSPG has been popularly used for assessing the hemodynamic severity of non-coronary arterial stenosis. We have recently developed a noninvasive functional assessment technique via ICHD [6,7,19] to assess the true hemodynamic severity, mild or severe, of arterial stenosis, together with a recommendation of yes (if severe) or no (if mild) for interventional treatment. In cases of severe stenosis, we can predict the potential outcomes of revascularization. Such a new non-invasive hemodynamic assessment will especially benefit the patient group with moderate arterial stenosis, avoiding under- and over-interventional treatment to promote public health. In ICHD, we have found [7] that the volume reduction (VR) of the arterial lumen is closer to TSPG than the diameter reduction of the cross-section. To demonstrate the applicability of the MCL for translational medical research, we studied how the degree of stenosis affects TSPG. As shown in Figure 9a, we parameterize the degree of stenosis using a real iliac stenosis case (Figure 1a(VI)) from the current (real) 50% of VR to 45%, 33%, and 0%. It is noted that 0% of VR corresponds to the removal of stenosis through a stenting treatment. We cropped a stenosed segment between the two pressure measurement locations for p_a and p_d to the stenosis. The VR is defined as the volume ratio of the reduced lumen volume due to a stenosis vs. the normal lumen volume without a stenosis. We experimentally measured p_a and p_d for each of the four stenosis cases, from which the $TSPG$ $(= p_a - p_d)$ is calculated. We plotted the relation between the TSPG of SBP (squares) and MAP (dots) and VR in Figure 9b. It is seen that TSPG increases when the stenosis progresses. The TSPG~VR curve is flat when VR is relatively small (<45%) but steep when VR is relatively high (45%). On the flat side, increasing VR from 0% to 45% results in TSPG increases of 9 mmHg and 2 mmHg for SBP and MAP, respectively. On the steep side, increasing VR from 45% to 50% results in TSPG increases from 10 to 37 mmHg for SBP and 3 to 14 mmHg for MAP. From the slope of the TSPG~VR curve, one can identify a threshold of VR, denoted as VR_{ms}, which separates mild ($VR < VR_{ms}$) and severe ($VR > VR_{ms}$) stenosis conditions. For the case studied, VR_{ms} = 45%. On the mild side ($VR < VR_{ms}$), a one percent increase in VR causes a 0.2 mmHg and 0.047 mmHg increase of TSPG for SBP and MAP, respectively. Whereas on the severe side ($VR > VR_{ms}$), a one percent increase in VR causes 5.4 mmHg and 2.2 mmHg increases in TSPG for SBP and MAP, respectively. Using VR_{ms} (=45%), one can assess that the existing stenosis of the case being studied, VR = 50%, is severe. The recommendation for stenting therapy is then made. Our recommendation based on the noninvasive assessment agrees with the clinical treatment for this case. In our recent study of ICHD, we applied the same technique to assess renal stenoses and achieved good agreement with clinical treatments [7,19]. Meanwhile, TSPG = 1 mmHg for both SBP and MAP at VR = 0 means that if the stenosis is interventionally stented, the TSPG will be back to a normal condition, implying that stenting therapy will benefit the patient. Similarly to Figure 8, the error bars for TSPG in Figure 9 span the ranges of 0.12 to 0.21 mmHg for MAP and 0.33 to 0.38 mmHg for SBP, respectively. Their small magnitudes make them visually indistinguishable when compared to the size of the data symbols. Therefore, we have also omitted them from the plot.

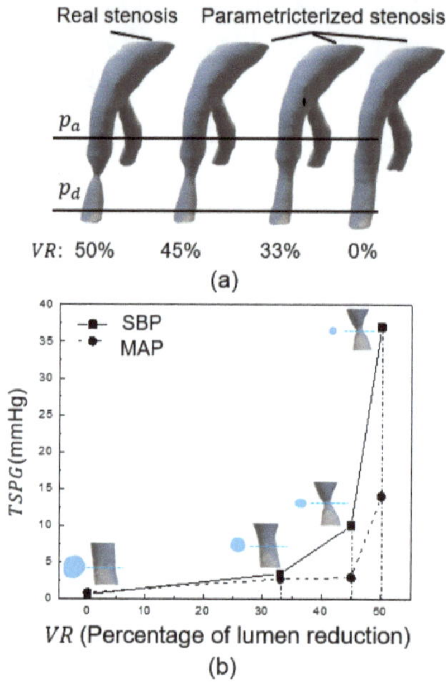

Figure 9. (**a**) A real (iliac case VI in Figure 1a) stenosis and its parameterization, characterized by VR. Distal (p_d) and proximal (p_a) pressure are measured at the indicated location. (**b**) $TSPG$~VR curves for SBP (squares) and MAP (dots). The cross-section area where the stenosis is located is shown in blue.

4. Summary and Discussion

We have developed and validated a new MCL that mimics the blood circulation in stenosed human systemic arterial systems and enables in vitro measurement of velocity and pressure waveforms in 3D-printed silicone arterial systems anatomically extracted from patient CTA imaging data. The MCL mainly consists of a human blood pump, a test section, a reservoir, Windkessel chambers, and resistance valves. Its measurement instruments include a DAQ system, a Doppler ultrasound machine, pressure transducers, and flowmeters. In this work, we focus on the in vitro measurement of blood pressure in the aortoiliac and aortorenal arterial systems. The objectives include reliably reproducing the computed blood pressures from ICHD using the same arterial system and the same flow conditions and revealing the pathophysiological properties that are essential to the ICHD modeling but are not readily available from current standard clinical measurements. Through this, we systematically studied five renal and six iliac cases to demonstrate the reliability of IEHD measurement and the applicability of IEHD for medical applications. The comparisons of the pressure data among the IEHD measurement, ICHD computation, and medical measurement indicate that both IEHD and ICHD are reliable for the noninvasive quantification of blood pressure in diseased human arteries. The IEHD-measured pressure waveforms agree well with those invasively measured and numerically simulated. A statistical analysis using all the iliac cases with 12 measurement samples and the renal cases with 13 measurement samples resulted in good correlations of SBP and DBP between experimental and medical measurements. The experimental measurement of SBP is more accurate than that of DBP. The reason might be that the signals the pressure transducers capture are stronger in systole than in diastole. We have also conducted two parametric studies using one renal and one iliac case. One is the cardiovascular response to MCL parameters. We found that the MAP in the aorta of an aortorenal arterial system has strong linear correlations with SV and HR, which are well understood. Another is the effect of VR of the arterial lumen on the TSPG of stenosis. Parametric variation of VR from real iliac stenosis and corresponding pressure measurements proximal and distal to the stenosis resulted in a TSPG~VR curve, which reveals a threshold of VR to assess the hemodynamic severity of the iliac stenosis. As the iliac stenosis was assessed as severe, we recommended interventional treatment for the stenosis, which agreed with the clinical treatment. Meanwhile, the TSPG at VR = 0 can predict the baseline pressure improvement after a potential revascularization, e.g., stenting, of the stenosis.

In addition to the in vitro pressure measurements, the MCL is being continuously developed for more research capabilities to support ICHD. We are interested in addressing the following open questions in the near future. First, how necessary is it to model the interaction between pulsatile blood flow and vessel deformation? Real arteries are deformable, but atherosclerosis (a common vascular disease in seniors) can significantly affect the elasticity of vessel walls during the cardiac cycle. Many ICHD computations use no-slip boundary conditions on arterial walls to avoid the complexity of modeling fluid-structure interaction and its demanding computation costs. IEHD can easily measure hemodynamics in deformable silicone arteries, but the deformation is affected by the 3D printing materials. The level of elasticity of a diseased artery segment is hard to determine due to a lack of available medical data. We plan to use three silicone vascular replicas (rigid, flexible, and elastic), medical artificial arteries, and cadaver arteries to study how the wall elasticity will affect the flow and pressure measurements. This study would be an important first-hand resource to guide the appropriate modeling of fluid-structure interactions in ICHD. Second, how can the patient-specific coefficients of vessel compliance and proximal and distal flow resistances in WK3 be determined? In current ICHD, these three coefficients are empirical, which significantly weakens the medical applications of ICHD. We are doing a systematic study for patient-specific vessel compliance and proximal and distal flow resistances, integrating engineering modeling in IEHD and ICHD by utilizing the available medical data. The study outcomes would advance the Windkessel model in ICHD for patient-specific applications.

The long-term goal of this research is twofold: (1) to establish a dependable resource that aids in the modeling of ICHD and validates the computational outcomes; and (2) to enable the comprehensive hemodynamic characterization of pathophysiological quantities, such as the coefficients in the WK3 model, for realistic blood flows in diseased human vessels. By leveraging the capabilities of the MCL, we can develop enhanced diagnostic tools, therapeutic strategies, and preventive measures using our powerful ICHD technique. These advancements play a critical role in improving patient care, alleviating the burden of arterial stenosis, and ultimately enhancing cardiovascular health outcomes.

Author Contributions: Conceptualization, H.Y., J.C., A.P.S. and W.H.; methodology, H.Y., J.C., W.H., J.T. and A.P.S.; formal analysis, W.H. and J.T.; investigation, H.Y., J.C., A.P.S., W.H. and J.T.; data curation, W.H. and J.T.; resources: A.P.S., D.M.R., X.F., J.L. and C.X.; writing—original draft preparation, H.Y. and W.H.; writing—review and editing, H.Y., J.C., A.P.S. and W.H.; visualization, W.H. and J.T.; supervision, H.Y., J.C. and A.P.S.; project administration, H.Y.; funding acquisition, H.Y. and A.P.S. All authors have read and agreed to the published version of the manuscript.

Funding: This research was partially supported by NSF grant CBET 1803845. This work used the Extreme Science and Engineering Discovery Environment (XSEDE), which is supported by the National Science Foundation Grant No. ACI-1548562. The first and corresponding author Yu would like to also acknowledge the IUPUI MEE Graduate Fellowship.

Data Availability Statement: The data presented in this study are available upon request from the corresponding authors. The data are not publicly available.

Conflicts of Interest: The data presented in this study are available on request from the corresponding author. The data are not publicly available.

Nomenclatures

CO	Cardiac output
CTA	Computed Tomography angiogram
DAQ	Data acquisition system
DBP	Diastolic blood pressure
DICOM	Digital Imaging and Communications in Medicine
DUS	Doppler ultrasound sonography
FFR	Fractional flow reserve
HR	Heart Rate
ICHD	Image-based computational hemodynamics
IEHD	Image-based experimental hemodynamics
IRB	Institutional Review Board
MAP	Mean arterial pressure
MCL	Mock circulation loop
SBP	Systolic blood pressure
STL	Standard Tessellation Language
SV	Stroke volume
TSPG	Trans-stenotic pressure gradient
V&V	Verification and validation
WK3	3-element Windkessel model

References

1. Pijls, N.H.; de Bruyne, B.; Peels, K.; van der Voort, P.H.; Bonnier, H.J.; Bartunek, J.; Koolen, J.J. Measurement of fractional flow reserve to assess the functional severity of coronary-artery stenoses. *N. Engl. J. Med.* **1996**, *334*, 1703–1708. [CrossRef] [PubMed]
2. Nørgaard, B.L.; Fairbairn, T.A.; Safian, R.D.; Rabbat, M.G.; Ko, B.; Jensen, J.M.; Nieman, K.; Chinnaiyan, K.M.; Sand, N.P.; Matsuo, H. Coronary CT angiography-derived fractional flow reserve testing in patients with stable coronary artery disease: Recommendations on interpretation and reporting. *Radiol. Cardiothorac. Imaging* **2019**, *1*, e190050. [CrossRef]
3. Ahmad, Y.; Götberg, M.; Cook, C.; Howard, J.P.; Malik, I.; Mikhail, G.; Frame, A.; Petraco, R.; Rajkumar, C.; Demir, O. Coronary hemodynamics in patients with severe aortic stenosis and coronary artery disease undergoing transcatheter aortic valve replacement: Implications for clinical indices of coronary stenosis severity. *JACC Cardiovasc. Interv.* **2018**, *11*, 2019–2031. [CrossRef] [PubMed]
4. Lu, M.T.; Ferencik, M.; Roberts, R.S.; Lee, K.L.; Ivanov, A.; Adami, E.; Mark, D.B.; Jaffer, F.A.; Leipsic, J.A.; Douglas, P.S. Noninvasive FFR derived from coronary CT angiography: Management and outcomes in the PROMISE trial. *JACC Cardiovasc. Imaging* **2017**, *10*, 1350–1358. [CrossRef]
5. Mangiacapra, F.; Trana, C.; Sarno, G.; Davidavicius, G.; Protasiewicz, M.; Muller, O.; Ntalianis, A.; Misonis, N.; Van Vlem, B.; Heyndrickx, G.R.; et al. Translesional pressure gradients to predict blood pressure response after renal artery stenting in patients with renovascular hypertension. *Circ. Cardiovasc. Interv.* **2010**, *3*, 537–542. [CrossRef]
6. Yu, H.; Khan, M.; Wu, H.; Zhang, C.; Du, X.; Chen, R.; Fang, X.; Long, J.; Sawchuk, A. Inlet and Outlet Boundary Conditions and Uncertainty Quantification in Volumetric Lattice Boltzmann Method for Image-Based Computational Hemodynamics. *Fluids* **2022**, *7*, 30. [CrossRef]
7. Yu, H.; Khan, M.; Wu, H.; Du, X.; Chen, R.; Rollins, D.M.; Fang, X.; Long, J.; Xu, C.; Sawchuk, A.P. A new noninvasive and patient-specific hemodynamic index for assessing the severity of renal arterial stenosis. *Int. J. Numer. Methods Biomed. Eng.* **2022**, *38*, e3611. [CrossRef] [PubMed]
8. Heinen, S.G.; van den Heuvel, D.A.; Huberts, W.; de Boer, S.W.; van de Vosse, F.N.; Delhaas, T.; de Vries, J.P. In Vivo Validation of Patient-Specific Pressure Gradient Calculations for Iliac Artery Stenosis Severity Assessment. *J. Am. Heart Assoc.* **2017**, *6*, e007328. [CrossRef]
9. Kinney, T.B.; Rose, S.C. Intraarterial pressure measurements during angiographic evaluation of peripheral vascular disease: Techniques, interpretation, applications, and limitations. *Am. J. Roentgenol.* **1996**, *166*, 277–284. [CrossRef]
10. de Boer, S.; Heinen, S.; van den Heuvel, D.; van de Vosse, F.; de Vries, J. How to define the hemodynamic significance of an equivocal iliofemoral artery stenosis: Review of literature and outcomes of an international questionnaire. *J. Vascular.* **2017**, *25*, 598–608. [CrossRef]
11. Marshall, R.S.; Pavol, M.A.; Cheung, Y.K.; Asllani, I.; Lazar, R.M. Cognitive Impairment Correlates Linearly with Mean Flow Velocity by Transcranial Doppler below a Definable Threshold. *Cerebrovasc. Dis. Extra* **2020**, *10*, 21–27. [CrossRef] [PubMed]
12. Koo, B.-K. The present and future of fractional flow reserve. *Circ. J.* **2014**, *78*, 1048–1054. [CrossRef] [PubMed]

13. Marsden, A.L.; Esmaily-Moghadam, M. Multiscale Modeling of Cardiovascular Flows for Clinical Decision Support. *Appl. Mech. Rev.* **2015**, *67*, 030804. [CrossRef]
14. Les, A.S.; Shadden, S.C.; Figueroa, C.A.; Park, J.M.; Tedesco, M.M.; Herfkens, R.J.; Dalman, R.L.; Taylor, C.A. Quantification of hemodynamics in abdominal aortic aneurysms during rest and exercise using magnetic resonance imaging and computational fluid dynamics. *Ann. Biomed. Eng.* **2010**, *38*, 1288–1313. [CrossRef]
15. Antiga, L.; Piccinelli, M.; Botti, L.; Ene-Iordache, B.; Remuzzi, A.; Steinman, A.D. An image-based modeling framework for patient-specific computational hemodynamics. *Med. Biol. Eng. Comput.* **2008**, *46*, 1097–1112. [CrossRef] [PubMed]
16. Nguyen, T.D.; Kadri, O.E.; Voronov, R.S. An introductory overview of image-based computational modeling in personalized cardiovascular medicine. *Front. Bioeng. Biotechnol.* **2020**, *8*, 529365. [CrossRef] [PubMed]
17. Yu, H.; Zhao, Y.; Lin, C. Unified Computational Method and System for in vivo Patient-Specific Hemodynamics. U.S. Patent 10482215, 19 November 2019.
18. Kim, H.; Vignon-Clementel, I.; Coogan, J.; Figueroa, C.; Jansen, K.; Taylor, C. Patient-specific modeling of blood flow and pressure in human coronary arteries. *Ann. Biomed. Eng.* **2010**, *38*, 3195–3209. [CrossRef] [PubMed]
19. Yu, H. Non-invasive Functional Assessment Technique for Determining Hemodynamics Severity of an Arterial Stenosis. U.S. Patent 11538153, 19 November 2022.
20. Murayama, Y.; Fujimura, S.; Suzuki, T.; Takao, H. Computational fluid dynamics as a risk assessment tool for aneurysm rupture. *Neurosurg. Focus* **2019**, *47*, E12. [CrossRef]
21. Liang, L.; Steinman, D.A.; Brina, O.; Chnafa, C.; Cancelliere, N.M.; Pereira, V.M. Towards the Clinical utility of CFD for assessment of intracranial aneurysm rupture—A systematic review and novel parameter-ranking tool. *J. Neurointerventional Surg.* **2019**, *11*, 153–158. [CrossRef]
22. Can, A.; Du, R. Association of hemodynamic factors with intracranial aneurysm formation and rupture: Systematic review and meta-analysis. *Neurosurgery* **2015**, *78*, 510–520. [CrossRef]
23. Formaggia, L.; Lamponi, D.; Quarteroni, A. One-dimensional models for blood flow in arteries. *J. Eng. Math.* **2003**, *47*, 251–276. [CrossRef]
24. Vignon-Clementel, I.E.; Figueroa, C.; Jansen, K.; Taylor, C. Outflow boundary conditions for 3D simulations of non-periodic blood flow and pressure fields in deformable arteries. *Comput. Methods Biomech. Biomed. Eng.* **2010**, *13*, 625–640. [CrossRef] [PubMed]
25. Vignon-Clementel, I.E.; Figueroa, C.A.; Jansen, K.E.; Taylor, C.A. Outflow boundary conditions for three-dimensional finite element modeling of blood flow and pressure in arteries. *Comput. Methods Appl. Mech. Eng.* **2006**, *195*, 3776–3796. [CrossRef]
26. Gallo, D.; De Santis, G.; Negri, F.; Tresoldi, D.; Ponzini, R.; Massai, D.; Deriu, M.; Segers, P.; Verhegghe, B.; Rizzo, G. On the use of in vivo measured flow rates as boundary conditions for image-based hemodynamic models of the human aorta: Implications for indicators of abnormal flow. *Ann. Biomed. Eng.* **2012**, *40*, 729–741. [CrossRef] [PubMed]
27. Vignon-Clementel, I.E.; Marsden, A.L.; Feinstein, J.A. A primer on computational simulation in congenital heart disease for the clinician. *Prog. Pediatr. Cardiol.* **2010**, *30*, 3–13. [CrossRef]
28. Alastruey, J.; Parker, K.; Peiró, J.; Sherwin, S. Lumped parameter outflow models for 1-D blood flow simulations: Effect on pulse waves and parameter estimation. *Commun. Comput. Phys.* **2008**, *4*, 317–336.
29. Stergiopulos, N.; Young, D.; Rogge, T. Computer simulation of arterial flow with applications to arterial and aortic stenoses. *J. Biomech.* **1992**, *25*, 1477–1488. [CrossRef]
30. Reymond, P.; Merenda, F.; Perren, F.; Rufenacht, D.; Stergiopulos, N. Validation of a one-dimensional model of the systemic arterial tree. *Am. J. Physiol. Heart Circ. Physiol.* **2009**, *297*, H208–H222. [CrossRef] [PubMed]
31. Bonfanti, M.; Balabani, S.; Greenwood, J.P.; Puppala, S.; Homer-Vanniasinkam, S.; Díaz-Zuccarini, V. Computational tools for clinical support: A multi-scale compliant model for haemodynamic simulations in an aortic dissection based on multi-modal imaging data. *J. R. Soc. Interface* **2017**, *14*, 20170632. [CrossRef]
32. Pirola, S.; Cheng, Z.; Jarral, O.; O'Regan, D.; Pepper, J.; Athanasiou, T.; Xu, X. On the choice of outlet boundary conditions for patient-specific analysis of aortic flow using computational fluid dynamics. *J. Biomech.* **2017**, *60*, 15–21. [CrossRef]
33. Morbiducci, U.; Gallo, D.; Massai, D.; Consolo, F.; Ponzini, R.; Antiga, L.; Bignardi, C.; Deriu, M.A.; Redaelli, A. Outflow conditions for image-based hemodynamic models of the carotid bifurcation: Implications for indicators of abnormal flow. *J. Biomech. Eng.* **2010**, *132*, 091005. [CrossRef] [PubMed]
34. Antonuccio, M.N.; Mariotti, A.; Fanni, B.M.; Capellini, K.; Capelli, C.; Sauvage, E.; Celi, S. Effects of uncertainty of outlet boundary conditions in a patient-specific case of aortic coarctation. *Ann. Biomed. Eng.* **2021**, *49*, 3494–3507. [CrossRef] [PubMed]
35. Westerhof, N.; Elzinga, G.; Sipkema, P. An artificial arterial system for pumping hearts. *J. Appl. Physiol.* **1971**, *31*, 776–781. [CrossRef] [PubMed]
36. Ohashi, Y.; De Andrade, A.; Nosé, Y. Hemolysis in an electromechanical driven pulsatile total artificial heart. *Artif. Organs* **2003**, *27*, 1089–1093. [CrossRef] [PubMed]
37. Nestler, F.; Bradley, A.P.; Wilson, S.J.; Timms, D.L.; Frazier, O.H.; Cohn, W.E. A hybrid mock circulation loop for a total artificial heart. *Artif. Organs* **2014**, *38*, 775–782. [CrossRef] [PubMed]
38. Gräf, F.; Finocchiaro, T.; Laumen, M.; Mager, I.; Steinseifer, U. Mock circulation loop to investigate hemolysis in a pulsatile total artificial heart. *Artif. Organs* **2015**, *39*, 416–422. [CrossRef]
39. Koenig, S.C.; Pantalos, G.M.; Gillars, K.J.; Ewert, D.L.; Litwak, K.N.; Etoch, S.W. Hemodynamic and pressure–volume responses to continuous and pulsatile ventricular assist in an adult mock circulation. *ASAIO J.* **2004**, *50*, 15–24. [CrossRef]

40. Pantalos, G.M.; Koenig, S.C.; Gillars, K.J.; Giridharan, G.A.; Ewert, D.L. Characterization of an adult mock circulation for testing cardiac support devices. *ASAIO J.* **2004**, *50*, 37–46. [CrossRef]
41. Liu, Y.; Allaire, P.; Wood, H.; Olsen, D. Design and initial testing of a mock human circulatory loop for left ventricular assist device performance testing. *Artif. Organs* **2005**, *29*, 341–345. [CrossRef] [PubMed]
42. Timms, D.; Hayne, M.; McNeil, K.; Galbraith, A. A complete mock circulation loop for the evaluation of left, right, and biventricular assist devices. *Artif. Organs* **2005**, *29*, 564–572. [CrossRef]
43. Colacino, F.M.; Moscato, F.; Piedimonte, F.; Danieli, G.; Nicosia, S.; Arabia, M. A modified elastance model to control mock ventricles in real-time: Numerical and experimental validation. *ASAIO J.* **2008**, *54*, 563–573. [CrossRef] [PubMed]
44. Schampaert, S.; Pennings, K.; Van de Molengraft, M.; Pijls, N.; Van de Vosse, F.; Rutten, M. A mock circulation model for cardiovascular device evaluation. *Physiol. Meas.* **2014**, *35*, 687. [CrossRef] [PubMed]
45. Gregory, S.D.; Pauls, J.P.; Wu, E.L.; Stephens, A.; Steinseifer, U.; Tansley, G.; Fraser, J.F. An advanced mock circulation loop for in vitro cardiovascular device evaluation. *Artif. Organs* **2020**, *44*, E238–E250. [CrossRef] [PubMed]
46. Rezaienia, M.; Paul, G.; Avital, E.; Mozafari, S.; Rothman, M.; Korakianitis, T. In-vitro investigation of the hemodynamic responses of the cerebral, coronary and renal circulations with a rotary blood pump installed in the descending aorta. *Med. Eng. Phys.* **2017**, *40*, 2–10. [CrossRef]
47. Gehron, J.; Zirbes, J.; Bongert, M.; Schäfer, S.; Fiebich, M.; Krombach, G.; Böning, A.; Grieshaber, P.; EMPACS (Exploration of the Mixing Phenomena during Interaction of Internal and External circulations) Study Group. Development and Validation of a Life-Sized Mock Circulatory Loop of the Human Circulation for Fluid-Mechanical Studies. *ASAIO J.* **2019**, *65*, 788–797. [CrossRef]
48. Ruiz, P.; Rezaienia, M.A.; Rahideh, A.; Keeble, T.R.; Rothman, M.T.; Korakianitis, T. In vitro cardiovascular system emulator (bioreactor) for the simulation of normal and diseased conditions with and without mechanical circulatory support. *Artif. Organs* **2013**, *37*, 549–560. [CrossRef] [PubMed]
49. Sénage, T.; Février, D.; Michel, M.; Pichot, E.; Duveau, D.; Tsui, S.; Trochu, J.N.; Roussel, J.C. A mock circulatory system to assess the performance of continuous-flow left ventricular assist devices (LVADs): Does axial flow unload better than centrifugal LVAD? *ASAIO J.* **2014**, *60*, 140. [CrossRef]
50. Pantalos, G.M.; Ionan, C.; Koenig, S.C.; Gillars, K.J.; Horrell, T.; Sahetya, S.; Colyer, J.; Gray, L.A., Jr. Expanded pediatric cardiovascular simulator for research and training. *ASAIO J.* **2010**, *56*, 67–72. [CrossRef]
51. Giridharan, G.A.; Ising, M.; Sobieski, M.A.; Koenig, S.C.; Chen, J.; Frankel, S.; Rodefeld, M.D. Cavopulmonary assist for the failing Fontan circulation: Impact of ventricular function on mechanical support strategy. *ASAIO J.* **2014**, *60*, 707. [CrossRef]
52. Giridharan, G.A.; Koenig, S.C.; Kennington, J.; Sobieski, M.A.; Chen, J.; Frankel, S.H.; Rodefeld, M.D. Performance evaluation of a pediatric viscous impeller pump for Fontan cavopulmonary assist. *J. Thorac. Cardiovasc. Surg.* **2013**, *145*, 249–257. [CrossRef]
53. Van Poucke, S.; Stevens, K.; Kicken, C.; Simons, A.; Marcus, A.; Lancé, M. Platelet function during hypothermia in experimental mock circulation. *Artif. Organs* **2016**, *40*, 288–293. [CrossRef]
54. Bleilevens, C.; Hill, A.; Grzanna, T.; Fechter, T.; Bohnen, M.; Weber, H.-J.; Beckers, C.; Borosch, S.; Zayat, R.; Benstoem, C. In vitro head-to-head comparison of anticoagulation properties of two heparin brands in a human blood miniature mock loop. *Interact. Cardiovasc. Thorac. Surg.* **2019**, *28*, 120–127. [CrossRef] [PubMed]
55. Knoops, P.G.; Biglino, G.; Hughes, A.D.; Parker, K.H.; Xu, L.; Schievano, S.; Torii, R. A mock circulatory system incorporating a compliant 3D-printed anatomical model to investigate pulmonary hemodynamics. *Artif. Organs* **2017**, *41*, 637–646. [CrossRef] [PubMed]
56. Conijn, M.; Wintermans, L.; Metselaar, R.; Ruisch, J.; Bax, E.; van Egmond, C.; Nieuwenstein, B.; Warmerdam, E.; Krings, G.J.B.P.; Express, E. A 3D printed pulmonary mock loop for hemodynamic studies in congenital heart disease. *Biomed. Phys. Eng. Express* **2022**, *8*, 065003. [CrossRef] [PubMed]
57. Vignali, E.; Gasparotti, E.; Mariotti, A.; Haxhiademi, D.; Ait-Ali, L.; Celi, S. High-versatility left ventricle pump and aortic mock circulatory loop development for patient-specific hemodynamic in vitro analysis. *ASAIO J.* **2022**, *68*, 1272–1281. [CrossRef] [PubMed]
58. Biglino, G.; Cosentino, D.; Steeden, J.A.; De Nova, L.; Castelli, M.; Ntsinjana, H.; Pennati, G.; Taylor, A.M.; Schievano, S. Using 4D cardiovascular magnetic resonance imaging to validate computational fluid dynamics: A case study. *Front. Pediatr.* **2015**, *3*, 107. [CrossRef]
59. Shepard, L.; Sommer, K.; Izzo, R.; Podgorsak, A.; Wilson, M.; Said, Z.; Rybicki, F.J.; Mitsouras, D.; Rudin, S.; Angel, E. Initial simulated FFR investigation using flow measurements in patient-specific 3D printed coronary phantoms. In Proceedings of the Medical Imaging 2017: Imaging Informatics for Healthcare, Research, and Applications, Orlando, FL, USA, 15–16 February 2017; pp. 192–203.
60. Sommer, K.N.; Shepard, L.; Karkhanis, N.V.; Iyer, V.; Angel, E.; Wilson, M.F.; Rybicki, F.J.; Mitsouras, D.; Rudin, S.; Ionita, C.N. 3D Printed Cardiovascular Patient Specific Phantoms Used for Clinical Validation of a CT-derived FFR Diagnostic Software. In Proceedings of the Medical Imaging 2018: Biomedical Applications in Molecular, Structural, and Functional Imaging, Houston, TX, USA, 11–13 February 2018; Volume 10578, p. 105780J.
61. Rong, C.; Chen, R.; Yan, W.; Yu, H.; Xu, Y. Hemodynamic analysis of external iliac artery based on VLBM. *J. Zhejiang Univ. Sci. Technol.* **2022**, *34*, 7–16.
62. Zhang, X.; Gomez-Paz, J.; Chen, X.; McDonough, J.M.; Islam, M.; Andreopoulos, Y.; Zhu, L.; Yu, H. Volumetric lattice Boltzmann method for wall stresses of image-based pulsatile flows. *Sci. Rep.* **2022**, *12*, 1697. [CrossRef]

63. An, S.; Yu, H.; Yao, J. GPU-accelerated Volumetric Lattice Boltzmann Method for Porous Media Flow. *J. Petro. Sci. Eng.* **2017**, *156*, 546–552. [CrossRef]
64. An, S.; Yu, H.; Wang, Z.; Chen, R.; Kapadia, B.; Yao, J. Unified Mesoscopic Modeling and GPU-accelerated Computational Method for Image-based Pore-scale Porous Media Flows. *Int. J. Heat Mass Trans.* **2017**, *115*, 1192–1202. [CrossRef]
65. Wang, Z.; Zhao, Y.; Sawchuck, A.P.; Dalsing, M.C.; Yu, H. GPU acceleration of Volumetric Lattice Boltzmann Method for patient-specific computational hemodynamics. *Comput. Fluids* **2015**, *115*, 192–200. [CrossRef]
66. Yu, H.; Chen, X.; Wang, Z.; Deep, D.; Lima, E.; Zhao, Y.; Teague, D.S. Mass-conserved volumetric lattice Boltzmann method for complex flows with willfully moving boundaries. *Phys. Rev. E* **2014**, *89*, 063304. [CrossRef] [PubMed]
67. Yu, H.; Rong, C.; Jin, X.; Xu, Y.; Murphy, M.; Motaganahalli, R.; Sawchuk, A. Fast and Noninvasive Evaluation of In Vivo Pressure in Stenosed Aortoiliac Arteries. *J. Vasc. Surg.* **2020**, *72*, e308–e309. [CrossRef]
68. Monnet, X.; Letierce, A.; Hamzaoui, O.; Chemla, D.; Anguel, N.; Osman, D.; Richard, C.; Teboul, J.-L. Arterial pressure allows monitoring the changes in cardiac output induced by volume expansion but not by norepinephrine. *Crit. Care Med.* **2011**, *39*, 1394–1399. [CrossRef] [PubMed]
69. Kluckow, M.; Evans, N. Relationship between blood pressure and cardiac output in preterm infants requiring mechanical ventilation. *J. Pediatr.* **1996**, *129*, 506–512. [CrossRef] [PubMed]
70. Chapman, J.H.; Elliott, P. Cardiovascular effects of static and dynamic exercise. *Eur. J. Appl. Physiol. Occup. Physiol.* **1988**, *58*, 152–157. [CrossRef]
71. Lind, A.; McNicol, G. Muscular factors which determine the cardiovascular responses to sustained and rhythmic exercise. *Can. Med. Assoc. J.* **1967**, *96*, 706.
72. Lind, A.R. Cardiovascular responses to static exercise (Isometrics, anyone?). *Circulation* **1970**, *41*, 173–176. [CrossRef]

fluids

MDPI

Article

V Flow Measurements of Pulsatile Flow in Femoral-Popliteal Bypass Proximal Anastomosis Compared with CFD Simulation

Andrey Yukhnev [1,*], Ludmila Tikhomolova [1], Yakov Gataulin [1], Alexandra Marinova [1], Evgueni Smirnov [1], Andrey Vrabiy [2], Andrey Suprunovich [2] and Gennady Khubulava [2]

1 Higher School of Applied Mathematics and Computational Physics, Peter the Great St. Petersburg Polytechnic University, 29 Polytechnicheskaya Str., 195251 St. Petersburg, Russia; ludmila060495@mail.ru (L.T.); yakov_gataulin@mail.ru (Y.G.); sanmarinova@gmail.com (A.M.); smirnov_em@spbstu.ru (E.S.)

2 Department of Faculty Surgery, Pavlov First St. Petersburg State Medical University, 6-8 L'va Tolstogo Str., 197022 St. Petersburg, Russia; dock1@yandex.ru (A.V.); doctoras@mail.ru (A.S.); ggkh07@rambler.ru (G.K.)

* Correspondence: a.yukhnev@mail.ru

Abstract: This paper presents the experience of using the V Flow high-frame-rate ultrasound vector imaging method to study the pulsatile velocity fields in the area of the proximal anastomosis for femoral popliteal bypass surgery in vitro and in vivo. A representative (average) anastomosis model and the experimental setup designed for in vitro studies covering forward and reverse flow phases throughout the cycle are described. The results of the measurements are presented for areas with a relatively uniform velocity distribution and for areas with pronounced spatial inhomogeneities due to the jet or recirculating nature of the flow. The results of ultrasonic studies of the velocity field of the three-dimensional pulsatile flow in vitro and in vivo are compared with the data of numerical simulations carried out for the average and personalized models based on the Navier–Stokes equations. Acceptable consistency between the results of experimental and numerical studies is demonstrated.

Keywords: pulsatile blood flow; 3D numerical simulation; 2D ultrasound vector imaging; femoral popliteal bypass surgery; proximal anastomosis; personalized model

Citation: Yukhnev, A.; Tikhomolova, L.; Gataulin, Y.; Marinova, A.; Smirnov, E.; Vrabiy, A.; Suprunovich, A.; Khubulava, G. V Flow Measurements of Pulsatile Flow in Femoral-Popliteal Bypass Proximal Anastomosis Compared with CFD Simulation. *Fluids* **2024**, *9*, 64. https://doi.org/10.3390/fluids9030064

Academic Editors: Fang-Bao Tian and Paolo Zunino

Received: 21 November 2023
Revised: 22 January 2024
Accepted: 21 February 2024
Published: 4 March 2024

1. Introduction

Computational fluid dynamics (CFD) is a powerful tool for studying blood flow in different parts of the vascular bed and can account for such factors as geometry complexity, vessel wall elasticity, blood rheology, and turbulence [1]. Compared to measurements, up-to-date numerical simulations of blood flow provide significantly more information about the velocity and pressure fields, as well as wall shear stresses. Obtained information in some cases is used for diagnosing blood flow disorders and for the planning of operations.

Modern technologies in CFD implement numerical methods that make it possible to simulate viscous fluid flows in areas of arbitrary, often rather complex, geometry; bifurcations of blood vessels are also in this row. This opportunity has arisen due to the intensive development of numerical methods known as the finite element method (FEM) and the finite volume method (FVM). Both of these methods belong to the class of mesh-based methods, but use different approaches to obtain algebraic systems that approximate the basic partial differential equations of continuum mechanics. The FEM method is especially in demand for solving problems of structural mechanics. Regarding fluid dynamic problems, the fundamental conservation property of the FVM makes it the preferable method in comparison to the FEM, and the FVM is currently the most widely used numerical technique for CFD studies.

The FVM is based on the approximate solution of the integral form of the conservation equations. The computational domain chosen for the investigated problem is divided into a set of non-overlapping control volumes referred to as finite volumes, where the variables of interest are assigned to the center of the finite volume. The finite volume is also referred to as a cell or element, and can be of different forms: tetrahedral, prismatic, hexahedral, and polyhedral. The governing equations are integrated over each finite volume, and some interpolation procedures are assumed, which serve to restore representative values of the variable of interest at the faces of the current cell by making up some combination of the central value and values from the centers of neighboring cells. The resulting discretization (algebraic) equation expresses with a specified order of accuracy the conservation principle for the variable inside the finite volume, and the system of such equations is to be solved using some algorithm, usually iterative. When solving time-dependent problems, the time derivative of the variable of interest is approximated for the center of the cell using one or another finite-difference scheme, and iterative calculations with discrete time steps are carried out sequentially at different time steps. In the case of 3D incompressible viscous fluid flows, the governing equations are the Navier–Stokes equations and the continuity equation, and the variables of interest are pressure and three components of the velocity vector. A detailed description of the numerical schemes and algorithms most widely used to solve problems of incompressible viscous fluid dynamics can be found elsewhere [2,3].

Blood flow studies in various vessels, healthy and damaged, as well as in vascular prostheses, are often carried out using simplified models, for the construction of which average statistical data are used. This approach makes it possible to conduct extended parametric studies, test new experimental methods for clinical diagnosis of blood flow in phantoms, and validate CFD methods under conditions of complex transient flow regimes. From the other side, the progress of computing tools has opened up the possibility of the widespread use of personalized vascular models in both pre- and post-operative settings. This approach allows for the individual selection of a prosthesis and surgical technique [4–6].

In the natural vessel bifurcations, a complex spatial structure of blood flow is formed, which in some cases is one of the causes of atherosclerotic changes in the vessel wall [7]. Significant disturbances in the physiological structure of blood flow are also observed in the areas of vessels' connection with synthetic prostheses of the "end-to-side" type, namely, in the areas of proximal and distal anastomoses. It can lead to neointimal hyperplasia, thrombosis of prostheses, and their complete blockage, requiring reoperation [8,9].

When studying the blood flow patterns in anastomoses, numerical modeling is wildly used, the results of which in some cases are compared with experimental data [10]. Most in vitro studies in anastomotic models for femoral popliteal bypass surgery were performed using particle image velocimetry (PIV) [11]. A recent experience in using the smoke image velocimetry (SIV) measurement technique for studying fluid motion with local turbulence in a proximal anastomosis model is presented in [12]. Ultrasound (US) methods for blood flow analysis are used mainly in clinical studies [13]. To obtain knowledge that is useful for the early prediction of dangerous postoperative complications, numerical calculations are performed for personalized anastomosis models based on magnetic resonance imaging (MRI) [14], US and multispiral computed tomography (MSCT) data [15–17].

In the case of numerical simulations of blood flow using personalized models for one or another part of the vascular bed, the problem of the incomplete definition of boundary conditions always arises, resulting from the difficulties of clinical blood flow measurements. Accordingly, it is necessary to introduce some assumptions. Sometimes, numerical results are supplemented by an uncertainty analysis, and for greater reliability, the calculation results are compared with US measurements. The authors of several works dealing with studies of complex blood flows clearly give preference to US methods in terms of accuracy and involve US data to validate the CFD results [18,19]. On the other side, the data reported in other works allows one to conclude that in the case of personalized blood flow studies, numerical modeling and ultrasound measurements provide results of comparable accuracy [20,21].

Nowadays, US methods remain the leader among clinical methods for blood flow diagnostics, despite their operator dependence. The development and introduction into clinical practice of new US methods for hemodynamics analysis were accompanied traditionally by their validation using PIV and laser Doppler anemometry (LDA) [22] in the last century, and MRI and CFD have been added to them [23] in recent decades. The increasing interest in choosing CFD as a validation method for US measurements is attributed to the fact that in the case of laminar flows, which include blood flow in most vessels, numerical modeling with well-defined boundary conditions provides results that are more accurate compared to the US measurement results. It is almost recognized today that in the case of comparative studies of laminar flows in blood flow phantoms under controlled experimental conditions (which is problematic at clinical measurements), the usage of CFD methods is an effective and reliable approach for validating US techniques [24,25].

Currently, among the numerous US methods of blood flow investigations, one can highlight modern developments that form the class of methods for high frame-rate vector flow imaging (VFI), detailed reviews of which can be found in [26,27]. Over the last decade, such methods have been developed using special US scientific equipment [28], and in recent years, they have begun to be used in commercial US scanners—Mindray, GE, BK Medical [29]. The main advantages of these methods are the independence of measurement results from the transducer installation angle and obtaining a dynamic 2D vector field of blood flow velocity; at that high time resolution (up to 600 fps) is achieved. Some scanners of this class, along with 2D vector velocity fields, produce velocity profiles in vessel sections [30], streamlines [31], as well as the magnitude and direction of near-wall velocity and wall shear stress at the user-defined monitoring points [32].

The measurement results presented in the present work were obtained using the V Flow vector visualization program implemented in the Mindray Resona 7 scanner (Shenzhen Mindray Bio-Medical Electronics Co., Ltd., Shenzhen, China). The V Flow technique is based on multi-directional Doppler US using an interleaved plane wave and focused wave transmissions [16]. The blood flow structure after short-term scanning at high frame rates and short post-processing is presented in the form of a non-stationary 2D vector velocity field in the scanning window. The color and length of the vectors correspond to the magnitude of the velocity module. The results do not depend on the angle of installation of the transducer relative to the blood vessel. The video clip of pulsatile blood flow can be viewed by increasing the cycle duration by 7–200 times. The clinical testing of this program was carried out both by the developers of the method and by some scientific groups in different countries [30,33]. Using the V Flow method, vector blood flow velocity fields were studied in the carotid and femoral arteries [34], arteriovenous fistulas [35], and veins of the lower extremities [36]. The authors of the report [37] tested the V Flow method in a "reference" flow using two commercial blood flow phantoms. When using a string phantom, the error in measuring the axial velocity was evaluated as 3.8%, and in the case of the Doppler flow phantom, the error in measuring the bulk velocity was 5.2%.

The authors of the present work are not aware of any studies devoted to comparison of the measurement data by the V Flow method with the results of the numerical simulation of three-dimensional blood flow. The method was compared with other clinical measurement methods only based on the integral parameters of blood flow. The purpose of this work is a comparative study in vitro and in vivo of the pulsatile blood flow velocity field in the area of the graft branch from the femoral artery during femoral popliteal bypass surgery (proximal anastomosis) using two methods, which are the high-frame-rate ultrasound vector imaging of V Flow and the finite-volume numerical simulation technique for solving the Navier-Stokes equations.

2. Facilities and Conditions for V Flow Measurements

2.1. Laboratory Model of Proximal Anastomosis and Experimental Setup

The simplified geometric model of a proximal anastomosis section after the femoral popliteal bypass surgery used in this work was developed based on the averaging of 15 personalized models obtained by slice-by-slice segmentation of multispiral computed angiography images [16]. For the processed personalized models, the diameter of the CFA was 7.5 ± 1.2 mm, the diameter of the graft was 9.5 ± 1.2 mm, the diameter of DFA was 4.2 ± 1.1 mm, the angle of the graft branch from the CFA was 45 ± 25°, and from the DFA was 44 ± 25°. The developed model is hereafter referred to as the average proximal anastomosis model. It includes the area of branch of the vascular graft from the common femoral artery (CFA), a short segment of the occluded surface femoral artery, and the deep femoral artery branch, as well as straight outlet sections of the graft and the deep femoral artery (Figure 1a). The central lines of the vessels forming the model lie in the plane of symmetry of the anastomosis area (plane YZ) in Figure 1a.

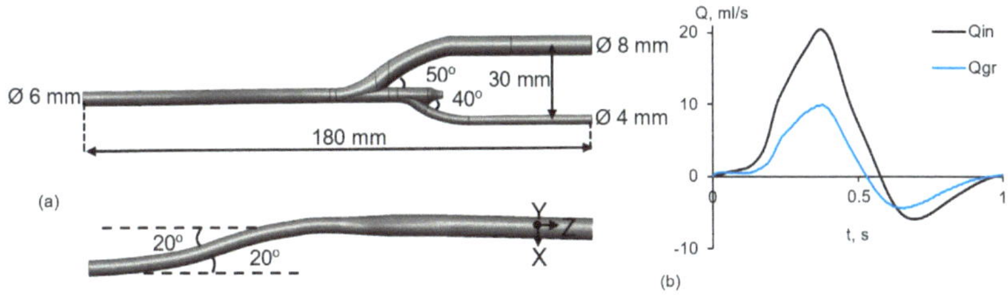

Figure 1. Average model of proximal anastomosis section for femoral popliteal bypass (**a**) and experimental flow rate curves (**b**).

For the developed proximal anastomosis model, the following dimensions were adopted: the internal diameters of the CFA, deep femoral artery, and graft are 6, 4, and 8 mm, respectively. The angle of the graft branch from the CFA is 50°, and from the DFA is 40°. The length of the inlet straight section of the femoral artery before the graft branch is 10 mm, and the total length of the model is 110 mm. The axes of the straight outlet sections of the graft and the deep femoral artery are spaced 30 mm from each other. To reduce the influence of uncertainties in setting inlet conditions for velocity, the proximal anastomosis model was supplemented with an inlet section (supply tube) with a length of 70 mm, modeling a moderate double bend of the femoral artery in a plane normal to the plane of symmetry of the anastomosis. A general displacement of the central line of this inlet section is 15 mm (Figure 1a). The model with a wall thickness of 1 mm was made by 3D printing using Tough 20003D plastic (Formlabs Co., Somerville, MS, USA).

To conduct an experimental study of fluid flow in the developed proximal anastomosis model, a laboratory setup was assembled (Figure 2). The setup consisted of a container with the model included in a hydrodynamic loop, a flow pump MaxiJet (Marineland Co., Blacksburg, VA, USA) with programmable timer ST4S-1p (Autonics Co., Mundelein, IL, USA), a damper that ensures the formation of the reverse flow phase, and two flow regulators that allow one to regulate the inlet flow amplitude and the ratio of flow rates in the outlet branches. To imitate the properties of biological tissue, the container was filled with tissue-mimicking material based on agar-agar with the addition of graphite. The material simulates the density (1050 kg/m^3), speed of sound (1550 m/s), and attenuation coefficient (0.7 dB/(cm × MHz)) of a biological tissue [38]. A water solution of glycerin (36% by volume) with the addition of salt (10 g/L) and fine pigment (5 g/L) was used as a blood-mimicking fluid with density ϱ = 1050 kg/m^3 and dynamic viscosity coefficient μ = 0.0035 Pa·s. The flow curve of a blood-mimicking fluid was modeled qualitatively in the setup (Figure 1b): the phases of forward and reverse flow had durations of 0.5 s and 0.4 s, respectively, in the CFA model [39]. For the accepted conditions, the maximum inlet Reynolds number was 1390, when calculated using the maximum flow rate and the internal diameter of the femoral artery. The measurement results presented below were obtained for the case in which the ratio of the maximum flow in the graft to the maximum inlet flow was equal to 0.48.

The measuring system consisted of a Nihon Kohden MF-46 electromagnetic flow meter (Nihon Kohden Corp., Tokyo, Japan) connected via an ADC to the computer, and a Mindray Resona 7 ultrasonic scanner (Shenzhen Mindray Bio-Medical Electronics Co., Ltd., Shenzhen, China), with a linear transducer installed above the model under study (Figure 2). The flow meter was equipped with two flow transducers, positioned before the model inlet and behind the graft outlet to measure flow rates $Q_{in}(t)$ and $Q_{gr}(t)$, correspondingly.

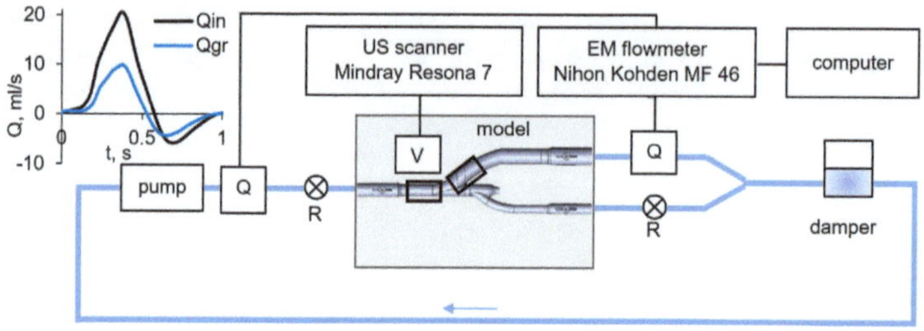

Figure 2. Diagram of the experimental setup: Q—electromagnetic flow transducer, R—flow regulator, and V—ultrasonic transducer.

When measuring the vector velocity field using the V Flow method, the ultrasonic transducer was fixed on a holder so that the scanning plane was located in the plane of symmetry of the anastomosis model. Two scanning windows (15 × 27 mm) selected to obtain vector velocity fields at the inlet section (CFA) and in the graft are shown in Figure 3; the location of the scanning windows is shown against the background of the calculated vector velocity field obtained for the symmetry plane as a result of the CFD simulation described below.

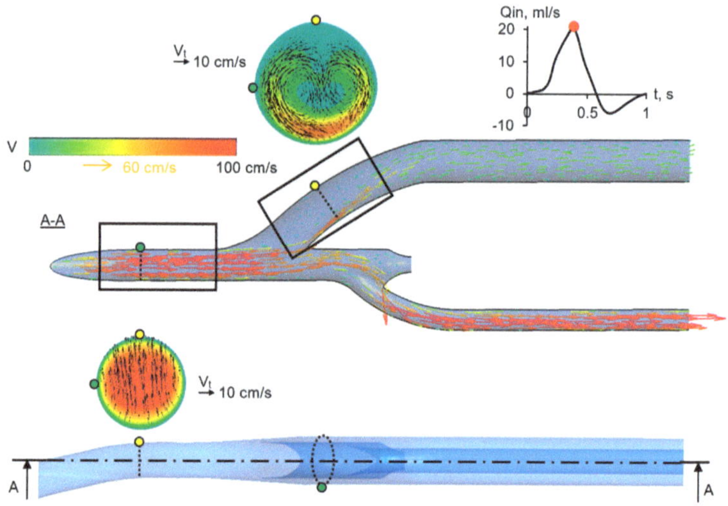

Figure 3. Locations of the ultrasonic scanning windows (rectangles) and lines in the window planes along which the streamwise velocity profiles were measured (dotted lines); for more information, the inlet flow curve $Q_{in}(t)$ and some results of the described-below numerical simulation are illustrated: the velocity vector distribution is shown for the longitudinal section A-A (velocities greater than 100 cm/s are painted red), and the streamwise velocity distributions supplemented by cross-flow patterns (view against the stream) are given for two cross-sections of the average anastomosis model.

For both scanning windows, the V Flow method provided recording video images of vector velocity fields at a frequency of 500 fps. The recording time was 3 s. Synchronization of the video images was carried out using a signal from the electromagnetic flow meter. The recorded velocity field information can be viewed in slow motion (from 7 to 200 times) compared to real time and some visualization parameters can be changed, such as vector density and vector length. Individual frames were processed to construct streamwise velocity profiles in cross-sections at different instants. When processing the video files for the construction of a velocity profile in the selected cross-section, the magnitude and angle of the velocity vector were measured by the V Flow program moving along the selected section with a step of 0.3 mm.

2.2. Clinical V Flow Measurements

Using the V Flow method, the vector field of blood flow velocity in the area of the proximal anastomosis of one of the patients was measured 7 months after the femoral popliteal bypass surgery performed in a clinic of the Pavlov First Saint Petersburg State Medical University. A synthetic graft was used for the operation.

During the measurements, the scanning plane of the ultrasound transducer coincided, as it was possible, with the middle longitudinal plane of the anastomosis areas under examination. Figure 4 shows the window of the V Flow program with a 2D vector velocity field directly at the site of the graft branch from the CFA. Here, the scanning window, highlighted with a white frame, has dimensions of 20 × 27 mm. The results of the vector field processing for the region of interest (ROI), highlighted with a white frame with dimensions of 2.7 × 8.1 mm, are visualized by the velocity curve (light green curve) at the point of maximum velocity in the ROI (green dot), being synchronized with the electrocardiogram signal shown at the window bottom. The upper part of the window displays the values of automatically calculated flow velocities for the ROI. The following notations are used here: Max—instantaneous maximum velocity, T-Max—maximum velocity during recording time (T1~T2), Mean—instantaneous average velocity over ROI, TAMean—average velocity over recording time (T1~T2), Median—instantaneous median velocity, Turbulen—velocity pulsation parameter, and Size—ROI size.

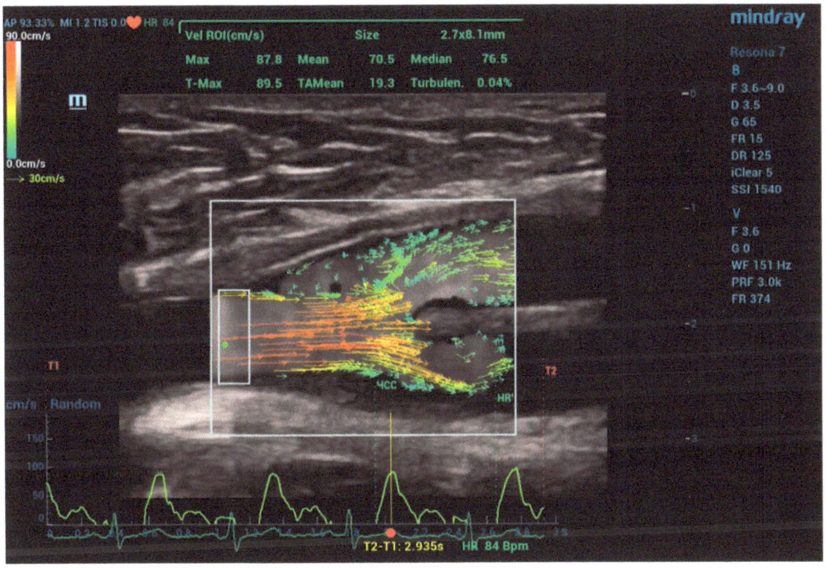

Figure 4. V Flow program window when placing the scanning window at the site of the graft branch from the patient's common femoral artery.

3. Numerical Simulation Technique

3.1. Computational Domains and Grids

The numerical modeling of pulsatile blood flow in the area of the proximal anastomosis was carried out based on Navier–Stokes equations using the average model and a personalized one.

In the case of the average model, the used computational domain fully corresponded to the 3D geometry of the laboratory model of the described-above anastomosis area, with the addition of the supply tube. The calculation results shown in the present paper were obtained with a basic computational grid that included 3.2 million elements, mainly tetrahedra (Figure 5). Near the wall, the grid contained prismatic layers for a better resolution of near-wall flow. The thickness of the elements closest to the wall was about one hundredth of the diameter of the vessel. At a preliminary stage of calculations aimed at the analysis of grid dependence, two more grids, characterized by decreased and increased characteristic sizes of the elements, were also constructed, maintaining a similar spatial distribution. The coarser and the refined grids contained 500 thousand and 11 million elements, respectively. Some results of the grid sensitivity analysis are presented in Section 3.3.

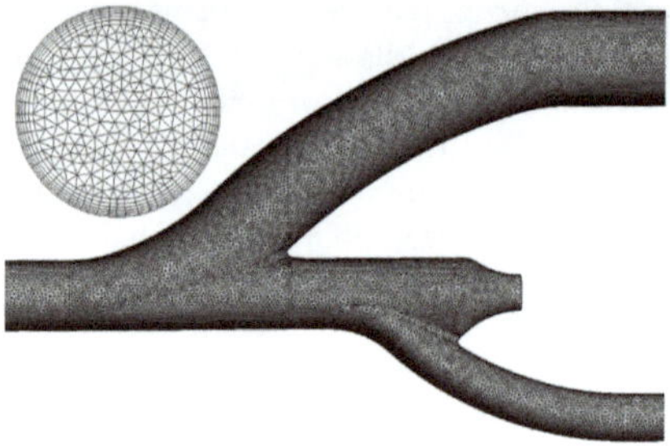

Figure 5. Computational grid for the average model of the anastomosis area.

Using the procedures described in our previous reports [15–17], a personalized geometric model of the proximal anastomosis area of the patient, for whom an examination was carried out in the University clinic 7 months after the operation, was built based on data of MSCT angiography. This personalized model is illustrated in Figure 6a. The diameter of the model section related to the common femoral artery (CFA) is evaluated as 7 mm, whereas the outlet diameter of the captured graft section is 9 mm. The computational grid generated for the personalized model consisted of 2 million elements.

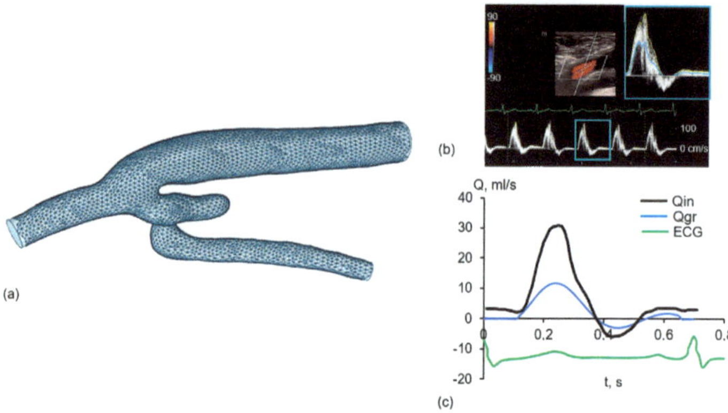

Figure 6. Personalized geometric model of the proximal anastomosis during femoral popliteal bypass (**a**), Doppler signals of blood flow in the CFA (**b**) and flow rate curves (**c**) for the computational domain inlet and for the outlet section of the graft model.

3.2. Boundary Conditions

Changes in the average flow rate over time were specified as boundary conditions at the inlet to the computational domain corresponding to the CFA, section S_{in}, and at one of the outlets corresponding to the graft, section S_{gr}. For the anastomosis average model, the temporal changes of the flow rates at sections S_{in} and S_{gr} were defined by the measurements performed on the experimental setup (Figure 1b). In the case of the personalized model, the flow rate curves $Q_{in}(t)$ and $Q_{gr}(t)$ shown in Figure 6b were obtained via processing the results of ultrasonic Doppler measurements at a section of the CFA and at a distal section of the graft. The ratio of the maximum flow in the graft to the maximum inlet flow was equal to 0.38. Data synchronization for the CFA and the graft sections was provided by the electrocardiogram signal. The patient had a biphasic blood flow rate curve measured in CFA. The duration of the systolic phase was 0.23 s with a peak velocity of 81 cm/s, the duration of the reverse flow was 0.14 s with the peak velocity of 16 cm/s, the cycle duration was 0.7 s, and the constant diastolic velocity—8 cm/s (Figure 6c).

The inlet velocity distribution was assumed to be uniform over the vessel cross-section. At the second outlet, the reduced pressure was set to zero. The no-slip condition was specified on all walls and treated as rigid. The measured radial displacements of the vessel wall were quite low: $0.05 \cdot D_{CFA}$ for CFA and $0.005 \cdot D_g$ for the graft.

3.3. Computational Aspects

For the simulations, the dynamic viscosity of the blood-mimicking fluid was assumed to be 0.0035 Pa·s, and the density was 1050 kg/m^3. With these properties of the fluid, the characteristic Reynolds number, based on the inlet flow rate at the maximum flow instant and on the diameter of the inlet section, is evaluated as 1390 for the anastomosis average model and 1640 for the personalized one.

The mean Reynolds numbers, evaluated with the time-averaged inlet velocity over the forward-flow phase, are about 500 for both models. It means that the flow regime is very far from regimes of developed turbulence, and one would expect only local (in space and time) manifestations of incipient turbulence. There are no known statistical turbulence models in the literature that could be applied to reliably capture the smoothing impact of this local incipient turbulence on the major large-scale components of the flow developing in the vessel models. The application of advanced approaches aimed at the resolution of a broad spectrum of vortices in the flow (direct numerical simulation or well-resolving large eddy simulation) is out of the scope of the present study. Apart from the need for enormous computational resources, the use of these approaches would require well-defined information about the spectrum of inlet disturbances, which is very problematic to extract from experiments.

Taking into account the above-mentioned circumstances, the present numerical simulation was carried out under the assumption that the flow is quasi-laminar. With this assumption, the physical smoothing action of small-scale eddies in potential areas with local turbulence is replaced by the dissipative properties of the numerical scheme used for solving the discretized Navies–Stokes equations.

Calculations of pulsatile flow fields for both the anastomosis average model and the personalized one were performed using the commercial finite-volume CFD package Ansys CFX 18.2 (Ansys Inc., Canonsburg, PA, USA). The "High Resolution" scheme was activated for the calculation of the convective fluxes, being, in fact, the second-order accuracy upwind scheme introducing a proper dissipation. The second-order backward Euler scheme was used for physical time stepping. The final calculations were performed with time step equal to 0.01 s. The convergence criterion was the reduction of normalized residuals to 10^{-6} at each time step. Typically, the calculations were carried out for three cycles, starting with the conditions corresponding to an initially quiescent fluid. It was established that the calculated flow fields in the second and third cycles practically coincided.

Figure 7 presents some results of the grid and time-step sensitivity analysis performed to assess the appropriateness of the basic computational grid characteristics, as well as the chosen time step of 0.01 s. This analysis included two series of computations dealing with the anastomosis average model. The first one covered a comparison of the steady-state solutions obtained with the basic and refined grids (N_{elm} = 3.2 and 11 million) for the inlet Reynolds number equal to 480, which is just the mean Reynolds number evaluated with the time-averaged inlet velocity over the forward-flow phase. A well agreement of the flow fields computed with two grids is illustrated by Figure 7a, where the streamwise velocity profiles related to the graft model are compared. The second series of the grid sensitivity analysis covered a comparison of the pulsatile flow solutions computed with two grids and different time steps for the time-dependent boundary conditions, as described in Section 3.2. Particular results of this comparison are demonstrated in Figure 7b, where three velocity profiles, also taken in the graft area, are shown for the maximum flow instant.

The performed analysis of the numerical solution sensitivity to changes in the grid size and time step allowed us to conclude that the basic grid and the chosen time step were quite acceptable for simulation in the case of the anastomosis average model. The structure and size of the base grid elements for the personalized model were chosen similarly to the average model case.

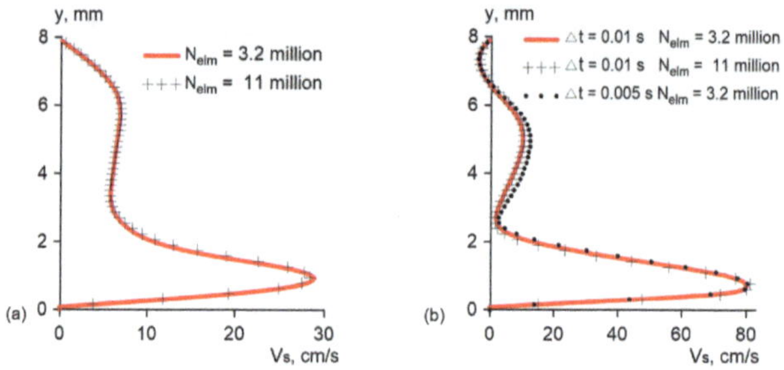

Figure 7. Grid and time-step influence on the distribution of the streamwise velocity in the model graft section indicated in Figure 3: profiles calculated with two grids for steady-state flow at Re_{in} = 480 (**a**), and instant profiles (at the maximum flow rate) predicted with two grids and two time steps under pulsatile flow conditions at $Re_{in,max}$ = 1390 (**b**).

4. Results and Discussion

To compare the experimental and computational results obtained for the average model of the proximal anastomosis two characteristic instants were chosen, the first of which corresponds to the maximum forward flow rate; the second corresponds to the maximum reverse flow rate.

Figures 8 and 9 present a comparison of CFD and V Flow data for the scanning window positioned before the model graft branch (Figure 3), whereas Figures 10 and 11 cover the compared data obtained with the window positioned in the model graft, just after the branching (Figure 3).

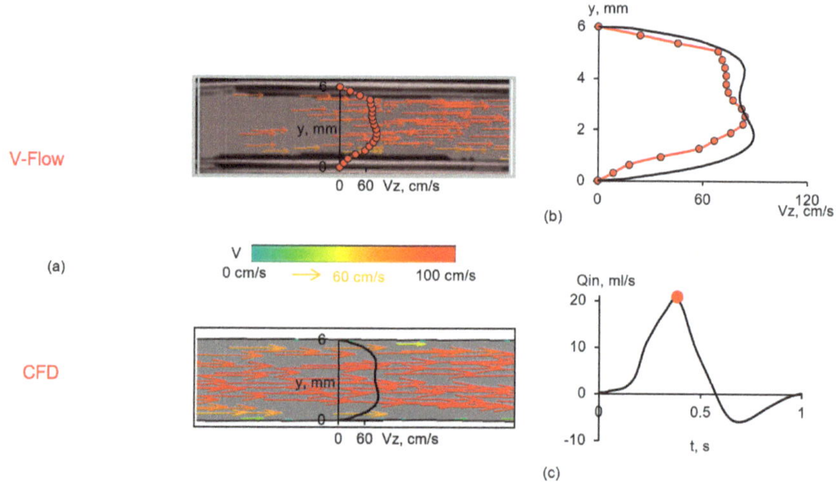

Figure 8. Vector fields (**a**) and streamwise velocity profiles (**b**) in the inlet CFA section of the anastomosis average model obtained by the V Flow (dots) and CFD (line) techniques for the marked instant of maximum forward-flow rate (**c**).

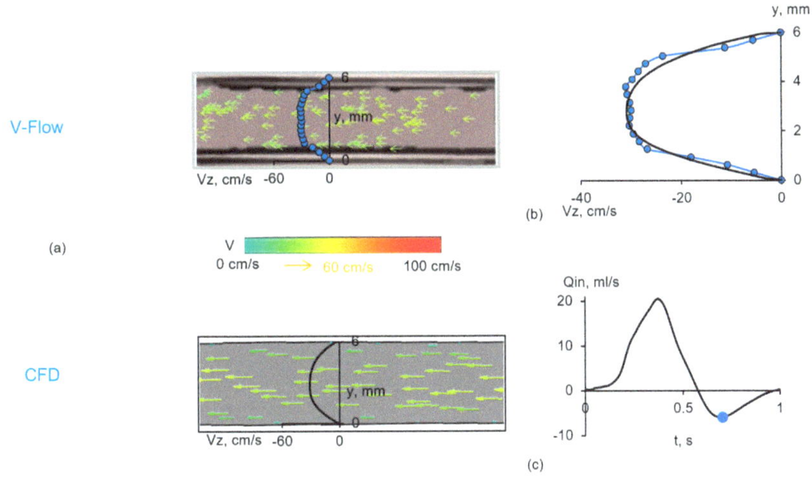

Figure 9. Vector fields (**a**) and streamwise velocity profiles (**b**) in the inlet CFA section of the anastomosis average model obtained by the V Flow (dots) and CFD (line) techniques for the marked instant of maximum reverse-flow rate (**c**).

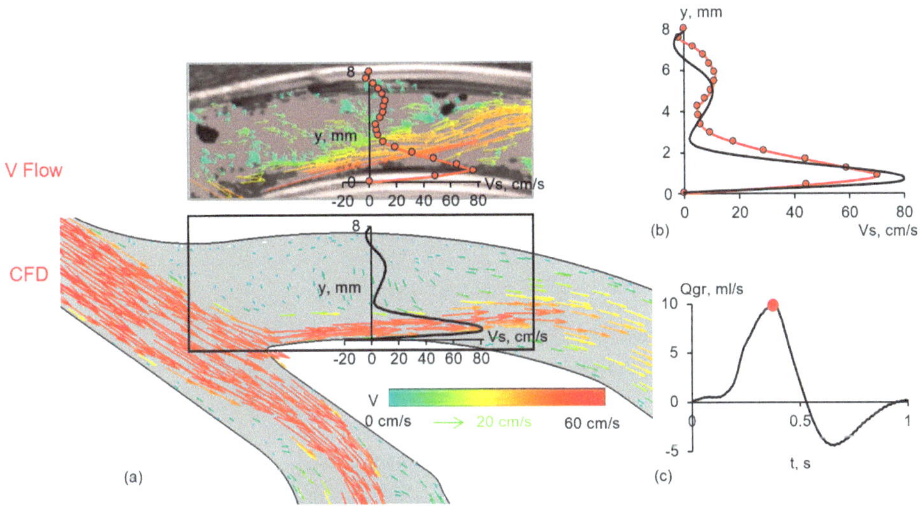

Figure 10. Vector fields (**a**) and streamwise velocity profiles (**b**) in the examined part of the anastomosis average model obtained by the V Flow (dots) and CFD (line) techniques for the marked instant of maximum forward-flow rate (**c**). Velocities greater than 60 cm/s are painted red.

For the section covered by the first window, both the V Flow technique and the simulation visualized the velocity field with nearly straight streamlines parallel to the vessel wall, even for the chosen instant of the reverse flow (Figures 8a and 9a). The measured and calculated profiles shape of the streamwise velocity, V_s, are in reasonable agreement (Figures 8b and 9b, $V_s = V_z$ in this case). The difference between the measured and calculated maximum velocity does not exceed 8%. The standard deviation of the experimental velocity profile points from the calculated one was 9% for the forward flow profile (Figure 8) and 1% for the reverse flow profile (Figure 9), with respect to the peak bulk velocity in the section where the velocity profile was measured.

Note that the experimental profiles were constructed by averaging the data obtained in five independent experiments to reduce the influence of random experimental errors.

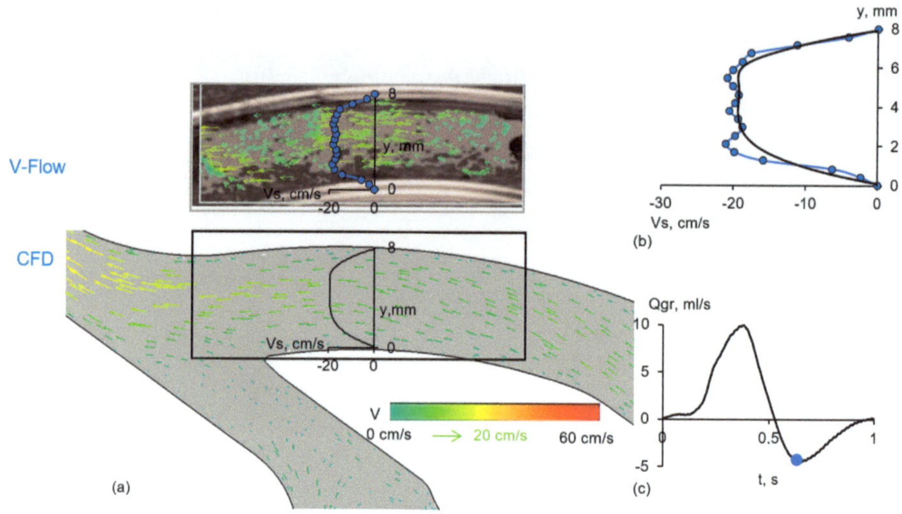

Figure 11. Vector fields (**a**) and streamwise velocity profiles (**b**) in the examined part of the anastomosis average model obtained by the V Flow (dots) and CFD (line) techniques for the marked instant of maximum reverse-flow rate (**c**).

The velocity vector patterns obtained by the two methods for the maximum flow rate instant in the case of the scanning window positioned in the graft are compared in Figure 10a. Both the experimental observations and the simulation show the formation of a large recirculation zone with relatively low velocities. The CFD pattern covering a larger area compared with the V Flow scanning window (see also Figure 3) shows that this recirculation zone, adjacent to the external side of the graft, originates at the beginning of the vessel bifurcation. On the internal side of the graft, the flow is a high-velocity jet. The profile shapes of the measured and calculated streamwise velocity, Vs, shown in Figure 10b, are in an acceptable agreement. The difference between the measured and calculated maximum values does not exceed 12%. The standard deviation of the experimental velocity profile points from the calculated ones was 13% for the forward flow profile (Figure 10) and 3% for the reverse flow profile (Figure 11), again with respect to the peak bulk velocity in the section where the velocity profile was measured.

The post-processing of the CFD solution allows one to better understand the flow structure in the graft, which is essentially three-dimensional for the forward flow phase. The latter is illustrated in Figure 3, which also covers a cross-flow vector pattern superimposed on the streamwise velocity map for a graft cross-section. One can observe that the secondary flow in the graft section under examination has the form of a vortex pair.

Figure 11 presents a comparison of CFD and V Flow data obtained for the examined part of the anastomosis average model at the instant of maximum reverse-flow rate. One can observe that, in this case, a relatively simple flow structure with a unidirectional fluid motion is formed both in the graft and in the femoral artery part. The velocity vector patterns obtained with the two methods are quite similar, and the measured and computed profiles of the streamwise velocity in the graft section are in reasonable agreement.

In Figure 12, the results of clinical measurements performed by the V Flow method for the graft branch from the femoral artery are compared with the results of the numerical simulation for the personalized model. In this case, a time instant corresponding to the phase of decreasing the forward flow (marked in Figure 12b) was chosen for comparison.

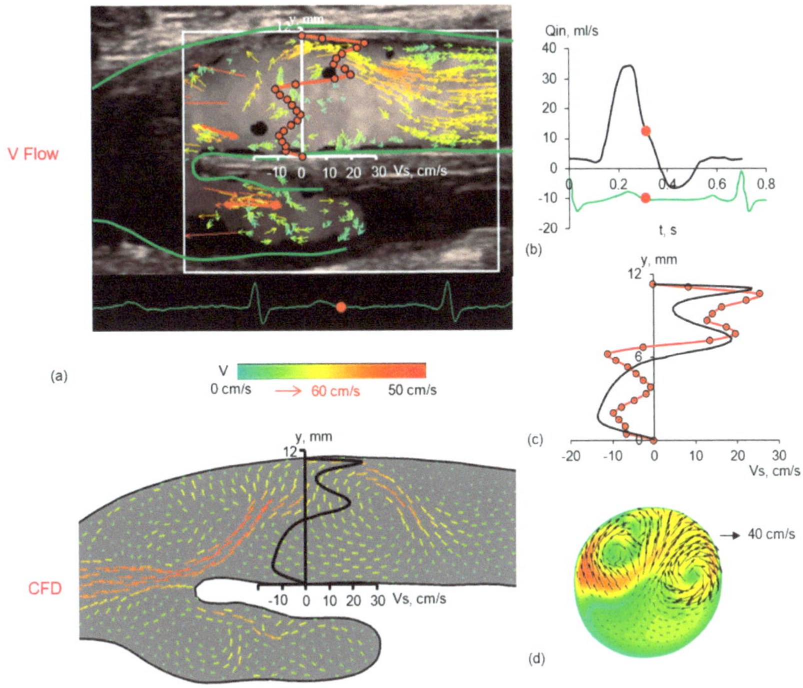

Figure 12. Comparison of the results of clinical V Flow measurements (dots) and CFD simulation (line) for the phase of decreasing forward-flow rate: measured and calculated vector velocity fields (**a**), time dependence of inlet flow with the marked time instant chosen for the comparison (**b**), measured and calculated profiles of the streamwise velocity in graft cross-section (**c**), and calculated flow structure in the graft cross-section (**d**).

One can state the similarity between the flow patterns, which were recorded in vivo and generated by the calculations using the personalized computational model. It should be emphasized that for the chosen instant, the flow structure in the scanning window is especially complicated and characterized by the presence of multiple vortices and an intensive cross flow directed from the inner wall of the graft to the external one (Figure 12a). It is also remarkable that the instant cross-flow pattern predicted by CFD for the chosen section (Figure 12d) looks like a vortex pair, but the direction of its circulation is opposite compared to that illustrated for the instant of the maximum flow in the anastomosis average model (Figure 3). Our experience in the parametric numerical simulation of pulsatile blood flow in different proximal anastomosis models, presented partially in [15–17], allows us to conclude that the circulation direction in the formed vortex pair depends not only on the examined phase of the cycle but also on the graft-to-CFA flow rate ratio and the individual geometry. The profiles of the streamwise velocity, V_s, measured and calculated for a graft section, are shown in Figure 12c (here, the experimental profile is a result of phase-averaging over five cycles of blood flow recorded in vivo). The difference between the measured and calculated maximum velocity of forward and reverse flow does not exceed 15%. The standard deviation of the experimental velocity profile points from the calculated one was 13% (Figure 12), with respect to the peak bulk velocity in the section where the velocity profile was measured.

In general, the plots presented in Figure 12 indicate that the results of the clinical V Flow measurements and the CFD simulation are in reasonable agreement. It should be noted here that due to a number of uncertainties, both arising when constructing a personalized computational model and inherent in the V Flow method itself, one can hardly expect complete consistency in the results obtained using the two methods. General limitations of the used methods/approaches can be summarized as follows. The V Flow method has the following limitations: (i) the impossibility of obtaining data for 3D vector flow visualization, (ii) the size of the scanning window is 20 × 27 mm only, (iii) the depth of the scanning window is limited by 50 mm, and (iv) recording time is limited by 3 s. The limitations of the applied numerical simulation technique are (i) neglecting pulsations of the vessel walls, and (ii) the assumption of the quasi-laminar flow regime. The main difficulty when comparing the results obtained with the two methods is attributed to the choice of a longitudinal section of the personalized model with the calculated 2D vector velocity pattern that would be as close as possible to the position of the scanning plane during the clinical study.

5. Conclusions

The V Flow high frame rate ultrasound vector flow imaging method and a CFD technique have been used for a coordinated study of the pulsatile velocity fields in the area of the proximal anastomosis for femoral popliteal bypass surgery in vitro and in vivo. Obtained by two methods, the dynamic vector images of the velocity field and constructed streamwise velocity profiles in the scanning plane of the flow in the area of the graft branch are in reasonable agreement, both in the case of the anastomosis average model developed for laboratory studies and in the case of the personalized model constructed to compare with clinic measurements.

Although the flow at the site of the graft branch from the artery depends significantly on the individual geometry, some general patterns can still be noted. At the beginning of the acceleration phase of the forward flow, as well as during the reverse flow phase, the fluid motion in the graft is unidirectional. In the accelerated forward flow, a separation zone appears at the external wall of the graft, rapidly increasing in size, with a significantly three-dimensional flow character and the determining role of the developed cross flow. During the deceleration phase of the forward flow, a relatively narrow zone of high flow velocities, initially adjacent to the internal wall of the graft, moves away from the wall, resulting in the formation of a large zone occupied by several vortex structures with opposite circulation.

The attractive advantages of the applied ultrasound high frame rate vector imaging technique are (i) the ability to carry out measurements of blood flow in real vessels, (ii) a high rate of obtaining a large volume of recorded data on pulsatile blood flow in complex areas of the vascular bed, and (iii) the possibility of the subsequent repeated slow-motion viewing for the purpose of processing and analysis of the received data. At the same time, the ultrasonic vector flow visualization method provides very limited opportunities for identifying three-dimensional vortex formations in the area under study, which, on the contrary, is fully provided by up-to-date CFD technologies; however, they are for flow in the models of real vessels only.

The complementary use of the V Flow method and three-dimensional numerical simulation has great potential for an in-depth study of the dynamics of pulsatile blood flow in branching sections of the vascular bed.

Author Contributions: Conceptualization, A.Y. and E.S.; Data curation, G.K. and E.S.; Investigation, L.T., Y.G., A.M., A.V. and A.S.; Methodology, L.T., Y.G. and A.V.; Writing—original draft, A.Y. and E.S.; Writing—review and editing, E.S. All authors have read and agreed to the published version of the manuscript.

Funding: The work was carried out with the support of the Russian Science Foundation, grant No. 20-65-47018.

Institutional Review Board Statement: The study was conducted according to the guidelines of the Declaration of Helsinki and approved by the local Institutional Review Board of the Pavlov First St. Petersburg State Medical University, protocol No. 3, 13 February 2023.

Informed Consent Statement: Written informed consent was obtained from all subjects involved in the study.

Data Availability Statement: Data are contained within the article.

Acknowledgments: Ultrasound measurements were performed with technical support and assistance from Mindray Medical Rus LTD and Sonar-Medical LTD. Simulations were performed on the Polytechnic RSC Tornado cluster of the Polytechnic Supercomputer Center http://www.scc.spbstu.ru (accessed on 20 February 2024).

Conflicts of Interest: The authors declare no conflicts of interest.

References

1. Morris, P.D.; Narracott, A.; von Tengg-Kobligk, H.; Silva Soto, D.A.; Hsiao, S.; Lungu, A.; Evans, P.; Bressloff, N.W.; Lawford, P.V.; Hose, D.R. Computational fluid dynamics modelling in cardiovascular medicine. *Heart* **2016**, *102*, 18–28. [CrossRef]
2. Ferziger, J.H.; Perić, M. *Computational Methods for Fluid Dynamics*; Springer: Berlin/Heidelberg, Germany, 2002; p. 426.
3. Blazek, J. *Computational Fluid Dynamics: Principles and Applications*, 3rd ed.; Elsevier: Kidlington, UK, 2015; p. 447.
4. Taylor, C.A.; Figueroa, C. Patient-specific modeling of cardiovascular mechanics. *Annu. Rev. Biomed. Eng.* **2009**, *11*, 109–134. [CrossRef]
5. Lopes, D.; Puga, H.; Teixeira, J.; Lima, R. Blood flow simulations in patient-specific geometries of the carotid artery: A systematic review. *J. Biomech.* **2020**, *111*, 110019. [CrossRef]
6. Vassilevski, Y.; Olshanskii, M.; Simakov, S.; Kolobov, A.; Danilov, A. *Personalized Computational Hemodynamics: Models, Methods, and Applications for Vascular Surgery and Antitumor Therapy*; Academic Press: London, UK, 2020; p. 280.
7. Dolan, J.M.; Kolega, J.; Meng, H. High wall shear stress and spatial gradients in vascular pathology: A review. *Ann. Biomed. Eng.* **2012**, *41*, 1411–1427. [CrossRef]
8. Chiu, J.-J.; Chien, S. Effects of Disturbed Flow on Vascular Endothelium: Pathophysiological Basis and Clinical Perspectives. *Physiol. Rev.* **2011**, *91*, 327–387. [CrossRef]
9. Donadoni, F.; Bonfanti, M.; Pichardo-Almarza, C.; Homer-Vanniasinkam, S.; Dardik, A.; Díaz-Zuccarini, V. An in silico study of the influence of vessel wall deformation on neointimal hyperplasia progression in peripheral bypass grafts. *Med. Eng. Phys.* **2019**, *74*, 137–145. [CrossRef]
10. Rivera, J.; van der Graaf, G.B.; Escudero, J.R.; Bellmunt, S.; van de Vosse, F. A computational fluid dynamics study on hemodynamics for different locations of the distal anastomosis of a bypass nearby a collateral vessel in the femoropopliteal area. *Int. J. Numer. Methods Biomed. Eng.* **2014**, *30*, 1263–1277. [CrossRef]
11. Leong, C.M.; Nackman, G.B.; Wei, T. Flow patterns through vascular graft models with and without cuffs. *PLoS ONE* **2018**, *13*, e0193304. [CrossRef]
12. Gataulin, Y.A.; Smirnov, E.M.; Molochnikov, V.M.; Mikheev, A.N. The structure of a 3D flow with local turbulence in the branching juncture of a circular-section channel. *St. Petersburg Polytech. Univ. J.-Phys. Math.* **2022**, *15*, 81–94.
13. Du, Y.; Ding, H.; He, L.; Yiu, B.Y.S.; Deng, L.; Yu, A.C.H.; Zhu, L. Quantitative blood flow measurements in the common carotid artery: A comparative study of high-frame-rate ultrasound vector flow imaging, pulsed wave Doppler, and phase contrast magnetic resonance imaging. *Diagnostics* **2022**, *12*, 690. [CrossRef]
14. He, Y.; Northrup, H.; Le, H.; Cheung, A.K.; Berceli, S.A.; Shiu, Y.T. Medical image-based computational fluid dynamics and fluid-structure interaction analysis in vascular diseases. *Front. Bioeng. Biotechnol.* **2020**, *10*, 855791. [CrossRef]
15. Ivanova, Y.F.; Tikhomolova, L.G.; Yukhnev, A.D.; Gataulin, Y.A.; Smirnov, E.M.; Kalmikova, R.V.; Morozov, A.N.; Suprunovich, A.A.; Vrabiy, A.A.; Khubulava, G.G. Patient-specific simulation of the blood flow in a proximal femoral artery-graft junction. *J. Phys. Conf. Ser.* **2021**, *1959*, 012025. [CrossRef]
16. Sherstneva, M.A.; Smirnov, E.M.; Yukhnev, A.D.; Vrabiy, A.A. Numerical simulation of the branching blood flow in a model of the femoral artery-graft junction. *J. Phys. Conf. Ser.* **2021**, *2088*, 012043. [CrossRef]
17. Ivanova, Y.; Yukhnev, A.; Tikhomolova, L.; Smirnov, E.; Vrabiy, A.; Suprunovich, A.; Morozov, A.; Khubulava, G.; Vavilov, V. Experience of patient-specific CFD simulation of blood flow in proximal anastomosis for femoral popliteal bypass. *Fluids* **2022**, *7*, 314. [CrossRef]
18. Tsang, A.C.O.; Lai, S.S.M.; Chung, W.C.; Tang, A.Y.S.; Leung, G.K.K.; Poon, A.K.K.; Hang, Y.A.C.; Chow, K.W. Blood flow in intracranial aneurysms treated with pipeline embolization devices: Computational simulation and verification with Doppler ultrasonography on phantom models. *Ultrasonography* **2015**, *34*, 98–108. [CrossRef]
19. Dol, A.V.; Ivanov, D.V.; Bakhmetev, A.S.; Maystrenko, D.N.; Edinova, M.V.; Rykova, A.Y. Boundary conditions at the outlets for numerical modeling of carotid artery hemodynamics. *Russ. J. Biomech.* **2021**, *25*, 20–31.
20. Hoskins, P.R. Simulation and validation of arterial ultrasound imaging and blood flow. *Ultrasound Med. Biol.* **2008**, *34*, 693–717. [CrossRef]
21. Caddy, H.T.; Thomas, H.J.; Kelsey, L.J.; Smith, K.J.; Doyle1, B.J.; Green, D.J. Comparison of computational fluid dynamics with transcranial Doppler ultrasound in response to physiological stimuli. *Biomech. Model. Mechanobiol.* **2023**. [CrossRef]
22. Weigand, C.J.; Liepsch, D.W. Color Doppler velocity measurements compared with laser Doppler anemometry under pulsatile flow conditions. *Instrum. Sci. Technol.* **1999**, *27*, 255–266. [CrossRef]
23. Swillens, A.; Degroote, J.; Vierendeels, J.; Lovstakken, L.; Segers, P. A simulation environment for validating ultrasonic blood flow and vessel wall imaging based on fluid-structure interaction simulations: Ultrasonic assessment of arterial distension and wall shear rate. *Med. Phys.* **2010**, *37*, 4318–4330. [CrossRef]

24. Zhou, H.; Meng, L.; Zhou, W.; Xin, L.; Xia, X.; Li, S.; Zheng, H.; Niu, L. Computational and experimental assessment of infuences of hemodynamic shear stress on carotid plaque. *BioMed. Eng. OnLine* **2017**, *16*, 92. [CrossRef]
25. Kotmakova, A.A.; Gataulin, Y.A.; Yukhnev, A.D. Swirling flow in a model of the carotid artery: Numerical and experimental study. *AIP Conf. Proc.* **2018**, *1959*, 090006.
26. Poelma, C. Ultrasound imaging velocimetry: A review. *Exp. Fluids* **2017**, *58*, 3–28. [CrossRef]
27. Au, J.; Hughson, R.; Yu, A. Riding the plane wave: Considerations for in vivo study designs employing high frame rate ultrasound. *Appl. Sci.* **2018**, *8*, 286–298. [CrossRef]
28. Yiu, B.Y.S.; Lai, S.S.M.; Yu, A.C.H. Vector projectile imaging: Time-resolved dynamic visualization of complex flow patterns. *Ultrasound Med. Biol.* **2014**, *40*, 2295–2309. [CrossRef]
29. Hansen, K.L.; Nielsen, M.B.; Jensen, J.A. Vector velocity estimation of blood flow—A new application in medical ultrasound. *Ultrasound* **2017**, *25*, 189–199. [CrossRef]
30. Du, Y.; Goddi, A.; Bortolotto, C.; Shen, Y.; Dell'Era, A.; Calliada, F.; Zhu, L. Wall shear stress measurements based on ultrasound vector flow imaging. Theoretical studies and clinical examples. *J. Ultrasound Med.* **2020**, *39*, 649–1664. [CrossRef]
31. Sengupta, P.P.; Pedrizzetti, G.; Kilner, P.J.; Kheradvar, A.; Ebbers, T.; Tonti, G.; Narula, J. Emerging trends in CV flow visualization. *JACC Cardiovasc. Imaging* **2012**, *5*, 305–316. [CrossRef]
32. Leow, C.H.; Tang, M.-X. Spatio-temporal flow and wall shear stress mapping based on incoherent ensemble-correlation of ultrafast contrast enhanced ultrasound images. *Ultrasound Med. Biol.* **2018**, *44*, 134–152. [CrossRef]
33. Ding, J.; Dub, Y.; Zhao, R.; Yang, Q.; Zhu, L.; Tong, Y.; Wen, C.; Wang, M. Detection of abnormal wall shear stress and oscillatory shear index via ultrasound vector flow imaging as possible indicators for arteriovenous fistula stenosis in hemodialysis. *Ultrasound Med. Biol.* **2023**, *49*, 1830–1836. [CrossRef]
34. Qiu, Y.; Dong, Y.; Mao, F.; Zhang, Q.; Yang, D.; Chen, K.; Shi, S.; Zuo, D.; Tian, X.; Yu, L.; et al. High-frame rate vector flow imaging technique: Initial application in evaluating the hemodynamic changes of carotid stenosis caused by atherosclerosis. *Front. Cardiovasc. Med.* **2021**, *8*, 617391. [CrossRef]
35. Fiorina, I.; Raciti, M.V.; Goddi, A.; Cantisani, V.; Bortolotto, C.; Chu, S.; Calliada, F. Ultrasound vector flow imaging—Could be a new tool in evaluation of arteriovenous fistulas for hemodialysis? *J. Vasc. Access* **2017**, *18*, 284–289. [CrossRef]
36. Kammerer, S.; Stroszczynski, C.; Jung, E.M. Functional ultrasound imaging of the venous valve of the great saphenous vein in the area around the crosse using the novel vector flow technique. *Clin. Hemorheol. Microcirc.* **2020**, *76*, 211–219. [CrossRef]
37. Du, Y.; Shen, Y.; Yiu, B.Y.S.; Yu, A.C.H.; Zhu, L. High frame rate vector flow imaging: Development as a new diagnostic mode on a clinical scanner. In Proceedings of the 2018 IEEE International Ultrasonics Symposium (IUS), Kobe, Japan, 22–25 October 2018; IEEE: New York, NY, USA, 2018; pp. 1–4.
38. McGarry, C.K.; Grattan, L.J.; Ivory, A.M.; Leek, F.; Liney, G.P.; Liu, Y.; Miloro, P.; Rai, R.; Robinson, A.P.; Shih, A.J. Tissue mimicking materials for imaging and therapy phantoms: A review. *Phys. Med. Biol.* **2020**, *65*, 23TR01. [CrossRef]
39. Hashimoto, J.; Ito, S. Pulse pressure amplification, arterial stiffness, and peripheral wave reflection determine pulsatile flow waveform of the femoral artery. *Hypertension* **2010**, *56*, 926–933. [CrossRef]

fluids

Review

Advancements and Opportunities in Characterizing Patient-Specific Wall Shear Stress Imposed by Coronary Artery Stenting

John F. LaDisa, Jr. [1,2,3,4,*], Arash Ghorbannia [2], David S. Marks [3], Peter Mason [3] and Hiromasa Otake [5]

1 Department of Pediatrics, Division of Cardiology, Herma Heart Institute, Children's Wisconsin and the Medical College of Wisconsin, Milwaukee, WI 53226, USA
2 Department of Biomedical Engineering, The Medical College of Wisconsin and Marquette University, Milwaukee, WI 53226, USA
3 Department of Medicine, Division of Cardiovascular Medicine, Medical College of Wisconsin, Milwaukee, WI 53226, USA
4 Department of Physiology, Medical College of Wisconsin, Milwaukee, WI 53226, USA
5 Graduate School of Medicine, Kobe University, Kobe 657-8501, Japan
* Correspondence: jladisa@mcw.edu

Abstract: The success of drug-eluting stents (DES) is limited by restenosis and, to a lesser extent, late stent thrombosis. Mechanical stimuli have been implicated in these outcomes, with indices of wall shear stress (WSS) determined from computational simulations being reported most frequently. The current work summarizes state-of-the-art computational approaches applicable to patient-specific models aimed at further understanding changes in WSS indexes imposed by stent implantation. We begin with a review of best practices involved in the process and then summarize the literature related to stent-induced WSS alterations. Image-based reconstruction methods are also discussed, along with the latest generation boundary conditions that replicate cardiac physiology and downstream vasculature in the setting of coronary artery disease. The influence of existing material property data on WSS results obtained with geometries reconstructed from finite element modeling and fluid structure interaction (FSI) simulations is reviewed, along with the novel approaches being used to provide coronary artery plaque data that are currently missing from the literature. We also consider the use of machine learning tools that have the potential for impact when assessing the role of adverse stent-induced WSS in suboptimal clinical outcomes. We conclude by focusing on challenging cases that involve DES implantation, which may benefit from recent advancements in patient-specific computational modeling.

Keywords: drug-eluting stents; computational fluid dynamics; simulation; fluid structure interaction

Citation: LaDisa, J.F., Jr.; Ghorbannia, A.; Marks, D.S.; Mason, P.; Otake, H. Advancements and Opportunities in Characterizing Patient-Specific Wall Shear Stress Imposed by Coronary Artery Stenting. *Fluids* **2022**, *7*, 325. https://doi.org/10.3390/fluids7100325

Academic Editor: Huidan (Whitney) Yu

Received: 30 June 2022
Accepted: 5 October 2022
Published: 11 October 2022

1. Background

Cardiovascular disease (CVD) remains the leading cause of mortality in the USA, with greater than 1000 daily CVD deaths and approximately 8 million procedures annually. A total of 126 million Americans (49.2% of adults) have CVD, and ~50% suffer from coronary artery disease (CAD) [1]. Bare metal (BMS) and drug-eluting stents (DES) have revolutionized CAD treatment. Stenting is now the most common method of revascularization and percutaneous coronary intervention (PCI). Despite notable advances, current DES are still limited by in-stent restenosis (ISR) and, to a lesser extent, stent thrombosis (ST). Restenosis is defined by a diameter reduction via angiography [2] or intravascular imaging modalities such as optical coherency tomography (OCT) [3] from excessive neointimal hyperplasia (NH; new tissue growth) [4]. Early DES decreased restenosis [5] relative to BMS but were prone to ST [6] as they inhibited endothelialization [7–9]. Newer DES improved the outcomes, but restenosis can still persist. Stent thrombosis is classified as early, late, and very late ST (>1 year after stenting). Early DES were associated with the increased incidence of late ST (LST) and very late ST. Although far less common than restenosis, LST also remains a concern for its high rates of morbidity and mortality [10–14] and the potential for delayed healing long after DES implantation [15]. Despite the use of current DES, some studies have estimated annual restenosis rates to be around 5–10% of cases, while LST occurs in 0.2–1.0% of cases [16–18]. These issues are accentuated for challenging lesions [19] and may only be partially alleviated by next-generation DES as their use extends to other vessels in the body, including some areas without extensive experience [20].

The precise cause of poor clinical outcomes of DES in some patients remains unknown. Recent reports have pointed to procedural and biological contributors, with at least one recent classification system proposed to better differentiate contributors on a patient-specific basis [21]. The malapposition of stent struts is likely a contributor within the procedural aspect of the proposed scoring system cited above. Malapposition is defined as a portion of the stent present at a distance greater than the strut's thickness from the artery surface [22]. Malapposition can occur late as a result of positive vessel remodeling due to an eluted drug [23], or early as the result of local plaque burden [24]. Coronary anatomy such as bifurcation lesions can also impact the deformation and positioning of stent struts [25], leading to the unintentional underexpansion or undersizing of stents. Malapposition is a prominent contributor, and hence a marker, of LST and possibly NH [26,27].

Mechanical stimuli have also been implicated in poor clinical outcomes following DES implantation, with indices of wall shear stress (WSS) calculated from computational simulations being reported most frequently. Hence, the current review primarily focuses on indices of WSS and aspects of the patent-specific computational modeling process that can influence these indices. Distributions of WSS indices (see Section 3) are locally influenced by strut positioning after stenting and can also modulate artery response to the stent implantation process [28]. Previous studies and reviews [28–30] have suggested adverse flow patterns and that the potential for flow stagnation are accentuated in the setting of malapposed struts. Stenting has also been noted to cause endothelial denudation [31] and arterial damage as part of the process of restoring artery patency. Studies going back decades have convincingly shown how WSS indices can determine where the neointimal tissue establishes in response to localized injury [32–35]. The over-expansion of stents can therefore create the undesirable situation of more prominent arterial injury along with a greater percentage of the stented artery being exposed to adverse distributions of WSS.

WSS patterns correlate with sites of NH in BMS and can inhibit the endothelialization of stents [36], which prevents LST. In contrast to BMS, the data relating WSS to NH or LST for DES can be conflicting [23,37,38]. This observation underscores the need for additional studies that link detailed wall shear stress distributions after stenting, as can be revealed by patient-specific computational modeling, to the local arterial response. Despite advances in modeling capabilities, there is a paucity of important data upon which advanced simulation methods rely for the accurate assessment of WSS distributions imposed by stenting as well as the vascular response to it. For example, the materials often used with computational models of stents employ hyperelastic constitutive relationships from a limited range of plaques [39–41] despite the literature underscoring a need for more representative data [42–44]. Similarly, some reports have suggested that the mechanism of action for a DES may differentially impact the relationship between NH and WSS seen in BMS [23,37,38], but such hypotheses have not been conclusively studied in vivo or with newer DES.

With the above background in mind, the current work summarizes state-of-the-art computational approaches that are applicable to patient-specific models, aimed at further understanding WSS indices imposed by stent implantation. We begin with a general review of the steps involved in computational simulations and review the literature related to WSS indices that have historically been reported most commonly. Considering WSS alterations in response to stenting requires a brief summary of the reconstruction methods and the latest generation boundary conditions that replicate cardiac physiology and downstream vasculature in the setting of coronary artery disease. The roles of existing material property data as well as of those that are currently missing from the literature are also reviewed. We also consider the use of current tools from mathematical and statistical sciences, such as machine learning, that have the potential for impact when assessing the role of WSS from stents in the clinical outcomes of restenosis and/or stent thrombosis. Lastly, we focus on clinical sequelae with particularly poor clinical outcomes after DES implantation that may benefit from advancements in patient-specific computational modeling.

2. Requirements and Best Practices

Patterns of fluid flow can be described by partial differential equations representing conservation laws for quantities that include mass and momentum [45]. Predicting the effect of such flows in biomedical applications and other disciplines that prominently study fluid flow is costly and time-consuming when the user does not employ computational techniques. Computational fluid dynamics (CFD) is one commonly used approach in the simulation of fluid that passes within or around objects. CFD solves the governing equations of fluid motion through numerical methods (as finite element or finite volume methods) and provides an estimation of the velocity vector in a discretized computational domain, which is represented by a collection of organized mesh elements. The current work focuses on blood vessels, where CFD allows the user to replace the partial differential equations with algebraic equations that can be solved numerically within the discretized computational domain (i.e., mesh) via digital computers. Commercial and open-source CFD software packages are available to facilitate the completion of these calculations. Most of these current-generation software packages have user-friendly interfaces that accept several types of medical imaging data. The pipeline for each software package then allows a user to generate hemodynamic results by leveraging the governing mathematical equations. Despite the apparent ease with which CFD simulations can seemingly be conducted, it is worth mentioning that there are several important considerations that the user should be cognizant of, as discussed in more detail below.

The general requirements related to studying blood flow alterations using CFD include the creation of a model of vessel geometry from medical imaging data. In addition, CFD requires the user to prescribe flow and/or pressure information at the entrance of the vessel. It is also necessary to prescribe the hemodynamic state beyond the outlets of the model created in order to obtain realistic results (e.g., imposing downstream resistance to obtain a physiologic range of pressure). Rheological properties, including the density and viscosity of blood, should then be assigned. The last step in the process involves using a powerful computer or a high-performance cluster of computers to solve the governing equations for fluid flow throughout the vessel geometry, which is represented as a computational mesh.

More precisely, the first step in performing CFD simulations involves the creation of a computer aided design (CAD) model within the arterial regions of interest, often from medical imaging data. For coronary arteries implanted with stents, this is typically angiography data from computed tomographic (CTA) or fluoroscopic systems. These data are readily available and provide a clear definition of anatomy. While CTA data sets are inherently volumetric, geometries for CFD modelling from fluoroscopic imaging systems are also sometimes reconstructed using two or more planes under certain assumptions, and higher resolution information from the stented region is then obtained via registration with an intravascular imaging modality, as discussed in more detail below. CFD geometries can also be created from other imaging modalities (e.g., magnetic resonance imaging). The models created can yield geometries on a patient-specific basis when it is desirable to focus on clinical questions for a specific patient [46] or for a group of patients with a similar pathology. Alternatively, idealized models are sometimes created to study blood flow that is generally reflective of a patient population. In such situations, the geometry within the model is typically informed by measurements taken from data within a collection of patients. The models may also offer guidance related to a specific clinical issue using a representative patient data set [47]. To date, our laboratory has primarily used SimVascular (simvascular.github.io), but there are other software packages that also permit the import and segmentation of medical imaging data, including ITK-SNAP (www.itksnap.org, accessed on 29 June 2022), Cardiovascular Integrated Modeling and Simulation (CRIMSON, www.crimson.software, accessed on 29 June 2022), Vascular Modeling Toolkit (VMTK, www.vmtk.org, accessed on 29 June 2022), Mimics (www.materialise.com/en/healthcare/mimics-innovation-suite, accessed on 29 June 2022, Plymouth, MI, USA), and others. Each of these programs facilitates the discretization of a CAD model after it has been created from imaging data by interacting with some type of meshing software. The parameters selected during the segmentation and meshing steps can have a large influence on the results obtained. For example, the accuracy of WSS distributions (see below for details on specific indices of WSS) substantially depends on the artery radius. Therefore, the care and reproducibility with which segments or 3D boundaries for the CAD model are created from the available imaging data is important in resulting WSS distributions.

In its simplest form (e.g., plane Couette flow), WSS is calculated as the product of viscosity and the rate of deformation (i.e., near-wall velocity gradient; Figure 1). For CFD, this change in velocity from the wall of an artery to the next nearest location is dictated by the arrangement of elements and local details of the computational mesh that has been created. The velocity on the wall is often zero due to a no-slip condition that is applicable to the interaction between the flowing blood and the vessel. Unfortunately, the computational costs (time and compute resources) of obtaining CFD results increases as a function of mesh density. This trade-off is often balanced in CFD studies through the use of adaptive meshing approaches [48–51], which can produce greater mesh density in spatial locations where it is most needed as a result of complex flow patterns. Such locations include the area near the vessel wall where, as mentioned above, improved accuracy is important for determining indices of WSS. As alluded to above, WSS indices are calculated from the velocity field and therefore cannot be directly quantified using transducers or non-invasive approaches in vivo. Hence, in the setting of careful reconstruction methods and results independent of the computational mesh (see below), CFD simulations allow for the computation of hemodynamic quantities such as WSS that are clinically difficult or even impossible to assess with a high degree of accuracy.

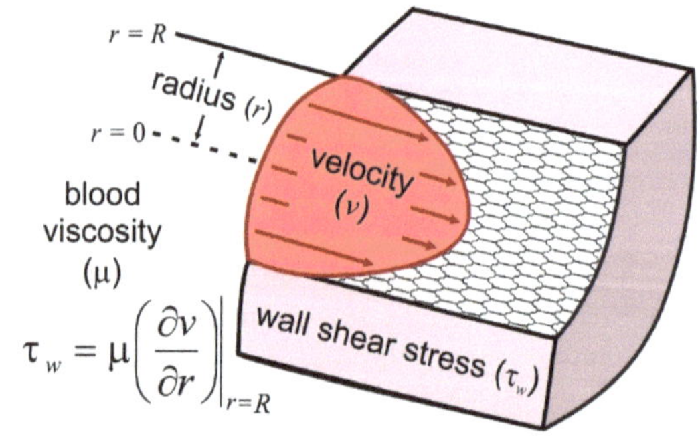

Figure 1. The figure provides an illustration of the velocity profile experienced by endothelial cells lining an artery as a result of blood flow, and a general expression for wall shear stress (τ_w; WSS). In its simplest form, WSS can be defined as the frictional force imposed on the blood vessel wall. For this simple case, WSS is the product of viscosity (μ) and the shear rate, which is also known as the rate of deformation or near-wall velocity gradient ($\partial v/\partial r$). Adapted from Samyn and LaDisa [52].

In practice, a value for the density of blood is typically selected from the literature. A Newtonian assumption (i.e., constant blood viscosity) is also most often employed. Although blood is a shear-thinning fluid, approximating its behavior as Newtonian may be reasonable, depending on the range of shear rates experienced by the portion of the vasculature being studied and the indices being reported [53]. Newtonian and non-Newtonian (e.g., Carreau) models have been implemented [54–56], which may provide more realistic results at certain times during the cardiac cycle and near stent struts when other sources of realism have been included and when sources of uncertainty in the modeling process have been mitigated; for example, in patient-specific simulations where reconstruction accuracy or physiologic boundary conditions have been implemented [57].

One unique aspect of CFD software programs specifically designed for blood flow in vessels lies in their ability to impose boundary conditions that replicate normal and pathologic physiology. Methods imposed for patient-specific inlet boundary conditions include velocity-encoded (VENC; i.e., phase-contrast) flow data from magnetic resonance imaging (PC-MRI). The patient-specific inlet flow rate can also be obtained from angiography [58]. When such data are not available, other approaches such as waveforms from the literature are sometimes made to approximate patient-specific conditions by scaling according to the body surface area. In terms of outlet boundary conditions, the impedance spectra (i.e., time-varying opposition to blood flow) can be determined from flow and pressure measurements if they are obtained at the same location in the vasculature. Given the impracticality of the necessary measurements within a clinical setting, Windkessel models are often used as an approximation of the impedance spectra [59] at the outlets of computational models. It is increasingly common in CFD modelling of applications involving non-coronary arteries to employ Windkessel representations derived from blood pressure measurements and PC-MRI flow data for outlet boundary conditions. The use of these more recent outlet boundary conditions is often associated with backflow stabilization methods [60,61], which limit the need for the extension of outlets, as had been commonly applied in the past. Pronounced resistance introduced by ventricular contraction results in a non-linear and time-varying system for the coronary arteries [62]. Recent advancements related to boundary conditions also account for this physiology by including the ability to mimic cardiac function through the use of closed-loop lumped-parameter networks (LPNs) with CFD models. Although initially developed to model single ventricle physiology, closed-loop LPNs are now being used to characterize flow patterns in the coronary arteries and other vascular regions [63]. The parameters used with closed-loop LPN models are typically tuned based on clinically obtained data (e.g., cardiac output, stroke volume, blood pressure, and ejection fraction). Diameter-based scaling laws have also been widely adopted in cases where the coronary branches are included, and the specific approaches implemented to date have been nicely summarized in a recent book chapter by Chiastra, Dubini, and Migliavacca [64].

As mentioned above, upon setting boundary conditions, patient-specific simulations then use specialized computers to solve for the conservation of mass and the balance of fluid momentum. In some cases, the elastodynamic equations [65] are also solved, allowing for fluid–structure interaction (FSI) simulations that replicate local deformation. FSI simulations therefore represent an extension beyond CFD modelling that considers the elastic nature and pulsatility of the arterial system. FSI modelling has the potential to introduce more clinically relevant features when determining WSS indices by including realistic local deformations. For example, when again considering the simple case of WSS calculated as the product of the near-wall velocity gradient and viscosity, the movement of the vessel wall, as it occurs in vivo, will impact this calculation [66]. Most coronary artery stents are relatively rigid to prevent recoil with the stented region [67], but including local deformations in WSS calculations has the potential to provide more realistic results and strains within intra-strut regions [66]. While the impact of FSI simulations on distributions of WSS in the literature has so far been dependent on available material properties, FSI studies conducted in idealized arteries with computational replicas of several commercially available stents show the advantages of circumferentially oriented sinusoidal strut patterns with limited connector elements and large open-cell designs, which have the smallest deviations in Von Mises stress and displacement when exposed to dynamic curvature [68]. Such studies provide support for design aspects that have been implemented by manufactures for decades. As with all the considerations above, adding the realism of deformable walls often leads to increased computational time. Hence, the need for this advancement should be considered relative to its importance, in relation to the quantities being reported. Conversely, simulating pulsatile blood flow for transient portions of the vasculature such as the coronary arteries, and best practices such as mesh independence, the presentation of results that are independent of the time step employed during a simulation, and the assurance that the results are periodic all represent considerations that are usually important regardless of the indices being studied.

3. Intravascular Reconstruction for Computational Simulations

Given the importance of geometry in the simulation process and WSS calculations reviewed above, it is not surprising that multiple approaches have been developed for the reconstruction of coronary arteries used in computational simulations [69–73], many of which using OCT data have recently been reviewed by Chiastra et al. [74]. We previously developed methods [73] to create patient-specific coronary artery reconstructions by combining conventional extravascular and high-resolution intravascular imaging. Our work in this area has featured the fusion of CTA with OCT data, but the conventional imaging data source could also be MRI or biplane angiography. Biplane (i.e., orthogonal) angiographic images are often used to determine the intravascular imaging pathway and have also been used to reconstruct a vessel lumen. Although biplane angiography is a common clinical process that is performed during PCI, non-orthogonal views, differences in data acquisition between planes, movement resulting from cardiac and/or respiratory dynamics, visibility of the imaging device, and general image quality are potential issues that may arise with this approach. Our approach (Figure 2) uses the conventional extravascular imaging data to establish boundaries for an optimization routine that determines the mathematical location of the intravascular imaging wire during pullback. The intravascular images provide high-resolution information from within the stented region, thereby theoretically permitting the use of volumes generated from biplane imaging planes despite the limitations mentioned above. Besides OCT, the high-resolution data can also come from IVUS. Both modalities can capture stent linkages. OCT has 10x the resolution of IVUS but requires contrast, and therefore some interventional cardiologists prefer IVUS. Our reconstruction and stenting approaches work with data from either modality. Briefly, intravascular images are processed to isolate lumen versus stent contours (i.e., segments) by thresholding. The pathway traversed by the wire during pullback is calculated by minimizing the total bending energy [73] within the volume delineated by the extravascular imaging data. Lumen segments are then registered longitudinally on the wire pathway using landmarks and according to the intravascular image spacing. These lumen segments are oriented such that they are orthogonal to the wire pathway and rotated so that the centroids are aligned with the vessel centerline. The aligned and oriented segments are then loaded into SimVascular for lofting and blending. The implanted DES from post-stenting OCT or IVUS is quickly replicated in reconstructions by a series of Boolean operations [66,73] or structural finite element modeling (FEM) [46]. Similar to full-scale structural FEM simulations that include stent recoil, the series of Boolean operations implemented for this approach can capture local stent malapposition.

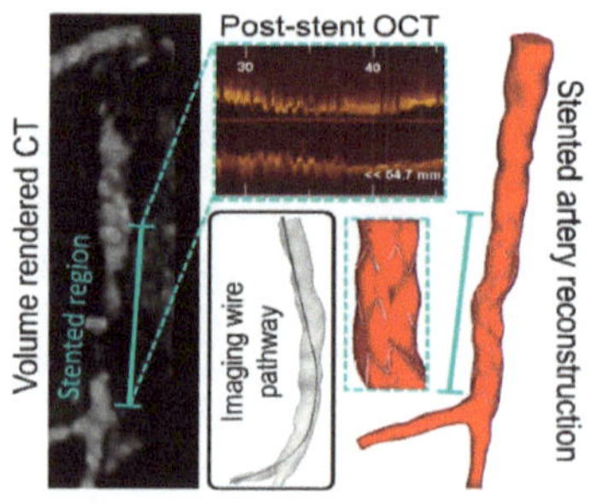

Figure 2. One method of patient-specific stented artery reconstruction. A coronary CTA is seeded with candidate points, and the OCT imaging wire path is calculated by minimizing total bending energy. The implanted stent is then replicated computationally. Adapted from Ellwein et al. [73].

Wu et al. recently created a reconstruction approach that is specifically tailored to coronary bifurcations [69]. Reproducibility and accuracy were assessed in five silicone bifurcation models created from patient-specific angiography, which then underwent OCT imaging and were compared to reference versions from contrast-enhanced microfocal computed tomography (μCT). The utility of the method was further scrutinized in seven diseased patient bifurcations of varying anatomical complexity. Agreement was high between reconstructions and reference morphology, as was reproducibility and the ability to conduct the reconstruction process within a reasonable amount of time (i.e., <1 h). This method has since been extended, with angiography and OCT data obtained directly from patients and with the addition of local linkages resulting from deployment through the optimization of several bifurcation stenting techniques [75].

Other approaches in the combination of conventional extravascular and high-resolution intravascular imaging data to reconstruct the stented flow domain are also presented with varying degrees of detail in the literature [70,76–79]. Recently, Li and colleagues described their process using 3D angiography and OCT imaging, which was featured for diseased coronary artery data sets [80] as well as those obtained after bioresorbable vascular scaffold (BVS) implantation [58]. Their approach pays particular attention to the inclusion of side branches that are likely to impact the local flow field. A similar approach was employed for BVS by Migliori et al., who further quantified restenosis and neointimal thickening [76]. Gogas et al. developed a similar process that is highlighted via a BVS case study [77,78]. Their approach is noted for including semiautomated lumen extraction and strut detection via shape recognition algorithms within each OCT image. These data are then aligned while accounting for patient curvature.

The choice of which intravascular imaging modality is to be used in each approach (e.g., OCT vs. IVUS) depends on many factors, such as those present from a clinical perspective that are related to the need for contrast, onboard versus offline plaque characterization protocols as well as clinical preference and the availability of each clinical imaging system within a given catheterization laboratory. While acknowledging these considerations, studies have suggested that patient-specific arteries created from OCT and IVUS runs of the same atherosclerotic arteries may be highly correlated [81]. In contrast, when conducted after stenting, it appears that there can be significant differences in clinical endpoints such as the thickness and area of neointimal tissue that impact local blood flow patterns manifesting in the WSS imposed on the endothelium [82]. These differences seem to be a result of improved resolution afforded by OCT.

It is important to note that to date, the inclusion of stents in computational models of reconstructed arteries has ranged from the general contour traced out by stent struts in intravascular images while ignoring the local struts and their associated perturbations, to the incorporation of residual stresses and strains experienced during balloon folding and pleating and stent crimping. Particularly noteworthy is the work of Chiastra et al., which demonstrated the importance of including local stent struts in patient-specific reconstructions exposed to steady inflow boundary conditions [55]. Local distributions of TAWSS are drastically different due to the geometry of the implanted stent, with 35% of the modelled left anterior descending coronary artery exposed to adverse TAWSS (i.e., <4 dyn/cm^2) for the stented case; this was only 2.6% when local struts and their perturbations were ignored [64]. The inclusion of residual stresses and strains experienced during balloon folding and pleating and stent crimping was shown to significantly increase the accuracy of deformations obtained from stent deployment simulations [83]. Although the authors used a 316 L stainless steel stent when many current stents are made of cobalt chromium, their point is well-taken and interesting for several reasons. First, the detailed geometry after implantation impacts the local flow disturbances manifested in indices of WSS and the response of the artery to DES-induced damage, but also because issues in the DES manufacturing process unique to a given DES may also influence the potential for local stent malapposition, which has been associated with LST and NH [26,27], as mentioned above.

4. WSS Findings to Date and Related Indices of Interest

Mechanical stimuli have been shown to influence the onset and progression of CVD. As alluded to above, WSS is of particular interest in response to stent implantation. In prior studies, sites of low time-averaged WSS (TAWSS) were thought to correlate with the localization of atherogenesis and inflammation [84–88]. These studies motivated the early hypotheses related to whether alterations in WSS manifesting from stenting may ultimately contribute to NH. Although WSS is represented by vectors that change instantaneously throughout the cardiac cycle, most studies report time-averaged representations along the wall for the region of interest. This is likely carried out for simplicity and because the mechanisms by which a each WSS index leads to NH are not yet fully known. There is some evidence indicating that spatial and temporal WSS alterations may also serve as stimuli for neointimal thickening. Many CFD studies also report the oscillatory shear index (OSI) [87], which is a measure of WSS directionality. Lower OSI values indicate that WSS is predominantly oriented in the primary direction of blood flow, whereas a value of 0.5 indicates bi-directional WSS with a time-averaged value of zero. Sites of the vasculature exposed to low WSS magnitude and high OSI are theoretically less likely to experience fluid forces that allow for the washout of noxious and potentially atherogenic materials in contact with the arterial surface (e.g., LDL). In general, adverse values for these indices are expressed as thresholds (e.g., TAWSS $\leq$4 dyn/cm^2 for the coronary arteries [86]). Data from prior idealized and preclinical stenting studies suggest that an OSI greater than ~0.1 may also be considered adverse [47], as well as spatial and temporal WSS gradients greater than 100 dyn/cm^3 and $\pm$200 dyn/cm^2/s, respectively [88–90].

New indices that attempt to better capture the stimuli exposed in an artery continue to be derived. One such index is the WSS exposure time (WSSET). The WSSET is determined from the vectorized form of WSS and generally represents the accumulated amount of time WSS trajectories visit a mesh element over the cardiac cycle [91]. When applied in computational simulations of DES, WSSET provides some measure of drug transport associated with each DES in the setting of local hemodynamics stimuli and its potential relationship to NH or resorption in a patient-specific manner. Additional indices of interest include helicity (sometimes reported as a normalized version that considers local velocity and vorticity magnitude) and relative residence time (RRT). Local normalized helicity describes the alignment of streamlines into spiral patterns within arteries [92]. RRT is considered an aggregate index of TAWSS and OSI, as both are included in its calculation. As its name implies, RRT is interpreted as a measure of particle residence time near the wall, with elevated RRT being linked to cellular proliferation in other vascular beds [93,94]. Results from patient-specific models exposed to steady boundary conditions have noted values for WSS and RRT that tend to be localized to regions of NH adjacent to stent struts, within bifurcations and in zones where stents are overlapping [55].

Over the past two decades, numerous studies have been conducted to determine if adverse distributions of WSS unique to the geometry of the stented portion of an artery correlate with NH. For example, we previously showed that areas of low WSS established after BMS implantation modulate the development of NH in rabbit iliac arteries [32]. As NH occurs within a stent implanted into an otherwise healthy artery, the geometry and associated WSS distributions change over time to progressively alleviate deleterious WSS distributions initially created by stenting. Additional studies by our lab and others [32,89,95–98] further demonstrated that the geometric properties of a stent may contribute to adverse indices (e.g., low TAWSS) associated with NH. Idealized studies have further shown that the number, width, and thickness of stent struts as well as the severity of shortening, local scaffolding, and the degree of curvature created by a stent can introduce potentially adverse flow disturbances that may lead to NH [32,89,97,99,100]. More specifically, the thickness and number of stent struts are believed to have a greater impact on the development of NH than their width [99]. The thickness of struts causes protrusion of the stent into the flow domain, which in turn causes disruptive flow patterns and increases in the area of the vessel exposed to adverse WSS. This theoretical finding is supported by numerous controlled preclinical and clinical studies [31,101,102]. Similarly, by increasing the number of stent struts, a greater scaffolding of the lumen is accomplished by limiting the protrusion of the vessel wall between linkages, which then leads to increased longitudinal and circumferential uniformity within the stented region [99]. However, since all stent struts disturb blood flow, the relative ratio of the stent-to-vessel area can cause deleterious flow patterns depending on the orientation of the struts in relation to the flow direction. These findings from idealized stented models have also served as a means of understanding similar computational studies using more recent stent geometries [103,104]. While the understanding recounted above does not necessarily remain true in the setting of diseased human coronary arteries implanted with current DES, it does provide the intuition with which flow disturbances resulting from more realistic clinical scenarios can be interpreted. The effect of the stent struts on the local hemodynamics, which results in the behavior recounted above, has been previously discussed in detail [89,105] and generally involves struts redirecting local velocity vectors in a converging or diverging manner; this is based on the orientation of the struts in relation to the primary direction of the blood flow and depends on the overall geometry of the stent model. Helical flow structures can also result from any curvature upstream of the stent and in bifurcation regions [55]. Such details have since been associated with an index called the WSS topological skeleton, which is calculated from the divergence of the normalized WSS vector field and provides an additional characterization of the forces that may be experienced by cells along the stented vessel surface [106].

Recent computational studies have also provided further details on altered indices of WSS and stent thrombosis. For example, Ng et al. studied the acute impact of strut malapposition and underexpansion of stents using a 2.75 mm inner diameter isotropic benchtop silicon model with elastic material properties (density 1110 kg/m^3; Young's modulus 1.2 MPa; Poisson's ratio 0.48, 0.45 mm thickness) and a 40% diameter stenosis [107]. Models implanted with fully apposed and malapposed sirolimus-eluting stents in the stenotic region were perfused with porcine blood for 1 h at 200 mL/min. The takeaway of their work was that underexpansion alone did not substantially impact thrombus formation, but thrombus assessment and quantification via OCT pullback was noted when underexpansion was co-localized with malapposition. Although malapposition in the absence of stent underexpansion was not studied, Gasior et al. conducted a related study in which a dedicated bifurcation stent was compared with other commercially available stents implanted in a bifurcation model [108]. Computational modeling with OCT and immunofluorescence following perfusion with porcine blood for the same duration and flow rate above revealed that indices of thrombus were more pronounced for stents in which malapposed struts and the associated WSS alterations were present at the side branch ostia.

When considering CFD simulations of stented human coronary arteries, the most common type of result reported to date has been the extraction of WSS distributions from steady simulations. There is merit to such studies as the vascular response to stimuli likely represents some integrated measure of exposure to mechanical stimuli over time, and it is reasonable to surmise that mean WSS provides a representative index of that exposure. However, data relating WSS to NH or LST for DES can be somewhat conflicting [23,37,38], which makes the details associated with computational modeling methods employed for a particular study important. Each study in this area has undoubtedly been useful and serves as a foundation for our understanding to date, but future work will expand on this work to include more detailed perturbations from the local contours of individual struts that impact our understanding of mean WSS as well as other indices that have recently been reported.

5. Optimizing the Stenting Procedure

The use of optimization in a mathematical sense allows for a systematic and unbiased approach to maximizing or minimizing a function relative to some design criteria. The use of optimization in stent design and model selection for patients is intriguing, given the idealized CFD findings reviewed above. The use of optimization in CFD is not new; Marsden and colleagues applied an optimization algorithm to surgically reconstruct vessels of children with a single functioning ventricle [109], while other notable work has also been conducted [110]. However, the application of optimization theory to stenting has so far been limited. Hence, in contrast to the abovementioned patient-specific work with realistic boundary conditions, in this section, we more generally review the optimization work related to the more idealized modeling of stents that has been performed. For example, our lab applied a surrogate management framework optimization algorithm for 3D CFD of stents using idealized BMS models [111,112]. Computational approaches such as this are not routinely implemented within stent development companies, possibly since a product can often be created and tested more rapidly than a computational analysis can be conducted. However, CFD is becoming more efficient [113] and can provide insights that cannot be obtained from bench-top testing or from the evaluation of prototypes using design matrices. Since any number of permutations can be modified computationally, optimization algorithms can facilitate the evaluation of theoretical and virtual approaches that could offer distinct advantages in a systematic and unbiased manner. Findings from an initial analysis showed that an existing optimization framework could be integrated with CFD of stents to determine the optimal angle of stent struts relative to the primary flow direction [111], and a follow-up study [112] recommended modified sizing matrices relative to those used by manufacturers of common stents for 2–5 mm diameter arteries (Figure 3). A more recent work applied multi-objective optimization to 12 mm stents within idealized cylinders to identify optimal strut width, thickness, angle, longitudinal spacing, and connector shape based on the percentage of vessel area exposed to low TAWSS, high TAWSS, and radial stiffness [114]. Results were generally consistent with those previously presented and with anecdotal stent design knowledge, but they also showed important differences based on whether strut cross sections were modelled as circular or rectangular.

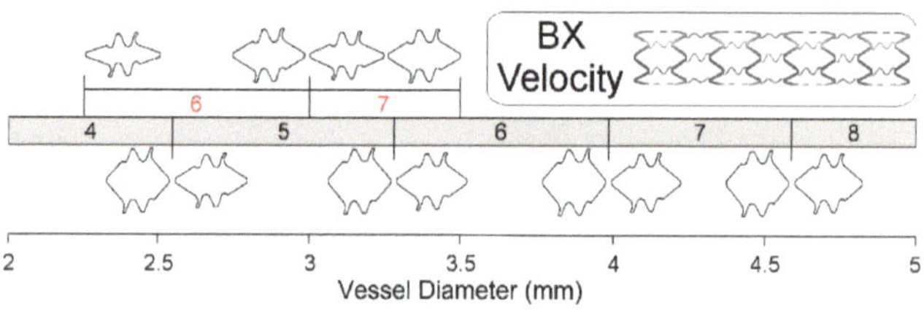

Figure 3. Comparing diameter ranges of a commercial stent vs. its diameter ranges for optimal WSS. The number of repeating units from the manufacturer's sizing matrix is shown (**red values**) with that from optimization using a TAWSS cost function (**black values**). The geometry of one unit for the limits of each range is also shown. Adapted from Gundert et al. [112].

The study mentioned above, which showed differences in optimal strut design based on its cross-sectional shape, is interesting, particularly considering a recent work that featured a parametric level set method to optimize the topology of struts [115]. This approach allows for the systematic computational modification of structural topologies to arrive at novel strut structures that may not have previously been devised. Although in its infancy relative to other current state-of-the-art CFD stimulations of stented arteries (e.g., steady flow, zero pressure outlet boundary conditions, sheet configuration), future work aimed at applying these approaches to patient-specific stent designs is exciting, given the existence of 3D printing/additive manufacturing techniques as well as current abilities to modify strut shapes through processes including wire electrical discharge machining.

Bench top studies from over a decade ago showed that the angle and orientation of stent struts can impact the potential for endothelialization [36]. More recent studies using intravascular imaging modalities have shown that stent malapposition can be associated with delayed neointimal healing via incomplete endothelialization. Malapposition also induces local disruptions in WSS indices that have previously been linked to neointimal proliferation and thrombus deposition in preclinical models as well as to clinical restenosis and stent thrombosis [116]. Hence, some subsequent studies have emerged to limit the likelihood of malapposition while also optimizing vessel wall stress (fluid and solid mechanics) and drug diffusion using surrogate modeling [117]. For example, by using constitutive parameters for a soft plaque [118] generally aimed at representing the combination of native vessel and plaque components [119], Ragkousis et al. were able to show [117] that the optimal deployment could be predicted computationally. Perhaps not surprisingly, malapposition was inversely proportional to stresses imposed on vascular tissue and the resulting drug diffusion, pointing to a delicate balance between implantation to optimize DES features upon delivery and local injury [120], which may adversely influence outcomes.

6. Limited Data from Atherosclerotic Arteries

There is some evidence in the literature to suggest that FSI simulations involving stents may not be necessary due to the rigidity imparted by many coronary stents on an artery [67,121]. However, the materials often used with FSI simulations involving stents as well as the flow domains obtained for CFD after the FEM of computationally deployed stents coincide with available hyperelastic constitutive relationships for a limited range of plaques [39–41] (e.g., lipid rich, fibrous, and calcified plaques [122,123]) and healthy arteries [124,125]. Our lab has also applied these methods [46], which can result in deformations with a reasonable level of agreement with reconstructed arteries. Strictly speaking, it therefore remains to be determined whether rigid wall simulations yield WSS results that are similar to FSI simulations following patient-specific assessment of local material properties. Put in another way, the utility of limited prior data falls short when the goal is to predict the likelihood of NH or LST using patient-specific computational models created with the use of imaging from the PCI period. The recent literature echoes this need for more data. Akyildiz et al. recounted how data were missing on the tensile and compressive properties of the coronary arteries because tissue was often obtained after autopsy, when mechanical properties had already degraded [42]. Chen and Kassab recently acknowledged the paucity of studies accounting for plaque substructure in constitutive models applied in coronary arteries [43]. McKittrick and colleagues further remind us that most models employed to date are from healthy vessels and do not capture the effects of disease, including the presence of atherosclerotic plaque [44]. Computational models of tissue prolapse [126] after stenting, which can impact WSS [90], have also used material properties and/or constants for their associated hyperplastic constitutive equations rooted in the references above. More clarity on material properties from a wider range of plaque types and combinations has the potential to expand prior work in order to create more accurate models of intra-strut prolapse that may be associated with restenosis on a patient-specific basis.

Recently Narayanan et al. used inverse finite element methods to estimate the material properties of arterial plaque components [127]. The authors leveraged intravascular ultrasound imaging data acquired during imaging wire pullback in vivo and an in silico (i.e., simulated) target geometry corresponding to 60 mmHg above the acquired geometry for three patients. Consistent with the work above, five material regions were identified, which included fibrous, lipid, calcium, mixed, and healthy wall tissue.

In an effort to address requests for additional material characterization studies related to atherosclerotic coronary arteries, our group has begun testing plaques obtained from fresh human coronary arteries, which were harvested after the untimely passing of patients. These arteries are obtained as part of an organ donation protocol at a partnering institution. OCT imaging was performed at 0–90 mmHg and the physiologic blood pressure noted in the associated documentation. In our initial protocol, the arteries were then dissected for uniaxial extension testing and characterized using a hyperelastic constitutive model [124]. Cauchy stress was expressed as a function of the Green–Lagrange Strain. To create a validated computational approach and to characterize plaques too small to test empirically, inverse finite element analysis simulations were performed using the displacement from OCT at each pressure (Figure 4A). There was good agreement within the physiological range between the stress–strain curves obtained experimentally and those from inverse finite element analysis simulations (green square within Figure 4B), but there were large deviations in the super-physiological (beyond physiological) range of loading imposed during stenting (red square in Figure 4B). To mitigate this issue, biaxial material testing is now implemented to better represent in vivo loading during stenting (Figure 4C). Moreover, to translate this ex vivo loading approach beyond the physiological range to the clinic, a specialized angioplasty balloon that accommodates the OCT imaging wire has been fabricated. Deformations throughout each artery are captured via OCT at pressures according to an associated compliance table, and inverse finite element analysis simulations are again performed, yielding a constitutive characterization of the tissue across the strain range applied during PCI. Results with these methods are helping fill the void in the literature for plaque types alone and in combination, which may ultimately permit a pre-stenting plaque morphology assessment that could influence the DES choice to mitigate poor outcomes by further exploiting imaging in vivo.

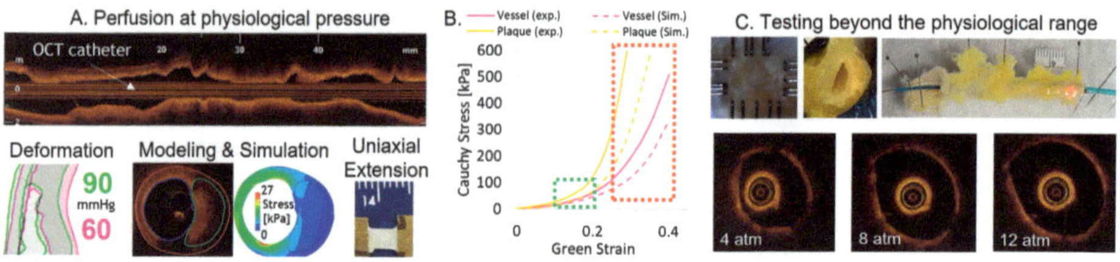

Figure 4. Current approaches to plaque characterization (**A**), results (**B**), and next-generation methods (**C**).

7. Application of Machine Learning (ML) and Artificial Intelligence (AI)

There is an increasing number of studies using ML and AI for several of the aspects related to the patient-specific stenting simulations reviewed above. The most common utility for such approaches seems to be the characterization of plaque components from IVUS or OCT images. As alluded to above, such advances are helpful for the field to expand beyond data currently available in the literature and to better characterize intra-strut deformations as well as changes in the curvature at the proximal and distal edges of an implanted stent, which can impact WSS indices. Kolluru et al. used a decision tree classifier approach with a database of 300 images to label each voxel from OCT runs conducted with cadaver coronary arteries as fibrotic, lipid-rich, calcified, or other [128]. Bae et al. recently published their work on the use of ML approaches to predict thin-cap fibroatheroma from IVUS results co-registered in OCT images [129]. Olender et al. also presented an alternative approach and extension to the characterization afforded by virtual histology with IVUS, which characterizes plaque components into dense calcium, necrotic core, fibrotic tissue, fibro-fatty tissue, or non-pathological features [130]. Other groups have also developed automated programs that leverage ML tools to detect stent linkages in IVUS [131] or OCT [132] images, delineate stent features and vessel areas in IVUS images [133,134], as well as automatically detect bifurcations within images from an OCT run [135].

Other intriguing work in this area has been carried out by Gharleghi et al., who recently applied deep learning techniques leveraging a 2 min steady simulation, artery geometry, and global features that included radii, curvature, and bifurcation angles to obtain TAWSS distributions representative of those obtained by transient CFD simulations [136]. This example study, along with related studies that have attempted to quantify [113] and reduce the time required to obtain patient-specific simulation results, represent important advancements that are critical for CFD/FSI simulations to be able to translate from the lab to clinical utility. While the initial simulations conducted with the end goal of obtaining a realistic patient-specific simulation and its associated WSS indices may be obtained in a ~1 day, it can easily take a month or more to tune boundary conditions, to adjust material properties until deformations are consistent with available measurements, to repeatedly quantify intermediary results, and to conduct mesh and time step independence analyses in order to generate WSS indices with a high level of confidence. It is also worth noting that although their findings were exciting and focused on the left main coronary bifurcation, Gharleghi et al. [136] and related work by Suk et al. [137] did not include stents and applied boundary conditions that were not patient-specific. Such advancements contribute substantially to the scale and complexity of the approach, thereby likely contributing the scarcity of such studies related to stenting.

Data reviewed in the Background section underscores the putative relationship between WSS indices and sites of restenosis after BMS implantation, while data relating such indices to outcomes for current-generation DES are still emerging. Groups have also applied ML tools to predict the likelihood of restenosis using angiographic and patient demographic data from the post-stenting period. Initial results suggest that such approaches may have improved predictive utility when considering the area under precision/recall curves relative to the existing restenosis scoring systems that used logistic regression, including Prevention of Restenosis With Tranilast and its Outcomes (PRESTO)-1, PRESTO-2, and Evaluation of Drug-Eluting Stents and Ischemic Events (EVENT) [138,139]. Interestingly, and to our knowledge, the ML classifiers tested identified features such as diabetes, multivessel disease, and post-PCI thrombolysis in myocardial infarction (PCI TIMI) flow as important features in the model, but did not consider stent-induced changes in WSS indices that can be revealed via CFD and/or FSI models.

8. Clinical Applications Using Patient-Specific Stenting

The best practices and considerations above have been implemented in several stenting studies with applications to specific clinical sequelae. One of the applications most poised for additional clinical guidance via computational modeling is the treatment of bifurcation lesions via provisional or two-stenting techniques. In particular, stenoses in the left main coronary artery (LMCA) can induce ischemia over a large fraction of the left ventricle and hence have the potential to result in substantial morbidity and/or mortality. One of the limitations to progress in this area is the fact that the PCI of LMCA lesions was considered harmful until recently (Class III in ACC/AHA guideline recommendations) and therefore often contraindicated during the clinical trials of DES [140,141]. Treatment with DES is now more reasonable with the recent lowering of the LMCA lesion classification, but without an abundance of available data for specific DES. Beyond being more limited in data and experience as compared to other coronary artery locations due to its prior classification, the LMCA has several unique considerations that also likely influence poor outcomes. Its derivation from the aorta makes it structurally unique, including a modest adventitia, a considerable smooth muscle content, and high elastic content proximally relative to distal coronary arteries [142]. The LMCA also has a larger caliber and shorter length [143], thus influencing its velocity profile and WSS relative to downstream arteries. LMCA lesions are present in 3–4% of the coronary catheterizations performed annually and often involve the LAD and/or LCX bifurcations. A wide range of bifurcation angles is also thought to contribute to flow disturbances and a high propensity for plaque in these bifurcations. Large-scale clinical trials are difficult due to the relatively low number of patients with LMCA lesions at most non-specialty centers and the vast number of DES designs now available. In contrast, patient-specific CFD and/or FSI simulations are well-suited to offer insight into the involvement of adverse blood flow disruptions without costly human trials or the potential for unfavorable clinical outcomes.

Recently, Samant et al. computationally and experimentally assessed the radial strength of the Synergy and several versions of the Megatron DES, which were specifically designed for the LMCA when deployed across a range of theoretical plaque types and four patient geometries [144]. Perhaps not surprisingly, the Megatron DES with 12 peaks circumferentially provided the best radial strength relative to those with fewer peaks. Nonetheless, this work highlights the balance and interplay between indices from solid mechanics relative to those from the fluid mechanics-based optimization studies [111,112] reviewed above, and differences in indices from studies using idealized arteries implanted with commercial stents [104]. In a follow-up study by Chatzizisis et al. [145], the authors showed the feasibility of a pre-interventional LMCA DES implantation planning procedure, whereby patient-specific models were created for three cases before a Megatron stent was computationally inserted into the 'optimal' position, as determined by clinical guidance through consideration of the stent positioning, sizing (length, diameter, inflation pressures), and strategy for each patient. There was good agreement in the mean diameters within the stented region following computational DES implantation as compared to the mean diameters obtained from IVUS conducted after the replication of the pre-interventional planning steps within the catheterization laboratory. Existing material properties from the literature were applied regionally, resulting in more uniform computational displacements relative to those obtained in vivo by visual inspection. Although extensions were used at the inlet and outlet of the models for the CFD analysis and only a snapshot of WSS results were presented, the work presents an undoubtedly exciting glimpse of what is possible through the clinical translation of computational tools.

Another recent work has also focused on bifurcations through the creation of a computational platform for patient-specific coronary bifurcation stenting [75]. As mentioned above, the methods employed most often include the assignment of material properties from available literature data of calcific, fibrous, and fibrolipid plaque components exposed to several stent platforms and techniques using one or two stents. There was good agreement along the length of the stented main vessel between the computational and clinical results for the five test cases implemented after the training cases for the platform. A visual inspection of OCT slices versus FEM results from spatially equivalent regions also showed good agreement. As pointed out by the authors, the goal was to create a framework to replicate clinical stenting steps using a trained computational stenting platform that informs the procedure conducted in the catheterization lab. Despite some limitations, the ability to conduct this work in real time and in a way that may ultimately further reduce restenosis, neoatherosclerosis, and/or LST is fascinating.

To illustrate the importance of the studies reviewed above relative to their potential for improved outcomes by optimizing the stenting procedure and the resulting stimuli for NH and/or LST, consider the examples shown in Figure 5. The examples use data from CTA and OCT imaging sessions obtained for patients with coronary artery lesions who subsequently underwent implantation with one of several DES commonly deployed within or outside the US: Xience Prime (Abbott) or Nobori (Terumo). Image-based CFD models were created, and the influence of the implanted stent on the distributions of WSS immediately after implantation were quantified. Post-stenting WSS results from the implanted stent were then also compared to those resulting from the virtual implantation of the other stent and quantified in terms of severity (i.e., total area of low WSS). When considering the stent-induced distributions of WSS for idealized representations of each stent, the CFD results suggest that the Xience Prime design, with its lower stent-to-artery area ratio and thinner struts, would be more favorable than the Nobori stent with its thick struts. However, geometric perturbations for a given patient influencing near-wall velocity patterns are superimposed on these theoretical results and can therefore factor into the observed results. While many patients likely receive what could be considered the optimal stent for their coronary artery, these examples using virtual stent implantation suggest that it is possible for some patients to benefit from the implantation of a DES other than that selected as a result of the strut thickness for the implanted stent, its amount of malapposition, and the geometric stent pattern adjacent to any malapposed regions. These findings demonstrate that geometric stent attributes, including strut thickness and stent-to-vessel area ratio alone, may not be predictive of adverse stent-induced WSS in patient-specific arteries, as had been previously thought based on results for idealized vessels. This example lends further support to the hypothesis that a certain DES may be more beneficial for a particular patient, thereby underscoring the excitement of computational planning tools for this purpose.

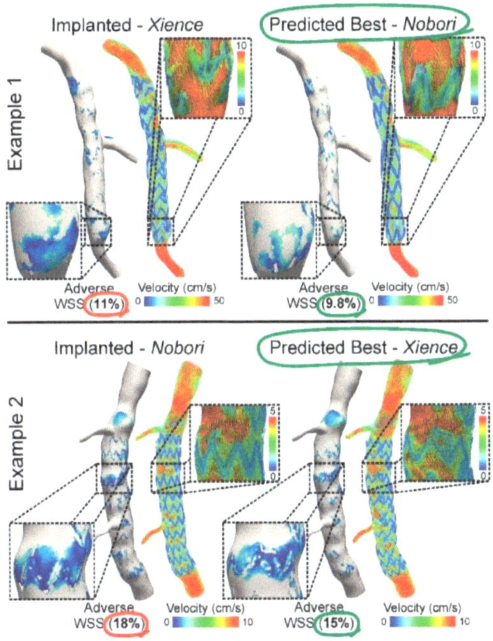

Figure 5. CFD simulations showing the amount of adverse WSS and near-wall velocity for two patients who may not have received the optimal DES according to CFD. Red circled percentages underscore more adverse areas of low WSS relative to green circled percentages for each example. These examples reveal how geometric perturbations for a given patient can factor into adverse WSS and may correlate with NH.

9. Future Directions

Among the most exciting computational advancements applicable to coronary artery disease management within the last decade is the development of a CT-derived fractional flow reserve (FFRct) index by HeartFlow [146,147]. Similar technology now also seems to be available from Toshiba [148], Siemens [149], and Canon Medical Systems Corporation. We are currently collaborating with HeartFlow for a novel research extension of their FFRct technology to noninvasively determine hemodynamically significant restenosis at follow-up, and similar approaches have been featured in at least one case report to date [150].

The ability to predict the likelihood of restenosis is intriguing, given the recent advancements in multiscale modeling [151]. For example, Zun et al. recently published an article describing a method of stent deployment and blood flow simulation, which also included tissue growth and was rooted in empirical observations of smooth muscle cell proliferation and extracellular matrix production. Their results showed good agreement with NH following BMS implantation via the use of a stent-to-artery deployment ratio of 1.4:1 in healthy curved and straight porcine coronary arteries. The adaption of this approach in order to include the inhibitory influence of a particular DES with validation in diseased human coronary arteries subjected to clinical deployment ratios would no doubt lead to substantial excitement in the field.

Corti et al. also recently conducted an extensive review focused on the multiscale analysis of arterial adaptation in stented and unstented vessels via agent-based models [152]. In contrast to the approaches discussed above, which used continuum models based on a system of differential equations, agent-based models can provide a natural description of biological systems using a set of rules that define agent activities such as apoptosis and mitosis [153]. This is particularly appealing when dealing with different spatiotemporal scales. Related studies to date have used such models to further elucidate the contributions of altered hemodynamics and arterial damage [154,155] in response to stenting as well as the impact of associated drugs eluted from stents to mitigate NH [156,157] and endothelialization related to LST based on WSS. Hemodynamic analyses have included idealized models (e.g., 3D with curvature) [151], and preliminary validation via histology in porcine arteries showed good agreement with model results. We eagerly await the additional extension of these approaches to patient-specific stented coronary arteries, as has been conducted in other stented vascular beds [158], which will especially contribute in an impactful way to the paucity of studies on stent thrombosis.

When considering the abovementioned details related to current technical abilities, with careful planning and in a setting with ample resources, it is possible to create truly patient-specific simulations of WSS indices today in response to an implanted DES. These models could consider local material properties extracted from pre-stent OCT conducted at multiple pressures (i.e., Figure 4C) in order to increase the likelihood of matching stent-induced deformations replicated though stent implantation via Boolean operations that leverage intravascular post-stent imaging [46] or full-scale FEM simulations of associated balloon folding, positioning, and elastoplastic expansion behavior of the stent [83]. Rheology could be included through the analysis of blood drawn from of the indwelling guide catheter, and boundary conditions that replicate the downstream pathogenic vasculature exerting pressure within the diseased coronary could be implemented using parameter estimates for coronary artery lumped parameter models [159–161]. Frequent (e.g., mean WSS) and less commonly reported stimuli (e.g., WSSET [91]) could be calculated within existing software packages or through post-processing after the convergence of simulations that consider time-step independence along with spatial mesh independence. Such simulations would be computationally expensive and time-consuming to conduct, but would provide the most comprehensive picture to date of the trade-off between the range of adverse WSS by stenting and the vascular response that is potentially mitigated by a given DES. Indeed, this is likely the future of such simulations that use current and next-generation DES as a further move toward personalized medicine, which seems to be limited only by computational ability and clinical approach. As noted above, both of these elements are rapidly progressing and becoming increasingly intertwined.

Author Contributions: Conceptualization, J.F.L.J.; investigation, J.F.L.J., A.G., D.S.M., P.M., H.O.; resources, J.F.L.J., A.G., D.S.M., P.M., H.O.; writing—original draft preparation, J.F.L.J., A.G.; writing—review and editing, J.F.L.J., A.G., D.S.M., P.M., H.O. All authors have read and agreed to the published version of the manuscript.

Funding: This research received no external funding.

Institutional Review Board Statement: Not applicable.

Informed Consent Statement: Not applicable.

Data Availability Statement: Not applicable.

Conflicts of Interest: The authors declare no conflict of interest.

References

1. Tsao, C.W.; Aday, A.W.; Almarzooq, Z.I.; Alonso, A.; Beaton, A.Z.; Bittencourt, M.S.; Boehme, A.K.; Buxton, A.E.; Carson, A.P.; Commodore-Mensah, Y.; et al. Heart Disease and Stroke Statistics-2022 Update: A Report From the American Heart Association. *Circulation* **2022**, *145*, e153–e639. [CrossRef]
2. Navarese, E.P.; Austin, D.; Gurbel, P.A.; Andreotti, F.; Tantry, U.; James, S.; Buffon, A.; Kozinski, M.; Obonska, K.; Bliden, K.; et al. Drug-coated balloons in treatment of in-stent restenosis: A meta-analysis of randomised controlled trials. *Clin. Res. Cardiol.* **2013**, *102*, 279–287. [CrossRef]
3. Lee, S.J.; Kim, B.K.; Kim, J.S.; Ko, Y.G.; Choi, D.; Jang, Y.; Hong, M.K. Evaluation of neointimal morphology of lesions with or without in-stent restenosis: An optical coherence tomography study. *Clin. Cardiol.* **2011**, *34*, 633–639. [CrossRef]

4. Hoffmann, R.; Mintz, G.S.; Dussaillant, G.R.; Popma, J.J.; Pichard, A.D.; Satler, L.F.; Kent, K.M.; Griffin, J.; Leon, M.B. Patterns and mechanisms of in-stent restenosis. A serial intravascular ultrasound study. *Circulation* **1996**, *94*, 1247–1254. [CrossRef] [PubMed]
5. Krone, R.J.; Rao, S.V.; Dai, D.; Anderson, H.V.; Peterson, E.D.; Brown, M.A.; Brindis, R.G.; Klein, L.W.; Shaw, R.E.; Weintraub, W.S. Acceptance, panic, and partial recovery the pattern of usage of drug-eluting stents after introduction in the, U.S. (a report from the American College of Cardiology/National Cardiovascular Data Registry). *JACC Cardiovasc. Interv.* **2010**, *3*, 902–910. [CrossRef] [PubMed]
6. Holmes, D.R., Jr.; Kereiakes, D.J.; Garg, S.; Serruys, P.W.; Dehmer, G.J.; Ellis, S.G.; Williams, D.O.; Kimura, T.; Moliterno, D.J. Stent thrombosis. *J. Am. Coll. Cardiol.* **2010**, *56*, 1357–1365. [CrossRef]
7. Finn, A.V.; Nakazawa, G.; Joner, M.; Kolodgie, F.D.; Mont, E.K.; Gold, H.K.; Virmani, R. Vascular responses to drug eluting stents: Importance of delayed healing. *Arterioscler. Thromb. Vasc. Biol.* **2007**, *27*, 1500–1510. [CrossRef]
8. Joner, M.; Finn, A.V.; Farb, A.; Mont, E.K.; Kolodgie, F.D.; Ladich, E.; Kutys, R.; Skorija, K.; Gold, H.K.; Virmani, R. Pathology of drug-eluting stents in humans: Delayed healing and late thrombotic risk. *J. Am. Coll. Cardiol.* **2006**, *48*, 193–202. [CrossRef]
9. Kotani, J.; Awata, M.; Nanto, S.; Uematsu, M.; Oshima, F.; Minamiguchi, H.; Mintz, G.S.; Nagata, S. Incomplete neointimal coverage of sirolimus-eluting stents: Angioscopic findings. *J. Am. Coll. Cardiol.* **2006**, *47*, 2108–2111. [CrossRef] [PubMed]
10. Kimura, T.; Morimoto, T.; Kozuma, K.; Honda, Y.; Kume, T.; Aizawa, T.; Mitsudo, K.; Miyazaki, S.; Yamaguchi, T.; Hiyoshi, E.; et al. Comparisons of baseline demographics, clinical presentation, and long-term outcome among patients with early, late, and very late stent thrombosis of sirolimus-eluting stents: Observations from the Registry of Stent Thrombosis for Review and Reevaluation (RESTART). *Circulation* **2010**, *122*, 52–61.
11. Ong, A.T.; Hoye, A.; Aoki, J.; van Mieghem, C.A.; Rodriguez Granillo, G.A.; Sonnenschein, K.; Regar, E.; McFadden, E.P.; Sianos, G.; van der Giessen, W.J.; et al. Thirty-day incidence and six-month clinical outcome of thrombotic stent occlusion after bare-metal, sirolimus, or paclitaxel stent implantation. *J. Am. Coll. Cardiol.* **2005**, *45*, 947–953. [CrossRef] [PubMed]
12. Van Werkum, J.W.; Heestermans, A.A.; de Korte, F.I.; Kelder, J.C.; Suttorp, M.J.; Rensing, B.J.; Zwart, B.; Brueren, B.R.; Koolen, J.J.; Dambrink, J.H.; et al. Long-term clinical outcome after a first angiographically confirmed coronary stent thrombosis: An analysis of 431 cases. *Circulation* **2009**, *119*, 828–834. [CrossRef]
13. Yeo, K.K.; Mahmud, E.; Armstrong, E.J.; Bennett, W.E.; Shunk, K.A.; MacGregor, J.S.; Li, Z.; Low, R.I.; Rogers, J.H. Contemporary clinical characteristics, treatment, and outcomes of angiographically confirmed coronary stent thrombosis: Results from a multicenter California registry. *Catheter. Cardiovasc. Interventions Off. J. Soc. Cardiac Angiogr. Interv.* **2012**, *79*, 550–556. [CrossRef] [PubMed]
14. Chechi, T.; Vecchio, S.; Vittori, G.; Giuliani, G.; Lilli, A.; Spaziani, G.; Consoli, L.; Baldereschi, G.; Biondi-Zoccai, G.G.; Sheiban, I.; et al. ST-segment elevation myocardial infarction due to early and late stent thrombosis a new group of high-risk patients. *J. Am. Coll. Cardiol.* **2008**, *51*, 2396–2402. [CrossRef] [PubMed]
15. De Cock, D.; Bennett, J.; Ughi, G.J.; Dubois, C.; Sinnaeve, P.; Dhooge, J.; Desmet, W.; Belmans, A.; Adriaenssens, T. Healing course of acute vessel wall injury after drug-eluting stent implantation assessed by optical coherence tomography. *Eur. Heart J. Cardiovasc. Imaging* **2014**, *15*, 800–809. [CrossRef]
16. Garg S and Serruys, P.W. Coronary stents: Current status. *J. Am. College Cardiol.* **2010**, *56*, S1–S42. [CrossRef]
17. Bonaa, K.H.; Mannsverk, J.; Wiseth, R.; Aaberge, L.; Myreng, Y.; Nygard, O.; Nilsen, D.W.; Klow, N.E.; Uchto, M.; Trovik, T.; et al. Drug-Eluting or Bare-Metal Stents for Coronary Artery Disease. *N. Engl. J. Med.* **2016**, *375*, 1242–1252. [CrossRef]
18. Byrne, R.A.; Joner, M.; Kastrati, A. Stent thrombosis and restenosis: What have we learned and where are we going? The Andreas Grüntzig Lecture ESC 2014. *Eur. Heart J.* **2015**, *36*, 3320–3331. [CrossRef]
19. Ragosta, M.; Dee, S.; Sarembock, I.J.; Lipson, L.C.; Gimple, L.W.; Powers, E.R. Prevalence of unfavorable angiographic characteristics for percutaneous intervention in patients with unprotected left main coronary artery disease. *Catheter. Cardiovasc. Interv.* **2006**, *68*, 357–362. [CrossRef]
20. Seyahi, N.; Altiparmak, M.R.; Tascilar, K.; Pekpak, M.; Serdengecti, K.; Erek, E. Ultrasonographic maturation of native arteriovenous fistulae: A follow-up study. *Ren. Fail.* **2007**, *29*, 481–486. [CrossRef]
21. Shlofmitz, E.; Iantorno, M.; Waksman, R. Restenosis of Drug-Eluting Stents: A New Classification System Based on Disease Mechanism to Guide Treatment and State-of-the-Art Review. *Circ. Cardiovasc. Interv.* **2019**, *12*, e007023, reprinted in *Circ. Cardiovasc. Interv.* **2019**, *12*, e000044. [CrossRef] [PubMed]
22. Nakagawa, M.; Otake, H.; Shinke, T.; Takaya, T.; Kozuki, A.; Hariki, H.; Inoue, T.; Osue, T.; Taniguchi, Y.; Iwasaki, M.; et al. Analysis by Optical Coherence Tomography of Long-term Arterial Healing After Implantation of Different Types of Stents. *Can. J. Cardiol.* **2014**, *30*, 904–911. [CrossRef] [PubMed]
23. Gijsen, F.J.; Oortman, R.M.; Wentzel, J.J.; Schuurbiers, J.C.; Tanabe, K.; Degertekin, M.; Ligthart, J.M.; Thury, A.; de Feyter, P.J.; Serruys, P.W.; et al. Usefulness of shear stress pattern in predicting neointima distribution in sirolimus-eluting stents in coronary arteries. *Am. J. Cardiol.* **2003**, *92*, 1325–1328. [CrossRef]
24. Celi, S.; Vaghetti, M.; Palmieri, C.; Berti, S. Superficial coronary calcium analysis by OCT: Looking forward an imaging algorithm for an automatic 3D quantification. *Int. J. Cardiol.* **2013**, *168*, 2958–2960. [CrossRef] [PubMed]
25. Mortier, P.; De Beule, M.; Dubini, G.; Hikichi, Y.; Murasato, Y.; Ormiston, J.A. Coronary bifurcation stenting: Insights from in vitro and virtual bench testing. *EuroIntervention* **2010**, *6*, 53–60. [CrossRef] [PubMed]

26. Moussa, I.; Di Mario, C.; Reimers, B.; Akiyama, T.; Tobis, J.; Colombo, A. Subacute stent thrombosis in the era of intravascular ultrasound-guided coronary stenting without anticoagulation: Frequency, predictors and clinical outcome. *J. Am. Coll. Cardiol.* **1997**, *29*, 6–12. [CrossRef]
27. Kuntz, R.E.; Safian, R.D.; Carrozza, J.P.; Fishman, R.F.; Mansour, M.; Baim, D.S. The importance of acute luminal diameter in determining restenosis after coronary atherectomy or stenting. *Circulation* **1992**, *86*, 1827–1835. [CrossRef]
28. Ng, J.; Bourantas, C.V.; Torii, R.; Ang, H.Y.; Tenekecioglu, E.; Serruys, P.W.; Foin, N. Local Hemodynamic Forces After Stenting: Implications on Restenosis and Thrombosis. *Arterioscler. Thromb. Vasc. Biol.* **2017**, *37*, 2231–2242. [CrossRef]
29. De Santis, G.; Conti, M.; Trachet, B.; De Schryver, T.; De Beule, M.; Degroote, J.; Vierendeels, J.; Auricchio, F.; Segers, P.; Verdonck, P.; et al. Haemodynamic impact of stent-vessel (mal)apposition following carotid artery stenting: Mind the gaps! *Comput. Methods Biomech. Biomed. Engin.* **2013**, *16*, 648–659. [CrossRef]
30. Rikhtegar, F.; Pacheco, F.; Wyss, C.; Stok, K.S.; Ge, H.; Choo, R.J.; Ferrari, A.; Poulikakos, D.; Muller, R.; Kurtcuoglu, V. Compound ex vivo and in silico method for hemodynamic analysis of stented arteries. *PLoS ONE* **2013**, *8*, e58147. [CrossRef]
31. Garasic, J.M.; Edelman, E.R.; Squire, J.C.; Seifert, P.; Williams, M.S.; Rogers, C. Stent and artery geometry determine intimal thickening independent of arterial injury. *Circulation* **2000**, *101*, 812–818. [CrossRef] [PubMed]
32. LaDisa, J.F., Jr.; Olson, L.E.; Molthen, R.C.; Hettrick, D.A.; Pratt, P.F.; Hardel, M.D.; Kersten, J.R.; Warltier, D.C.; Pagel, P.S. Alterations in wall shear stress predict sites of neointimal hyperplasia after stent implantation in rabbit iliac arteries. *Am. J. Physiol. Heart Circ. Physiol.* **2005**, *288*, H2465–H2475. [CrossRef]
33. Liu, S.Q.; Goldman, J. Role of blood shear stress in the regulation of vascular smooth muscle cell migration. *IEEE Trans. Biomed. Eng.* **2001**, *48*, 474–483. [CrossRef]
34. Liu, S.Q.; Tang, D.; Tieche, C.; Alkema, P.K. Pattern formation of vascular smooth muscle cells subjected to nonuniform fluid shear stress: Mediation by the gradient of cell density. *Am. J. Physiol.* **2003**, *285*, H1072–H1080.
35. Liu, S.Q.; Zhong, L.; Goldman, J. Control of the shape of a thrombus-neointima-like structure by blood shear stress. *J. Biomech. Eng.* **2002**, *124*, 30–36. [CrossRef] [PubMed]
36. Hamuro, M.; Palmaz, J.C.; Sprague, E.A.; Fuss, C.; Luo, J. Influence of stent edge angle on endothelialization in an in vitro model. *J. Vascular Interv. Radiol.* **2001**, *12*, 607–611. [CrossRef]
37. Papafaklis, M.I.; Bourantas, C.V.; Theodorakis, P.E.; Katsouras, C.S.; Naka, K.K.; Fotiadis, D.I.; Michalis, L.K. The effect of shear stress on neointimal response following sirolimus- and paclitaxel-eluting stent implantation compared with bare-metal stents in humans. *JACC Cardiovasc. Interv.* **2010**, *3*, 1181–1189. [CrossRef]
38. Suzuki, N.; Nanda, H.; Angiolillo, D.J.; Bezerra, H.; Sabate, M.; Jimenez-Quevedo, P.; Alfonso, F.; Macaya, C.; Bass, T.A.; Ilegbusi, O.J.; et al. Assessment of potential relationship between wall shear stress and arterial wall response after bare metal stent and sirolimus-eluting stent implantation in patients with diabetes mellitus. *Int. J. Cardiovasc. Imaging* **2008**, *24*, 357–364. [CrossRef]
39. Lee, R.T.; Grodzinsky, A.J.; Frank, E.H.; Kamm, R.D.; Schoen, F.J. Structure-dependent dynamic mechanical behavior of fibrous caps from human atherosclerotic plaques. *Circulation* **1991**, *83*, 1764–1770. [CrossRef]
40. Loree, H.M.; Grodzinsky, A.J.; Park, S.Y.; Gibson, L.J.; Lee, R.T. Static circumferential tangential modulus of human atherosclerotic tissue. *J. Biomech.* **1994**, *27*, 195–204. [CrossRef]
41. Kragel, A.H.; Reddy, S.G.; Wittes, J.T.; Roberts, W.C. Morphometric analysis of the composition of atherosclerotic plaques in the four major epicardial coronary arteries in acute myocardial infarction and in sudden coronary death. *Circulation* **1989**, *80*, 1747–1756. [CrossRef] [PubMed]
42. Akyildiz, A.C.; Speelman, L.; Gijsen, F.J. Mechanical properties of human atherosclerotic intima tissue. *J. Biomech.* **2014**, *47*, 773–783. [CrossRef] [PubMed]
43. Chen H and Kassab, G.S. Microstructure-based biomechanics of coronary arteries in health and disease. *J. Biomech.* **2016**, *49*, 2548–2559. [CrossRef]
44. McKittrick, C.M.; Kennedy, S.; Oldroyd, K.G.; McGinty, S.; McCormick, C. Modelling the Impact of Atherosclerosis on Drug Release and Distribution from Coronary Stents. *Ann. Biomed. Eng.* **2016**, *44*, 477–487. [CrossRef] [PubMed]
45. De Nevers, N. *Fluid Mechanics for Chemical Engineers*, 3rd ed.; McGraw Hill: New York, NY, USA, 2004.
46. Chiastra, C.; Wu, W.; Dickerhoff, B.; Aleiou, A.; Dubini, G.; Otake, H.; Migliavacca, F.; LaDisa, J.F., Jr. Computational replication of the patient-specific stenting procedure for coronary artery bifurcations: From OCT and CT imaging to structural and hemodynamics analyses. *J. Biomech.* **2016**, *49*, 2102–2111. [CrossRef] [PubMed]
47. Williams, A.R.; Koo, B.K.; Gundert, T.J.; Fitzgerald, P.J.; LaDisa, J.F., Jr. Local hemodynamic changes caused by main branch stent implantation and subsequent virtual side branch balloon angioplasty in a representative coronary bifurcation. *J. Appl. Physiol.* **2010**, *109*, 532–540. [CrossRef]
48. Muller, J.; Sahni, O.; Li, X.; Jansen, K.E.; Shephard, M.S.; Taylor, C.A. Anisotropic adaptive finite element method for modelling blood flow. *Comput. Methods Biomech. Biomed. Eng.* **2005**, *8*, 295–305. [CrossRef]
49. Sahni, O.; Muller, J.; Jansen, K.E.; Shephard, M.S.; Taylor, C.A. Efficient anisotropic adaptive discretization of the cardiovascular system. *Comput. Methods Appl. Mech. Eng.* **2006**, *195*, 5634–5655. [CrossRef]
50. Ellwein, L.; Samyn, M.M.; Danduran, M.; Schindler-Ivens, S.; Liebham, S.; LaDisa, J.F., Jr. Toward translating near-infrared spectroscopy oxygen saturation data for the non-invasive prediction of spatial and temporal hemodynamics during exercise. *Biomech. Modeling Mechanobiol.* **2017**, *16*, 75–96. [CrossRef]

51. Kwon, S.; Feinstein, J.A.; Dholakia, R.J.; LaDisa, J.F., Jr. Quantification of local hemodynamic alterations caused by virtual implantation of three commercially available stents for the treatment of aortic coarctation. *Pediatric Cardiol.* **2014**, *35*, 732–740. [CrossRef]
52. Samyn, M.M.; LaDisa, J.F., Jr. Novel Applications of Cardiovascular Magnetic Resonance Imaging-Based Computational Fluid Dynamics Modeling in Pediatric Cardiovascular and Congenital Heart Disease. *Assess. Cell. Organ Funct. Dysfunc. Using Direct Derived MRI Methodol.* **2016**, 27–56.
53. Benard, N.; Perrault, R.; Fau-Coisne, D.; Coisne, D. Computational approach to estimating the effects of blood properties on changes in intra-stent flow. *Ann. Biomed. Eng.* **2006**, *34*, 1259–1271. [CrossRef] [PubMed]
54. Arzani, A. Accounting for residence-time in blood rheology models: Do we really need non-Newtonian blood flow modelling in large arteries? *J. R. Soc. Interface* **2018**, *15*, 20180486. [CrossRef]
55. Chiastra, C.; Morlacchi, S.; Gallo, D.; Morbiducci, U.; Cardenes, R.; Larrabide, I.; Migliavacca, F. Computational fluid dynamic simulations of image-based stented coronary bifurcation models. *J. R. Soc. Interface* **2013**, *10*, 20130193. [CrossRef] [PubMed]
56. Thondapu, V.; Tenekecioglu, E.; Poon, E.K.W.; Collet, C.; Torii, R.; Bourantas, C.V.; Chin, C.; Sotomi, Y.; Jonker, H.; Dijkstra, J.; et al. Endothelial shear stress 5 years after implantation of a coronary bioresorbable scaffold. *Eur. Heart J.* **2018**, *39*, 1602–1609. [CrossRef]
57. Lee, S.W.; Steinman, D.A. On the relative importance of rheology for image-based CFD models of the carotid bifurcation. *J. Biomech. Eng.* **2007**, *129*, 273–278. [CrossRef]
58. Li, Y.; Li, Z.; Holck, E.N.; Xu, B.; Karanasos, A.; Fei, Z.; Chang, Y.; Chu, M.; Dijkstra, J.; Christiansen, E.H.; et al. Local flow patterns after implantation of bioresorbable vascular scaffold in coronary bifurcations- novel findings by computational fluid dynamics. *Circ. J.* **2018**, *82*, 1575–1583. [CrossRef]
59. Nichols, W.W.; O'Rourke, M.F. *McDonald's Blood Flow in Arteries: Theoretical, Experimental and Clinical Principles*, 5th ed.; Hodder Arnold Group: London, UK, 2005.
60. Kim, H.J.; Figueroa, C.A.; Hughes, T.J.R.; Jansen, K.E.; Taylor, C.A. Augmented Lagrangian method for constraining the shape of velocity profiles at outlet boundaries for three-dimensional Finite Element simulations of blood flow. *Comput. Methods Appl. Mech. Eng.* **2009**, *198*, 3551–3566. [CrossRef]
61. Esmaily Moghadam, M.; Bazilevs, Y.; Hsia, T.-Y.; Vignon-Clementel, I.E.; Marsden, A.L.; Modeling of Congenital Hearts Alliance. A comparison of outlet boundary treatments for prevention of backflow divergence with relevance to blood flow simulations. *Comput. Mech.* **2011**, *48*, 277–291. [CrossRef]
62. Bovendeerd, P.H.; Borsje, P.; Arts, T.; van De Vosse, F.N. Dependence of intramyocardial pressure and coronary flow on ventricular loading and contractility: A model study. *Ann. Biomed. Eng.* **2006**, *34*, 1833–1845. [CrossRef] [PubMed]
63. Sankaran, S.; Esmaily Moghadam, M.; Kahn, A.M.; Tseng, E.E.; Guccione, J.M.; Marsden, A.L. Patient-specific multiscale modeling of blood flow for coronary artery bypass graft surgery. *Ann. Biomed. Eng.* **2012**, *40*, 2228–2242. [CrossRef] [PubMed]
64. Chiastra, C.; Dubini, G.; Migliavacca, F. Hemodynamic perturbations due to the presence of stents. In *Biomechanics of Living Organs, Biomechanics of Coronary Atherosclerotic Plaque*; Jacques Ohayon, G.F., Pettigrew, R.I., Eds.; Academic Press: Cambridge, MA, USA, 2021; pp. 251–271.
65. Figueroa, C.A.; Vignon-Clementel, I.E.; Jansen, K.E.; Hughes, T.J.R.; Taylor, C.A. A coupled momentum method for modeling blood flow in three-dimensional deformable arteries. *Comput. Methods Appl. Mech. Eng.* **2006**, *195*, 5685–5706. [CrossRef]
66. Gundert, T.J.; Shadden, S.C.; Williams, A.R.; Koo, B.K.; Feinstein, J.A.; LaDisa, J.F., Jr. A rapid and computationally inexpensive method to virtually implant current and next-generation stents into subject-specific computational fluid dynamics models. *Ann. Biomed. Eng.* **2011**, *39*, 1423–1437. [CrossRef] [PubMed]
67. LaDisa, J.F., Jr.; Hettrick, D.A.; Olson, L.E.; Guler, I.; Gross, E.R.; Kress, T.T.; Kersten, J.R.; Warltier, D.C.; Pagel, P.S. Stent implantation alters coronary artery hemodynamics and wall shear stress during maximal vasodilation. *J. Appl. Physiol.* **2002**, *93*, 1939–1946. [CrossRef]
68. Bukač, M.; Čanić, S.; Tambača, J.; Wang, Y. Fluid–structure interaction between pulsatile blood flow and a curved stented coronary artery on a beating heart: A four stent computational study. *Comput. Methods Appl. Mech. Eng.* **2019**, *350*, 679–700. [CrossRef]
69. Wu, W.; Samant, S.; de Zwart, G.; Zhao, S.; Khan, B.; Ahmad, M.; Bologna, M.; Watanabe, Y.; Murasato, Y.; Burzotta, F.; et al. 3D reconstruction of coronary artery bifurcations from coronary angiography and optical coherence tomography: Feasibility, validation, and reproducibility. *Sci. Rep.* **2020**, *10*, 18049. [CrossRef] [PubMed]
70. Slager, C.J.; Wentzel, J.J.; Schuurbiers, J.C.; Oomen, J.A.; Kloet, J.; Krams, R.; von Birgelen, C.; van der Giessen, W.J.; Serruys, P.W.; de Feyter, P.J. True 3-dimensional reconstruction of coronary arteries in patients by fusion of angiography and IVUS (ANGUS) and its quantitative validation. *Circulation* **2000**, *102*, 511–516. [CrossRef]
71. Athanasiou, L.; Nezami, F.R.; Galon, M.Z.; Lopes, A.C.; Lemos, P.A.; de la Torre Hernandez, J.M.; Ben-Assa, E.; Edelman, E.R. Optimized Computer-Aided Segmentation and Three-Dimensional Reconstruction Using Intracoronary Optical Coherence Tomography. *IEEE J. Biomed. Health Inform.* **2018**, *22*, 1168–1176. [CrossRef]
72. Timmins, L.H.; Suo, J.; Eshtehardi, P.; Molony, D.S.; McDaniel, M.C.; Oshinski, J.N.; Giddens, D.P.; Samady, H. Comparison of angiographic and IVUS derived coronary geometric reconstructions for evaluation of the association of hemodynamics with coronary artery disease progression. *Int. J. Cardiovasc. Imaging* **2016**, *32*, 1327–1336. [CrossRef]

73. Ellwein, L.M.; Otake, H.; Gundert, T.J.; Koo, B.K.; Shinke, T.; Honda, Y.; Shite, J.; LaDisa, J.F., Jr. Optical coherence tomography for patient-specific 3D artery reconstruction and evaluation of wall shear stress in a left circumflex coronary artery. *Cardiovasc. Eng. Tech.* **2011**, *2*, 212–217. [CrossRef]
74. Chiastra, C.; Migliori, S.; Burzotta, F.; Dubini, G.; Migliavacca, F. Patient-Specific Modeling of Stented Coronary Arteries Reconstructed from Optical Coherence Tomography: Towards a Widespread Clinical Use of Fluid Dynamics Analyses. *J. Cardiovasc. Transl. Res.* **2018**, *11*, 156–172. [CrossRef] [PubMed]
75. Zhao, S.; Wu, W.; Samant, S.; Khan, B.; Kassab, G.S.; Watanabe, Y.; Murasato, Y.; Sharzehee, M.; Makadia, J.; Zolty, D.; et al. Patient-specific computational simulation of coronary artery bifurcation stenting. *Sci. Rep.* **2021**, *11*, 1–17. [CrossRef]
76. Migliori, S.; Rampat, R.; Bologna, M.; Montin, E.; Burzotta, F.; Hildick-Smith, D.; Dubini, G.; Mainardi, L.; Migliavacca, F.; Cockburn, J.; et al. A Patient-Specific Study Investigating the Relation between Coronary Hemodynamics and Neo-Intimal Thickening after Bifurcation Stenting with a Polymeric Bioresorbable Scaffold. *Appl. Sci.* **2018**, *8*, 1510. [CrossRef]
77. Gogas, B.D.; Yang, B.; Piccinelli, M.; Giddens, D.P.; King, S.B., 3rd; Kereiakes, D.J.; Ellis, S.G.; Stone, G.W.; Veneziani, A.; Samady, H. Novel 3-Dimensional Vessel and Scaffold Reconstruction Methodology for the Assessment of Strut-Level Wall Shear Stress After Deployment of Bioresorbable Vascular Scaffolds From the ABSORB III Imaging Substudy. *JACC Cardiovasc. Interv.* **2016**, *9*, 501–503. [CrossRef] [PubMed]
78. Gogas, B.D.; King, S.B., 3rd; Timmins, L.H.; Passerini, T.; Piccinelli, M.; Veneziani, A.; Kim, S.; Molony, D.S.; Giddens, D.P.; Serruys, P.W.; et al. Biomechanical assessment of fully bioresorbable devices. *JACC Cardiovasc. Interv.* **2013**, *6*, 760–761. [CrossRef] [PubMed]
79. Tu, S.; Holm, N.R.; Koning, G.; Huang, Z.; Reiber, J.H. Fusion of 3D QCA and IVUS/OCT. *Int. J. Cardiovasc. Imaging.* **2011**, *27*, 197–207. [CrossRef]
80. Li, Y.; Gutierrez-Chico, J.L.; Holm, N.R.; Yang, W.; Hebsgaard, L.; Christiansen, E.H.; Maeng, M.; Lassen, J.F.; Yan, F.; Reiber, J.H.; et al. Impact of Side Branch Modeling on Computation of Endothelial Shear Stress in Coronary Artery Disease: Coronary Tree Reconstruction by Fusion of 3D Angiography and, O.C.T. *J. Am. Coll. Cardiol.* **2015**, *66*, 125–135. [CrossRef] [PubMed]
81. Papafaklis, M.I.; Bourantas, C.V.; Yonetsu, T.; Vergallo, R.; Kotsia, A.; Nakatani, S.; Lakkas, L.S.; Athanasiou, L.S.; Naka, K.K.; Fotiadis, D.I.; et al. Anatomically correct three-dimensional coronary artery reconstruction using frequency domain optical coherence tomographic and angiographic data: Head-to-head comparison with intravascular ultrasound for endothelial shear stress assessment in humans. *EuroIntervention* **2015**, *11*, 407–415. [CrossRef]
82. Bourantas, C.V.; Papafaklis, M.I.; Lakkas, L.; Sakellarios, A.; Onuma, Y.; Zhang, Y.J.; Muramatsu, T.; Diletti, R.; Bizopoulos, P.; Kalatzis, F.; et al. Fusion of optical coherence tomographic and angiographic data for more accurate evaluation of the endothelial shear stress patterns and neointimal distribution after bioresorbable scaffold implantation: Comparison with intravascular ultrasound-derived reconstructions. *Int. J. Cardiovasc. Imaging* **2014**, *30*, 485–494.
83. Geith, M.A.; Swidergal, K.; Hochholdinger, B.; Schratzenstaller, T.G.; Wagner, M.; Holzapfel, G.A. On the importance of modeling balloon folding, pleating, and stent crimping: An FE study comparing experimental inflation tests. *Int. J. Numer. Method Biomed. Eng.* **2019**, *35*, e3249. [CrossRef]
84. Kleinstreuer, C.; Hyun, S.; Buchanan, J.R.; Longest, P.W.; Archie, J.P.; Truskey, G.A. Hemodynamic parameters and early intimal thickening in branching blood vessels. *Crit. Rev. Biomed. Eng.* **2001**, *29*, 1–64. [CrossRef] [PubMed]
85. Ku, D.N.; Giddens, D.P.; Zarins, C.K.; Glagov, S. Pulsatile flow and atherosclerosis in the human carotid bifurcation. Positive correlation between plaque location and low oscillating shear stress. *Arteriosclerosis* **1985**, *5*, 293–302. [CrossRef]
86. Malek, A.M.; Alper, S.L.; Izumo, S. Hemodynamic shear stress and its role in atherosclerosis. *JAMA* **1999**, *282*, 2035–2042. [CrossRef]
87. Moore, J.E.; Xu, C.; Glagov, S.; Zarins, C.K.; Ku, D.N. Fluid wall shear stress measurements in a model of the human abdominal aorta: Oscillatory behavior and relationship to atherosclerosis. *Atherosclerosis* **1994**, *110*, 225–240. [CrossRef]
88. Ojha, M. Spatial and temporal variations of wall shear stress within an end-to-side arterial anastomosis model. *J. Biomech.* **1993**, *26*, 1377–1388. [CrossRef]
89. LaDisa, J.F.; Olson, L.E.; Guler, I.; Hettrick, D.A.; Kersten, J.R.; Warltier, D.C.; Pagel, P.S. Circumferential vascular deformation after stent implantation alters wall shear stress evaluated with time-dependent 3D computational fluid dynamics models. *J. Appl. Physiol.* **2005**, *98*, 947–957. [CrossRef] [PubMed]
90. White, C.R.; Haidekker, M.; Bao, X.; Frangos, J.A. Temporal gradients in shear, but not spatial gradients, stimulate endothelial cell proliferation. *Circulation* **2001**, *103*, 2508–2513. [CrossRef] [PubMed]
91. Arzani, A.; Gambaruto, A.M.; Chen, G.; Shadden, S.C. Wall shear stress exposure time: A Lagrangian measure of near-wall stagnation and concentration in cardiovascular flows. *Biomech. Modeling Mechanobiol.* **2017**, *16*, 787–803. [CrossRef] [PubMed]
92. Morbiducci, U.; Ponzini, R.; Grigioni, M.F.; Redaelli, A. Helical flow as fluid dynamic signature for atherogenesis risk in aortocoronary bypass. A numeric study. *J. Biomech.* **2007**, *40*, 519–534. [CrossRef]
93. Himburg, H.A.; Grzybowski, D.M.; Hazel, A.L.; LaMack, J.A.; Li, X.; Friedman, M.H. Spatial comparison between wall shear stress measures and porcine arterial endothelial permeability. *Am. J. Physiol. Heart Circ. Physiol.* **2004**, *286*, H1916–H1922. [CrossRef] [PubMed]
94. Hoi, Y.; Zhou, Y.; Zhang, X.; Henkelman, R.M.; Steinman, D.A. Correlation between local hemodynamics and lesion distribution in a novel aortic regurgitation murine model of atherosclerosis. *Ann. Biomed. Eng.* **2011**, *39*, 1414–1422. [CrossRef] [PubMed]

95. Bashar, A.H.; Suzuki, K.; Kazui, T.; Okada, M.Y.; Suzuki, T.; Washiyama, N.; Terada, H.; Yamashita, K. Changes in cerebrospinal fluid and blood lactate concentrations after stent-graft implantation at critical aortic segment: A preliminary study. *Interact. Cardiovasc. Thorac. Surg.* **2008**, *7*, 262–266. [CrossRef] [PubMed]
96. He, Y.; Duraiswamy, N.; Frank, A.O.; Moore, J.E., Jr. Blood flow in stented arteries: A parametric comparison of strut design patterns in three dimensions. *J. Biomech. Eng.* **2005**, *127*, 637–647. [CrossRef] [PubMed]
97. Kawaguchi, R.; Sabate, M.; Angiolillo, D.J.; Jimenez-Quevedo, P.; Suzuki, N.; Corros, C.; Futamatsu, H.; Alfonso, F.; Hernandez-Antolin, R.; Macaya, C.; et al. Angiographic and 3D intravascular ultrasound assessment of overlapping bare metal stent and three different formulations of drug-eluting stents in patients with diabetes mellitus. *Int. J. Cardiovasc. Imaging* **2008**, *24*, 125–132. [CrossRef]
98. Murphy, J.B.; Boyle, F.J. A full-range, multi-variable, CFD-based methodology to identify abnormal near-wall hemodynamics in a stented coronary artery. *Biorheology* **2010**, *47*, 117–132. [CrossRef]
99. LaDisa, J.F., Jr.; Olson, L.E.; Guler, I.; Hettrick, D.A.; Audi, S.H.; Kersten, J.R.; Warltier, D.C.; Pagel, P.S. Stent design properties and deployment ratio influence indexes of wall shear stress: A three-dimensional computational fluid dynamics investigation within a normal artery. *J. Appl. Physiol.* **2004**, *97*, 424–430. [CrossRef]
100. LaDisa, J.F., Jr.; Olson, L.E.; Hettrick, D.A.; Warltier, D.C.; Kersten, J.R.; Pagel, P.S. Axial stent strut angle influences wall shear stress after stent implantation: Analysis using 3D computational fluid dynamics models of stent foreshortening. *Biomed. Eng. Online* **2005**, *4*, 59. [CrossRef]
101. Briguori, C.; Sarais, C.; Pagnotta, P.; Liistro, F.; Montorfano, M.; Chieffo, A.; Sgura, F.; Corvaja, N.; Albiero, R.; Stankovic, G.; et al. In-stent restenosis in small coronary arteries: Impact of strut thickness. *J. Am. Coll. Cardiol.* **2002**, *40*, 403–409. [CrossRef]
102. Kastrati, A.; Mehilli, J.; Dirschinger, J.; Dotzer, F.; Schuhlen, H.; Neumann, F.J.; Fleckenstein, M.; Pfafferott, C.; Seyfarth, M.; Schomig, A. Intracoronary stenting and angiographic results: Strut thickness effect on restenosis outcome (ISAR-STEREO) trial. *Circulation* **2001**, *103*, 2816–2821. [CrossRef]
103. Beier, S.; Ormiston, J.; Webster, M.; Cater, J.; Norris, S.; Medrano-Gracia, P.; Young, A.; Cowan, B. Hemodynamics in Idealized Stented Coronary Arteries: Important Stent Design Considerations. *Ann. Biomed. Eng.* **2016**, *44*, 315–329. [CrossRef]
104. Gundert, T.J.; Dholakia, R.J.; McMahon, D.; LaDisa, J.F. Computational fluid dynamics evaluation of equivalency in hemodynamic alterations between Driver, Integrity, and similar stents implanted into an idealized coronary artery. *J. Med. Devices* **2013**, *7*, 011004. [CrossRef]
105. LaDisa, J.F., Jr.; Guler, I.; Olson, L.E.; Hettrick, D.A.; Kersten, J.R.; Warltier, D.C.; Pagel, P.S. Three-dimensional computational fluid dynamics modeling of alterations in coronary wall shear stress produced by stent implantation. *Ann. Biomed. Eng.* **2003**, *31*, 972–980. [CrossRef] [PubMed]
106. Chiastra, C.; Mazzi, V.; Lodi Rizzini, M.; Calo, K.; Corti, A.; Acquasanta, A.; De Nisco, G.; Belliggiano, D.; Cerrato, E.; Gallo, D.; et al. Coronary Artery Stenting Affects Wall Shear Stress Topological Skeleton. *J. Biomech. Eng.* **2022**, *144*, 061002. [CrossRef] [PubMed]
107. Ng, J.C.K.; Lian, S.S.; Zhong, L.; Collet, C.; Foin, N.; Ang, H.Y. Stent malapposition generates stent thrombosis: Insights from a thrombosis model. *Int. J. Cardiol.* **2022**, *353*, 43–45. [CrossRef] [PubMed]
108. Gasior, P.; Lu, S.; Ng, C.K.J.; Toong, W.Y.D.; Wong, E.H.P.; Foin, N.; Kedhi, E.; Wojakowski, W.; Ang, H.Y. Comparison of overexpansion capabilities and thrombogenicity at the side branch ostia after implantation of four different drug eluting stents. *Sci. Rep.* **2020**, *10*, 20791. [CrossRef]
109. Katritsis, D.; Kaiktsis, L.; Chaniotis, A.; Pantos, J.; Efstathopoulos, E.P.; Marmarelis, V. Wall shear stress: Theoretical considerations and methods of measurement. *Prog. Cardiovasc. Dis.* **2007**, *49*, 307–329. [CrossRef]
110. Pizarro, C.; De Leval, M.R. Surgical variations and flow dynamics in cavopulmonary connections: A historical review. *Semin. Thorac. Cardiovasc. Surg. Pediatr. Card Surg. Annu.* **1998**, *1*, 53–60. [CrossRef]
111. Gundert, T.J.; Marsden, A.L.; Yang, W.; LaDisa, J.F., Jr. Optimization of cardiovascular stent design using computational fluid dynamics. *J. Biomech. Eng.* **2012**, *134*, 011002. [CrossRef]
112. Gundert, T.J.; Marsden, A.L.; Yang, W.; Marks, D.S.; LaDisa, J.F., Jr. Identification of hemodynamically optimal coronary stent designs based on vessel caliber. *IEEE Trans. Biomed. Eng.* **2012**, *59*, 1992–2002. [CrossRef]
113. LaDisa, J.F.; Bowers, M.; Harmann, L.; Prost, R.; Doppalapudi, A.V.; Mohyuddin, T.; Zaidat, O.; Migrino, R.Q. Time-efficient patient-specific quantification of regional carotid artery fluid dynamics and spatial correlation with plaque burden. *Med. Phys.* **2010**, *37*, 784–792. [CrossRef]
114. Gharleghi, R.; Wright, H.; Luvio, V.; Jepson, N.; Luo, Z.; Senthurnathan, A.; Babaei, B.; Prusty, B.G.; Ray, T.; Beier, S. A multi-objective optimization of stent geometries. *J. Biomech.* **2021**, *125*, 110575. [CrossRef] [PubMed]
115. Xue, H.; Saha, S.C.; Beier, S.; Jepson, N.; Luo, Z. Topological Optimization of Auxetic Coronary Stents Considering Hemodynamics. *Front. Bioeng. Biotechnol.* **2021**, *9*, 728914. [CrossRef] [PubMed]
116. Ozaki, Y.; Okumura, M.; Ismail, T.F.; Naruse, H.; Hattori, K.; Kan, S.; Ishikawa, M.; Kawai, T.; Takagi, Y.; Ishii, J.; et al. The fate of incomplete stent apposition with drug-eluting stents: An optical coherence tomography-based natural history study. *Eur. Heart J.* **2010**, *31*, 1470–1476. [CrossRef] [PubMed]
117. Ragkousis, G.E.; Curzen, N.; Bressloff, N.W. Multi-objective optimisation of stent dilation strategy in a patient-specific coronary artery via computational and surrogate modelling. *J. Biomech.* **2016**, *49*, 205–215. [CrossRef] [PubMed]

118. Ragkousis, G.E.; Curzen, N.; Bressloff, N.W. Simulation of longitudinal stent deformation in a patient-specific coronary artery. *Med. Eng. Phys.* **2014**, *36*, 467–476. [CrossRef] [PubMed]
119. Wong, H.C.; Cho, K.N.; Tang, W.C. Bending of a stented atherosclerotic artery. In *COMSOL Conference*; Comsol: Boston, MA, USA, 2009.
120. Welch, T.R.; Eberhart, R.C.; Banerjee, S.; Chuong, C.J. Mechanical Interaction of an Expanding Coiled Stent with a Plaque-Containing Arterial Wall: A Finite Element Analysis. *Cardiovasc. Eng. Technol.* **2016**, *7*, 58–68. [CrossRef]
121. Chiastra, C.; Migliavacca, F.; Martinez, M.A.; Malve, M. On the necessity of modelling fluid-structure interaction for stented coronary arteries. *J. Mech. Behav. Biomed. Mater.* **2014**, *34*, 217–230. [CrossRef]
122. Schroeder, S.; Kuettner, A.; Leitritz, M.; Janzen, J.; Kopp, A.F.; Herdeg, C.; Heuschmid, M.; Burgstahler, C.; Baumbach, A.; Wehrmann, M.; et al. Reliability of differentiating human coronary plaque morphology using contrast-enhanced multislice spiral computed tomography: A comparison with histology. *J. Comput. Assist. Tomogr.* **2004**, *28*, 449–454. [CrossRef]
123. Brodoefel, H.; Reimann, A.; Heuschmid, M.; Tsiflikas, I.; Kopp, A.F.; Schroeder, S.; Claussen, C.D.; Clouse, M.E.; Burgstahler, C. Characterization of coronary atherosclerosis by dual-source computed tomography and HU-based color mapping: A pilot study. *Eur. Radiol.* **2008**, *18*, 2466–2474. [CrossRef]
124. Holzapfel, G.A.; Sommer, G.; Gasser, C.T.; Regitnig, P. Determination of layer-specific mechanical properties of human coronary arteries with nonatherosclerotic intimal thickening and related constitutive modeling. *Am. J. Physiol. Heart Circ. Physiol.* **2005**, *289*, H2048–H2058. [CrossRef]
125. Pericevic, I.; Lally, C.; Toner, D.; Kelly, D.J. The influence of plaque composition on underlying arterial wall stress during stent expansion: The case for lesion-specific stents. *Med. Eng. Phys.* **2009**, *31*, 428–433. [CrossRef] [PubMed]
126. Hajiali, Z.; Dabagh, M.; Debusschere, N.; Beule, M.D.; Jalali, P. Tissue prolapse and stresses in stented coronary arteries: A computer model for multi-layer atherosclerotic plaque. *Comput. Biol. Med.* **2015**, *66*, 39–46. [CrossRef] [PubMed]
127. Narayanan, B.; Olender, M.L.; Marlevi, D.; Edelman, E.R.; Nezami, F.R. An inverse method for mechanical characterization of heterogeneous diseased arteries using intravascular imaging. *Sci. Rep.* **2021**, *11*, 22540. [CrossRef] [PubMed]
128. Kolluru, C.; Prabhu, D.; Gharaibeh, Y.; Wu, H.; Wilson, D.L. Voxel-based plaque classification in coronary intravascular optical coherence tomography images using decision trees. In *Medical Imaging 2018: Computer-Aided Diagnosis*; SPIE: Bellingham, WA, USA, 2018; Volume 10575, pp. 657–662.
129. Bae, Y.; Kang, S.J.; Kim, G.; Lee, J.G.; Min, H.S.; Cho, H.; Kang, D.Y.; Lee, P.H.; Ahn, J.M.; Park, D.W.; et al. Prediction of coronary thin-cap fibroatheroma by intravascular ultrasound-based machine learning. *Atherosclerosis* **2019**, *288*, 168–174. [CrossRef]
130. Olender, M.L.; Athanasiou, L.S.; Michalis, L.K.; Fotiadis, D.I.; Edelman, E.R. A Domain Enriched Deep Learning Approach to Classify Atherosclerosis Using Intravascular Ultrasound Imaging. *IEEE J. Sel. Top. Signal Process.* **2020**, *14*, 1210–1220. [CrossRef]
131. Ciompi, F.; Balocco, S.; Rigla, J.; Carrillo, X.; Mauri, J.; Radeva, P. Computer-aided detection of intracoronary stent in intravascular ultrasound sequences. *Med. Phys.* **2016**, *43*, 5616–5625. [CrossRef]
132. Zhao, W.; Jenkins, M.W.; Linderman, G.C.; Bezerra, H.G.; Fujino, Y.; Costa, M.A.; Wilson, D.L.; Rollins, A.M. 3-D Stent Detection in Intravascular OCT Using a Bayesian Network and Graph Search. *IEEE Trans. Med. Imaging* **2015**, *34*, 1549–1561.
133. Nishi, T.; Yamashita, R.; Imura, S.; Tateishi, K.; Kitahara, H.; Kobayashi, Y.; Yock, P.G.; Fitzgerald, P.J.; Honda, Y. Deep learning-based intravascular ultrasound segmentation for the assessment of coronary artery disease. *Int. J. Cardiol.* **2021**, *333*, 55–59. [CrossRef]
134. Shinohara, H.A.-O.; Kodera, S.; Ninomiya, K.A.-O.; Nakamoto, M.; Katsushika, S.A.-O.; Saito, A.; Minatsuki, S.; Kikuchi, H.; Kiyosue, A.; Higashikuni, Y.; et al. Automatic detection of vessel structure by deep learning using intravascular ultrasound images of the coronary arteries. *PLoS ONE* **2021**, *16*, e0255577. [CrossRef]
135. Macedo, M.M.; Guimarães, W.V.; Galon, M.Z.; Takimura, C.K.; Lemos, P.A.; Gutierrez, M.A. A bifurcation identifier for IV-OCT using orthogonal least squares and supervised machine learning. *Comput. Med. Imaging Graph.* **2015**, *46*, 237–248. [CrossRef]
136. Gharleghi, R.; Sowmya, A.; Beier, S. Transient wall shear stress estimation in coronary bifurcations using convolutional neural networks. *Comput. Methods Programs Biomed.* **2022**, *225*, 107013. [CrossRef] [PubMed]
137. Suk, J.; Haan, P.; Lippe, P.; Brune, C.; Wolterink, J.M. Mesh Convolutional Neural Networks for Wall Shear Stress Estimation in 3D Artery Models. In *Statistical Atlases and Computational Models of the Heart Multi-Disease, Multi-View, and Multi-Center Right Ventricular Segmentation in Cardiac MRI Challenge*; Springer: Cham, Switzerland, 2022; pp. 93–102.
138. Sampedro-Gomez, J.; Dorado-Diaz, P.I.; Vicente-Palacios, V.; Sanchez-Puente, A.; Jimenez-Navarro, M.; San Roman, J.A.; Galindo-Villardon, P.; Sanchez, P.L.; Fernandez-Aviles, F. Machine Learning to Predict Stent Restenosis Based on Daily Demographic, Clinical, and Angiographic Characteristics. *Can. J. Cardiol.* **2020**, *36*, 1624–1632. [CrossRef]
139. Avram, R.; Olgin, J.E.; Tison, G.H. The Rise of Open-Sourced Machine Learning in Small and Imbalanced Datasets: Predicting In-Stent Restenosis. *Can. J. Cardiol.* **2020**, *36*, 1574–1576. [CrossRef] [PubMed]
140. Kushner, F.G.; Hand, M.; Smith, S.C., Jr.; King, S.B., 3rd; Anderson, J.L.; Antman, E.M.; Bailey, S.R.; Bates, E.R.; Blankenship, J.C.; Casey, D.E., Jr.; et al. 2009 Focused Updates: ACC/AHA Guidelines for the Management of Patients With ST-Elevation Myocardial Infarction (updating the 2004 Guideline and 2007 Focused Update) and ACC/AHA/SCAI Guidelines on Percutaneous Coronary Intervention (updating the 2005 Guideline and 2007 Focused Update): A report of the American College of Cardiology Foundation/American Heart Association Task Force on Practice Guidelines. *Circulation* **2009**, *120*, 2271–2306. [PubMed]
141. Wijns, W.; Kolh, P.; Danchin, N.; Di Mario, C.; Falk, V.; Folliguet, T.; Garg, S.; Huber, K.; James, S.; Knuuti, J.; et al. Guidelines on myocardial revascularization. *Eur. Heart J.* **2010**, *31*, 2501–2555. [CrossRef] [PubMed]

142. El-Menyar, A.A.; Al Suwaidi, J.; Holmes, D.R., Jr. Left main coronary artery stenosis: State-of-the-art. *Curr. Probl. Cardiol.* **2007**, *32*, 103–193. [CrossRef] [PubMed]
143. Ellwein, L.; Marks, D.S.; Migrino, R.Q.; Foley, W.D.; Sherman, S.; LaDisa, J.F., Jr. Image-based quantification of 3D morphology for bifurcations in the left coronary artery: Application to stent design. *Catheter. Cardiovasc. Interv. Off. J. Soc. Card. Angiogr. Interv.* **2016**, *87*, 1244–1255. [CrossRef] [PubMed]
144. Samant, S.; Wu, W.; Zhao, S.; Khan, B.; Sharzehee, M.; Panagopoulos, A.; Makadia, J.; Mickley, T.; Bicek, A.; Boismier, D.; et al. Computational and experimental mechanical performance of a new everolimus-eluting stent purpose-built for left main interventions. *Sci. Rep.* **2021**, *11*, 8728. [CrossRef] [PubMed]
145. Chatzizisis Yiannis, S.; Makadia, J.; Zhao, S.; Panagopoulos, A.; Sharzehee, M.; Khan, B.; Samant, S.; Fayaz, M.; Pandya, J.; Akkad, H.; et al. First-in-Human Computational Preprocedural Planning of Left Main Interventions Using a New Everolimus-Eluting Stent. *JACC: Case Rep.* **2022**, *4*, 325–335. [CrossRef] [PubMed]
146. Chinnaiyan, K.M.; Akasaka, T.; Amano, T.; Bax, J.J.; Blanke, P.; De Bruyne, B.; Kawasaki, T.; Leipsic, J.; Matsuo, H.; Morino, Y.; et al. Rationale, design and goals of the HeartFlow assessing diagnostic value of non-invasive FFRCT in Coronary Care (ADVANCE) registry. *J. Cardiovasc. Comput. Tomogr.* **2017**, *11*, 62–67. [CrossRef]
147. Nørgaard, B.L.; Leipsic, J.; Gaur, S.; Seneviratne, S.; Ko, B.S.; Ito, H.; Jensen, J.M.; Mauri, L.; De Bruyne, B.; Bezerra, H.; et al. Diagnostic performance of noninvasive fractional flow reserve derived from coronary computed tomography angiography in suspected coronary artery disease: The NXT trial (Analysis of Coronary Blood Flow Using CT Angiography: Next Steps). *J. Am. Coll. Cardiol.* **2014**, *63*, 1145–1155. [CrossRef]
148. Ko, B.S.; Cameron, J.D.; Munnur, R.K.; Wong, D.T.L.; Fujisawa, Y.; Sakaguchi, T.; Hirohata, K.; Hislop-Jambrich, J.; Fujimoto, S.; Takamura, K.; et al. Noninvasive CT-Derived FFR Based on Structural and Fluid Analysis: A Comparison With Invasive FFR for Detection of Functionally Significant Stenosis. *JACC Cardiovasc. Imaging* **2017**, *10*, 663–673. [CrossRef]
149. Kruk, M.; Wardziak, L.; Demkow, M.; Pleban, W.; Pregowski, J.; Dzielinska, Z.; Witulski, M.; Witkowski, A.; Ruzyllo, W.; Kepka, C. Workstation-Based Calculation of CTA-Based FFR for Intermediate Stenosis. *JACC Cardiovasc. Imaging* **2016**, *9*, 690–699. [CrossRef] [PubMed]
150. Andreini, D.; Mushtaq, S.; Pontone, G.; Rogers, C.; Pepi, M.; Bartorelli, A.L. Severe in-stent restenosis missed by coronary CT angiography and accurately detected with FFRCT. *Int. J. Cardiovasc. Imaging* **2017**, 33, 119–120. [CrossRef] [PubMed]
151. Zun, P.S.; Narracott, A.J.; Chiastra, C.; Gunn, J.; Hoekstra, A.G. Location-specific comparison between a 3D in-stent restenosis model and micro-CT and histology data from porcine in vivo experiments. *Cardiovasc. Eng. Tech.* **2019**, *10*, 568–582. [CrossRef]
152. Corti, A.; Colombo, M.; Migliavacca, F.; Rodriguez Matas, J.F.; Casarin, S.; Chiastra, C. Multiscale computational modeling of vascular adaptation: A systems biology approach using agent-based models. *Front. Bioeng. Biotechnol.* **2021**, *9*, 744560. [CrossRef]
153. Hwang, M.; Garbey, M.; Berceli, S.A.; Tran-Son-Tay, R. Rule-based simulation of multi-cellular biological systems-a review of modeling techniques. *Cell Mol. Bioeng.* **2009**, *2*, 285–294. [CrossRef] [PubMed]
154. Boyle, C.J.; Lennon, A.B.; Prendergast, P.J. In silico prediction of the mechanobiological response of arterial tissue: Application to angioplasty and stenting. *J. Biomech. Eng.* **2011**, *133*, 081001. [CrossRef]
155. Zahedmanesh, H.; Van Oosterwyck, H.; Lally, C. A multi-scale mechanobiological model of in-stent restenosis: Deciphering the role of matrix metalloproteinase and extracellular matrix changes. *Comput. Methods Biomech. Biomed. Eng.* **2014**, *17*, 813–828. [CrossRef] [PubMed]
156. Caiazzo, A.; Evans, D.; Falcone, J.-L.; Hegewald, J.; Lorenz, E.; Stahl, B.; Wang, D.; Bernsdorf, J.; Chopard, B.; Gunn, J.; et al. A Complex Automata approach for in-stent restenosis: Two-dimensional multiscale modelling and simulations. *J. Comput. Sci.* **2011**, *2*, 9–17. [CrossRef]
157. Nolan, D.R.; Lally, C. An investigation of damage mechanisms in mechanobiological models of in-stent restenosis. *J. Comput. Sci.* **2018**, *24*, 132–142. [CrossRef]
158. Corti, A.; Colombo, M.; Rozowsky, J.M.; Casarin, S.; He, Y.; Carbonaro, D.; Migliavacca, F.; Rodriguez Matas, J.F.; Berceli, S.A.; Chiastra, C. A predictive multiscale model of in-stent restenosis in femoral arteries: Linking haemodynamics and gene expression with an agent-based model of cellular dynamics. *J. R. Soc. Interface* **2022**, *19*, 20210871. [CrossRef]
159. Razavi, A.; Sachdeva, S.; Frommelt, P.C.; LaDisa, J.F., Jr. Patient-Specific Numerical Analysis of Coronary Flow in Children With Intramural Anomalous Aortic Origin of Coronary Arteries. *Semin. Thorac. Cardiovasc. Surg.* **2021**, *33*, 155–167. [CrossRef] [PubMed]
160. Razavi, A.; Sachdeva, S.; Frommelt, P.C.; LaDisa, J.F. Computational Assessment of Hemodynamic Significance in Patients With Intramural Anomalous Aortic Origin of the Coronary Artery Using Virtually Derived Fractional Flow Reserve and Downstream Microvascular Resistance. *J. Biomech. Eng.* **2022**, *144*, 031005. [CrossRef] [PubMed]
161. Ghorbanniahassankiadeh, A.; Marks, D.S.; LaDisa, J.F. Correlation of computational instantaneous wave-fee ratio with fractional flow reserve for intermediate multivessel coronary disease. *J. Biomech. Eng.* **2021**, *143*, 051011. [CrossRef]

MDPI AG
Grosspeteranlage 5
4052 Basel
Switzerland
Tel.: +41 61 683 77 34

Fluids Editorial Office
E-mail: fluids@mdpi.com
www.mdpi.com/journal/fluids